Fluids and Surfaces

Fluids and Surfaces

Guest Editor
Manfredo Guilizzoni

Basel • Beijing • Wuhan • Barcelona • Belgrade • Novi Sad • Cluj • Manchester

Guest Editor
Manfredo Guilizzoni
Department of Energy
Politecnico di Milano
Milan
Italy

Editorial Office
MDPI AG
Grosspeteranlage 5
4052 Basel, Switzerland

This is a reprint of the Special Issue, published open access by the journal *Fluids* (ISSN 2311-5521), freely accessible at: https://www.mdpi.com/journal/fluids/special_issues/fluids_surfaces.

For citation purposes, cite each article independently as indicated on the article page online and as indicated below:

Lastname, A.A.; Lastname, B.B. Article Title. *Journal Name* **Year**, *Volume Number*, Page Range.

ISBN 978-3-7258-3133-3 (Hbk)
ISBN 978-3-7258-3134-0 (PDF)
https://doi.org/10.3390/books978-3-7258-3134-0

Cover image courtesy of Manfredo Guilizzoni

© 2025 by the authors. Articles in this book are Open Access and distributed under the Creative Commons Attribution (CC BY) license. The book as a whole is distributed by MDPI under the terms and conditions of the Creative Commons Attribution-NonCommercial-NoDerivs (CC BY-NC-ND) license (https://creativecommons.org/licenses/by-nc-nd/4.0/).

Contents

About the Editor vii

Manfredo Guilizzoni
Fluids and Surfaces
Reprinted from: *Fluids* 2025, 10, 8, https://doi.org/10.3390/fluids10010008 1

Motasem Y. D. Alazaiza, Tahra Al Maskari, Ahmed Albahansawi, Salem S. Abu Amr, Mohammed F. M. Abushammala and Maher Aburas
Diesel Migration and Distribution in Capillary Fringe Using Different Spill Volumes via Image Analysis
Reprinted from: *Fluids* 2021, 6, 189, https://doi.org/10.3390/fluids6050189 4

Dave Persaud, Mikhail Smirnov, Daniel Fong and Pejman Sanaei
Modeling of the Effects of Pleat Packing Density and Cartridge Geometry on the Performance of Pleated Membrane Filters
Reprinted from: *Fluids* 2021, 6, 209, https://doi.org/10.3390/fluids6060209 15

Pritam Kumar Roy, Shraga Shoval, Leonid A. Dombrovsky and Edward Bormashenko
Oscillatory Reversible Osmotic Growth of Sessile Saline Droplets on a Floating Polydimethylsiloxane Membrane
Reprinted from: *Fluids* 2021, 6, 232, https://doi.org/10.3390/fluids6070232 34

James Kofi Arthur
Turbulent Flow through and over a Compact Three-Dimensional Model Porous Medium: An Experimental Study
Reprinted from: *Fluids* 2021, 6, 337, https://doi.org/10.3390/fluids6100337 44

Bjørn Kvamme
Small Alcohols as Surfactants and Hydrate Promotors
Reprinted from: *Fluids* 2021, 6, 345, https://doi.org/10.3390/fluids6100345 79

Inna A. Nemirovskaya and Anastasia V. Khramtsova
Features of the Hydrocarbon Distribution in the Bottom Sediments of the Norwegian and Barents Seas
Reprinted from: *Fluids* 2021, 6, 456, https://doi.org/10.3390/fluids6120456 109

Abraham Medina, Abel López-Villa and Carlos A. Vargas
Functional Acrylic Surfaces Obtained by Scratching
Reprinted from: *Fluids* 2021, 6, 463, https://doi.org/10.3390/fluids6120463 123

Manfredo Guilizzoni and Giuseppe Frontera
Crater Depth after the Impact of Multiple Drops into Deep Pools
Reprinted from: *Fluids* 2022, 7, 50, https://doi.org/10.3390/fluids7020050 136

Viktor G. Grishaev, Ivan K. Bakulin, Alidad Amirfazli and Iskander S. Akhatov
Puncture of a Viscous Liquid Film Due to Droplet Falling
Reprinted from: *Fluids* 2022, 7, 196, https://doi.org/10.3390/fluids7060196 154

Roberta Caruana, Stefano De Antonellis, Luca Marocco, Paolo Liberati and Manfredo Guilizzoni
Experimental Characterization of the Wettability of Coated and Uncoated Plates for Indirect Evaporative Cooling Systems
Reprinted from: *Fluids* 2023, 8, 122, https://doi.org/10.3390/fluids8040122 163

About the Editor

Manfredo Guilizzoni

Manfredo Guilizzoni is an Associate Professor at the Department of Energy, Politecnico di Milano, where he is also the Head of the Fluid–Surface Interaction Research Area of the Multiphase Laboratory.

His main research activities deal with wettability and thermal interactions between fluids and complex or innovative surfaces, measurement techniques and data analysis for adiabatic multiphase flows, and thermal–fluid dynamic analyses of industrial and space devices and systems. His studies are carried out both via experiments and computational fluid dynamics. An additional research line concerns the field of wind energy. His teaching activities include applied thermodynamics, heat transfer and thermal analysis, and advanced measurement methods.

Editorial

Fluids and Surfaces

Manfredo Guilizzoni

Department of Energy, Politecnico di Milano, Via Lambruschini 4, 20156 Milan, Italy; manfredo.guilizzoni@polimi.it

Fluids is pleased to present a Special Issue named "Fluids and Surfaces", a curated collection of ten research articles focused on capillary phenomena and the interaction between fluids and surfaces. In fact, the latter play a key role in a broad set of circumstances, ranging from natural ones to industrial applications, environment protection, and cultural heritage conservation. A deep understanding of the fluid behavior in such scenarios has therefore acquired an increasing importance in recent years, also due to the growing number of devices and processes that are based on engineered surfaces, porous media, and miniaturization involving microfluidics. In addition, the design of bulk materials has reached a high level of maturity, so the effort to reach further levels of improvement is leading to a shift towards surface engineering too.

Fluid-surface interactions are certainly not a new research field, but despite more than two centuries of studies, many aspects are still not thoroughly clarified for both static and dynamic conditions, particularly on innovative surfaces (by chemistry and/or morphology), within porous media, for complex fluids and when fluid dynamics, capillarity, and heat transfer are coupled. Wettability and adhesion for complex surfaces and/or complex fluids, fluids in porous media and the relationship between external wettability and in-pore behavior, capillarity-driven flows, reactive wetting and electrowetting, evaporation, Marangoni, and thermocapillary convection, drop impact onto liquid pools and still and moving films, drop impact onto heterogeneous or rough/engineered dry surfaces (including porous, flexible, and textile surfaces) are all open research topics. Furthermore, the development of advanced measurement techniques, down to the microscale, opens new possibilities for experimental investigation, while the advent of computational fluid dynamics offers new tools for modeling and simulation.

The articles collected in this Special Issue cover very different topics, with the aim of giving an overview about the many different aspects of research in this field.

Alazaiza et al. [1] experimentally studied the migration of non-aqueous phase liquids in porous media, which is of interest for prevention of groundwater pollution, by image processing pictures of capillary migration of diesel in a laboratory-scale sand column. Their results prove that image processing is a suitable technique to study this phenomenon and offered insight about the dynamic behavior of these fluids in porous samples.

Persaud et al. [2] developed a mathematical model for the fluid flow in a pleated membrane filter, which is a type of filter widely used and that may offer significant advantages with respect to other types, but whose design must be very careful to avoid or reduce the points of weakness. Their analytical model allows to describe the fluid flow across a whole pleated filter cartridge.

Roy et al. [3] experimentally studied osmosis between a saline droplet and a pool of pure water, across a cured polydimethylsiloxane membrane. They performed cycles of growth and evaporation, a process that can be of interest for desalination and separation for batteries, evaluating the drop volume and contact angle and providing models for the involved phenomena.

Received: 29 December 2024
Accepted: 2 January 2025
Published: 6 January 2025

Citation: Guilizzoni, M. Fluids and Surfaces. *Fluids* **2025**, *10*, 8. https://doi.org/10.3390/fluids10010008

Copyright: © 2025 by the author. Licensee MDPI, Basel, Switzerland. This article is an open access article distributed under the terms and conditions of the Creative Commons Attribution (CC BY) license (https://creativecommons.org/licenses/by/4.0/).

Arthur [4] focused on coupled flow over and within a compact porous medium. Planar particle image velocimetry was used to acquire the fluid velocity field, from which all the quantities of interest were calculated. The results evidence how the compact porous medium induces large-scale anisotropy in the surrounding turbulent flow, which can be important in applications involving such scenario, from natural ones to heat transfer devices.

Kvamme [5] analyzed from the thermodynamic point of view the use of small alcohols as surfactants to promote hydrate formation, in addition to their more common use as hydrate inhibitors. The model results shed light on the thermodynamic stability of systems including water, methane, carbon dioxide, nitrogen, and ethanol.

Nemirovskaya and Khramtsova [6] analyzed the results of two cruise studies about aliphatic and polycyclic aromatic hydrocarbons in the bottom sediments within the Norwegian-Barents Sea basin, evaluating the anthropogenic input and the endogenous influence in the different regions, to explain the local anomalies in content and composition.

Medina et al. [7] used image processing to study the wetting behavior of functional acrylic surfaces obtained by scratching with sandpaper of different grit. Capillary rise and drop aspect were evaluated to evaluate the effect of the micro-grooves created by the scratching, which resembles those that are progressively created when using and cleansing commonly used surfaces.

Guilizzoni and Frontera [8] performed a computational fluid dynamics study of the impact of multiple water drops into large pools, to extend the classic results for single droplets. Specific focus was on the crater depth, which is one of the most important parameters to evaluate the influence of the drops on the pool, for example for pollution transport. The possible extension of classic models for single drop impact to the multiple drop impact scenario was also evaluated.

Grishaev et al. [9] performed experiments - using a high-speed camera—and created a semi-analytical model for film rupture due to drop impact, analyzing the needed energy and the viscous dissipation in the film and in the impinging drop. Their results offer new insight about the effect of the fluid viscosity and the dry spots that are formed after the drop impact.

Caruana et al. [10] studied the wettability of plates used in the recuperators of Indirect Evaporative Cooling systems. In fact, creating a uniform film of water on the plates significantly improves the performances, thus static and dynamic contact angles were experimentally investigated for three commercially used surfaces: one aluminum uncoated surface and the same surface covered with a standard epoxy coating and with a hydrophilic lacquer.

Finally, I would like to thank the authors and the reviewers that built this Special Issue with their articles and comments, and Ms. Xiaochun Peng and Ms. Wing Wang for all their help and assistance.

Conflicts of Interest: The author declares no conflicts of interest.

References

1. Alazaiza, M.Y.D.; Al Maskari, T.; Albahansawi, A.; Amr, S.S.A.; Abushammala, M.F.M.; Aburas, M. Diesel Migration and Distribution in Capillary Fringe Using Different Spill Volumes via Image Analysis. *Fluids* **2021**, *6*, 189. [CrossRef]
2. Persaud, D.; Smirnov, M.; Fong, D.; Sanaei, P. Modeling of the Effects of Pleat Packing Density and Cartridge Geometry on the Performance of Pleated Membrane Filters. *Fluids* **2021**, *6*, 209. [CrossRef]
3. Roy, P.K.; Shoval, S.; Dombrovsky, L.A.; Bormashenko, E. Oscillatory Reversible Osmotic Growth of Sessile Saline Droplets on a Floating Polydimethylsiloxane Membrane. *Fluids* **2021**, *6*, 232. [CrossRef]
4. Arthur, J.K. Turbulent Flow through and over a Compact Three-Dimensional Model Porous Medium: An Experimental Study. *Fluids* **2021**, *6*, 337. [CrossRef]
5. Kvamme, B. Small Alcohols as Surfactants and Hydrate Promotors. *Fluids* **2021**, *6*, 345. [CrossRef]

6. Nemirovskaya, I.A.; Khramtsova, A.V. Features of the Hydrocarbon Distribution in the Bottom Sediments of the Norwegian and Barents Seas. *Fluids* **2021**, *6*, 456. [CrossRef]
7. Medina, A.; López-Villa, A.; Vargas, C.A. Functional Acrylic Surfaces Obtained by Scratching. *Fluids* **2021**, *6*, 463. [CrossRef]
8. Guilizzoni, M.; Frontera, G. Crater Depth after the Impact of Multiple Drops into Deep Pools. *Fluids* **2022**, *7*, 50. [CrossRef]
9. Grishaev, V.G.; Bakulin, I.K.; Amirfazli, A.; Akhatov, I.S. Puncture of a Viscous Liquid Film Due to Droplet Falling. *Fluids* **2022**, *7*, 196. [CrossRef]
10. Caruana, R.; De Antonellis, S.; Marocco, L.; Liberati, P.; Guilizzoni, M. Experimental Characterization of the Wettability of Coated and Uncoated Plates for Indirect Evaporative Cooling Systems. *Fluids* **2023**, *8*, 122. [CrossRef]

Disclaimer/Publisher's Note: The statements, opinions and data contained in all publications are solely those of the individual author(s) and contributor(s) and not of MDPI and/or the editor(s). MDPI and/or the editor(s) disclaim responsibility for any injury to people or property resulting from any ideas, methods, instructions or products referred to in the content.

Article

Diesel Migration and Distribution in Capillary Fringe Using Different Spill Volumes via Image Analysis

Motasem Y. D. Alazaiza [1],*, Tahra Al Maskari [1], Ahmed Albahansawi [2], Salem S. Abu Amr [3], Mohammed F. M. Abushammala [4] and Maher Aburas [5]

[1] Department of Civil and Environmental Engineering, College of Engineering, A'Sharqiyah University, Ibra 400, Oman; tahra.almaskari@asu.edu.om
[2] Department of Environmental Engineering-Water Center (SUMER), Gebze Technical University, Kocaeli 41400, Turkey; ahmedalbahnasawi@gmail.com
[3] Faculty of Engineering, Demir Campus, Karabuk University, Karabuk 78050, Turkey; sabuamr@hotmail.com
[4] Department of Civil Engineering, Middle East College, Knowledge Oasis Muscat, Al Rusayl 124, Oman; mabushammala@mec.edu.om
[5] The High Institute for Engineering Proffessions, ALmajurie, Benghazi 16063, Libya; boras222@yahoo.com
* Correspondence: my.azaiza@gmail.com

Citation: Alazaiza, M.Y.D.; Al Maskari, T.; Albahsawi, A.; Amr, S.S.A.; Abushammala, M.F.M.; Aburas, M. Diesel Migration and Distribution in Capillary Fringe Using Different Spill Volumes via Image Analysis. *Fluids* **2021**, *6*, 189. https://doi.org/10.3390/fluids6050189

Academic Editor: Manfredo Guilizzoni

Received: 19 March 2021
Accepted: 6 May 2021
Published: 17 May 2021

Publisher's Note: MDPI stays neutral with regard to jurisdictional claims in published maps and institutional affiliations.

Copyright: © 2021 by the authors. Licensee MDPI, Basel, Switzerland. This article is an open access article distributed under the terms and conditions of the Creative Commons Attribution (CC BY) license (https://creativecommons.org/licenses/by/4.0/).

Abstract: Laboratory-scale column experiments were conducted to assess the impact of different LNAPL volumes on LANPL migration behavior in capillary zone in porous media. Three different volumes of diesel (50 mL, 100 mL, and 150 mL) were released in different experiments using a 1D rectangular column filled with natural sand. The water table was set at 29 cm from the bottom of the column. The image analysis results provided quantitative time-dependent data on the LNAPL distribution through the duration for the experiments. Results demonstrated that the higher diesel volume (150 mL) exhibited the faster LNAPL migration through all experiments. This observation was due to the high volume of diesel as compared to other cases which provides high pressure to migrate deeper in a short time. In all experiments, the diesel migration was fast during the first few minutes of observation and then, the velocity was decreased gradually. This is due to pressure exerted by diesel in order to allow the diesel to percolate through the sand voids. Overall, this study proved that the image analysis can be a good and reliable tool to monitor the LNAPL migration in porous media.

Keywords: LNAPL; groundwater; soil; image analysis; water table

1. Introduction

Organic compound pollution of groundwater and soil in the form of non-aqueous phase liquids (NAPLs) is a widespread environmental problem [1,2]. These pollutants have a negative effect on groundwater quality, rendering it unsuitable for human use and irrigation [3]. Because of their low aqueous solubility, many NAPLs are insoluble in water. As a result, they have the potential to persist for several years, contaminating large parts of groundwater [4]. Because of the complex nature of the water–air–NAPL multiphase system, it is difficult to predict the behavior of NAPLs in porous media, especially in the vadose region. NAPLs are divided into two categories based on their density in relation to water: light non-aqueous phase liquids (LNAPLs) and dense non-aqueous phase liquids (DNAPLs). Chlorinated solvents like tetrachloroethylene and tetrachloroethylene are common examples of DNAPL, while petroleum compounds like benzene, toluene, and xylene (BTEX) are common examples of LNAPL [5–7]. LNAPLs migrate downward through unsaturated soil due to gravity when a NAPL is released into the subsurface, leaving small ganglia along their way [8]. Because of the wide density difference between LNAPLs and air, the vadose zone provides little resistance, allowing the LNAPL to enter and accumulate on top of the water table. DNAPLs, on the other hand, will join the saturated zone and continue to migrate downward until they encountered hydraulic or capillary barrier [3]. Understanding the fate and transport of NAPLs in subsurface structures is critical for identifying NAPL source zone geometries as well as designing and evaluating remediation schemes [9,10].

Several studies have shown that NAPL migration in the subsurface is influenced by complex processes that are influenced by a variety of factors, including the NAPL's viscosity, density, saturation, immiscibility, capillary pressure, wettability, and polar nature, as well as the porous medium's permeability, porosity, pore size distribution, and surface properties [11–14].

The rise of water through a soil due to the fluid property known as surface tension is also known as capillarity. Because of the surface tension of pore water acting on the meniscus produced in void spaces between soil particles, pore water pressures are less than atmospheric due to capillarity.

Experiments on the behavior of immiscible fluids in soils have been conducted by a number of researchers [15–18]. For monitoring fluid properties and movement in soil and groundwater, non-invasive techniques such as gamma ray, X-ray attenuation, and electrical conductivity probes have been used [19,20]. Although these studies have provided useful information on the fate and transport of NAPLs in porous media, they have some drawbacks, the most notable of which is that they do not allow for the observation of fluid migration under dynamic conditions and across the entire domain of interest [2,7]. Image analysis approaches such as light propagation visualization, light reflection method, spectral image analysis method, and simplified image analysis method (SIAM) have gained more attention from researchers because of the limitations [16,18,21].

As previously mentioned, NAPLs released into the subsurface environment will last for a very long time. Though capillary forces may keep the NAPL entrapped in the soil, it may undergo some redistribution that affects its fate and transport. In the vadose zone, water percolation into the subsurface can influence the migration of entrapped NAPL ganglia as well as the NAPL mass accumulated on the water table. Alazaiza et al. [14] conducted a 2-D experimental study to observe the influence of rainfall on the migration of LNAPL in double-porosity soil structure. The authors reported that water table fluctuations caused the LNAPL that was entrapped in the porous media to be pushed further downward. In addition, they observed that the capillary fringe thickness was depressed due to the influence of both infiltration and LNAPL volume. Alazaiza et al. [18] applied the simplified image analysis method (SIAM) to investigate the saturation distributions of the LNAPL and water in the entire domain the influence of water table fluctuations. The authors concluded that the SIAM technique is an accurate and cost-effective tool for the visualization of the time-dependent NAPL/water movement in laboratory-scale experiments and quantifying dynamic changes in fluid saturation in porous media.

The present study aims to investigate the influence of LNAPL spill volume on the migration behavior in porous media. Specifically, SIAM is applied to produce detailed time-dependent LNAPL migration in a 1-D sand column. Diesel was used as a representative of LNAPL, and the sand used was a natural river sand.

2. Materials and Methods

2.1. Materials Characteristics

A natural sand collected from a river was used as a porous media. General characterization was conducted to investigate the main properties of the sand prior to experiments. Particle size analysis indicated that the pore grain size varied between 0.4 mm and 1.22 mm with an average size of 0.71 mm. The basic physical characteristics of the sand are presented in Table 1. LNAPL was represented using a commercial diesel fuel with a density of 0.83 g/cm^3 and a viscosity at 20 °C of 5.8 mm^2/s. The thermal expansion coefficient was 88 kg/m^3, flash point at 62 °C, and vapor pressure was 10 kPa at 15 °C [22]. The LNAPL was mixed with Red Sudan III dye while water was dyed using a Brilliant Blue FCF dye with 0.1% of volume to enhance the visualization.

Table 1. The basic physical properties of sand.

Parameter	Test	Value
Coefficient of permeability (m/s)	Constant head permeability	4×10^{-10}
Average diameter (mm)	Sieve analysis	0.30
Coefficient of uniformity, C_u		2.19
Density (g/cm^3)	Small pyknometer	2.66
Porosity	Calculated	0.40

2.2. Experimental Setup

For all experiments, a 1-D square acrylic glass column with dimensions of 30 mm width × 30 mm depth × 500 mm height and a 10 mm thick wall was used. The acrylic glass was translucent so that fluid migration within the sand column could be observed. Transparent glue was used to stack the column joints together. Two digital cameras (Nikon D5100 and Nikon D5300) were mounted 1 m in front of the column with two separate narrow bandpass filters (BN-650 and BN-470) attached to capture images during the experiments. In the column test, the BN-650 bandpass filter was used to assess the light intensity of the diesel, while the BN-470 bandpass filter was used to detect the light intensity of the water. In addition, two LED floodlights were installed as the primary source of illumination in the dark space. As a white color guide, a GretaMacbeth white balance board was fixed beside the column. To avoid displacement and vibration during image processing, a Nikon Camera Control Pro Software was used to capture images automatically. Manual changes were made to the camera settings to improve the dynamic range of the cameras by using exposure times of just a few seconds. The aperture of the lenses used was set at f-16 for all photographs, and the exposure time was 2.5 s, yielding a resolution of f/16 (0.030 by 0.035 cm^2). According to the procedure defined by Bob et al. [16], the temporal variation in light intensity was corrected. A reference image was produced that reflected the image captured when the column was fully saturated with water. On the reference image, two small squares (2 cm × 2 cm) were identified as "correction zones".

The correction coefficients were determined using the ratio of the reference image's average light intensity of the correction zones to the image's average light intensity of the correction zones that needed to be processed. To correct a particular image, for example, the correction coefficient of the image to be corrected was multiplied by the image's light intensities. Three subsequent images were taken for each image, and the average value of the light intensities of these three images was used in the calculations. MATLAB was used to analyze all the images. All tests were carried out in duplicate inside a dark room that was held at a temperature of 23 ± 1 °C. Figure 1 portrays the experimental setup.

2.3. Experimental Procedure

Before starting the experiments, the sand was washed with distilled water to eliminate any fine residuals and then oven dried for 48 h at 45 °C to remove any moisture [15,23]. Following that, the sand was gradually packed into the column in 2 cm layers. The sand column was fixed on a mechanical vibrator with a vibration rate of 50 hertz to ensure uniform compaction and identical void ratios in all column experiments. It was found that the porosity of the sand was 0.39, which is equal to a pore volume of 295 mL. After that, the sand was saturated with blue-dyed water and put in a vacuum chamber for 24 h to eliminate any trapped air and ensure that the sand was fully saturated with water. The drainage valve was then opened to drain the water into the tank, resulting in the water table being set at 29 cm from the column's bottom. Three different diesel volumes were used to investigate the effect of different diesel spill volumes on LNAPL migration in sand column. These volumes were 50 mL, 100 mL, and 150 mL. Digital cameras were set to capture an image each 1 min in order to monitor the changes in diesel behavior.

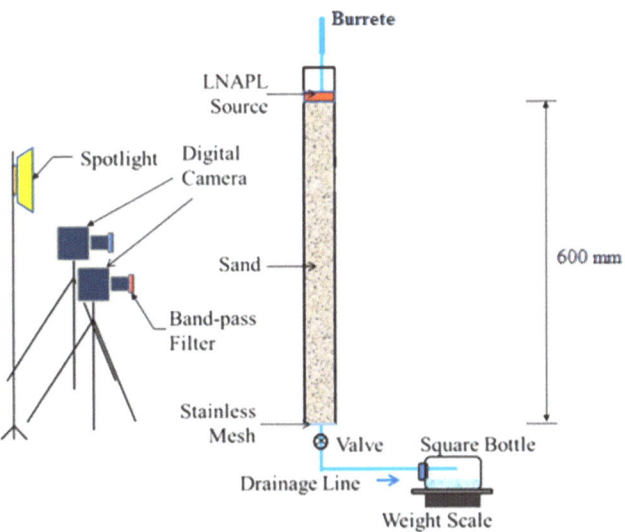

Figure 1. Experimental setup.

2.4. Simplified Image Analysis

The definition of NAPL zones needs precise NAPL saturation measurements. Additionally, in controlled laboratory environments, collecting precise quantitative data on saturation is a difficult task. Since soil samples must be separated from a soil sample, the gravimetric sampling process, though reliable, cannot provide continuous saturation data. Furthermore, sampling only offers information from one or a few chosen locations. Alazaiza et al. [18] developed the SIAM to observe the saturation distributions of LNAPL and water in the partially saturated sand column under complex conditions in this study. The SIAM's theory is based on the Beer-Lambert law of light transmittance, which assumes a linear relationship between the average optical density, LNAPL, and water saturation (S_o and S_w, respectively). The following formula can be used to measure the average optical density:

$$D_i = \frac{1}{N}\sum_{j=1}^{N} d_{ji} = \frac{1}{N}\sum_{j=1}^{N}\left[-\log_{10}\left(\frac{I^r_{ji}}{I^0_{ji}}\right)\right] \quad (1)$$

where D_i is the average optical density, i is the spectral band (i = 470 or 650), N is the number of pixels in the area of interest, d_{ji} is the optical density of the individual pixels, I^r_{ji} is the light intensity of the reflected light by individual pixel value, and I^0_{ji} is the light intensity of the reflected light by an ideal white surface.

From the calculated D_i, SIAM can estimate the saturation distribution matrices $[S_w]$ and $[S_o]$ by performing a comparison between the average optical densities of the matrix element (D_i) and the three calibrated images (D^w_i, D^0_i, D^d_i) for sand fully saturated with water (S_w = 100%, S_o = 0%), sand fully saturated with LNAPL (S_w = 0%, S_o = 100%), and dry sand (S_w = 0%, S_o = 0%), respectively, for the same range. The average optical density values for each cell of the studied range can be calculated and compared to the corresponding ones for all three calibration images. A matrix of correlation equation sets can also be obtained, with each one corresponding to each cell as follows:

$$\begin{bmatrix} D_i \\ D_j \end{bmatrix}_{mn} = \begin{bmatrix} (D^w_i - D^d_i)S_w + (D^o_i - D^d_i)S_o + D^d_i \\ (D^w_j - D^d_j)S_w + (D^o_j - D^d_j)S_o + D^d_j \end{bmatrix}_{mn} \quad (2)$$

where m and n are matrix dimensions, $[D_i]_{mn}$ and $[D_j]_{mn}$ are the average values of the optical density of each mesh element for wavelengths i and j, $[D_{di}]_{mn}$ and $[D_{dj}]_{mn}$ are the average optical density of each mesh element for dry sand; $[D_{wi}]_{mn}$ and $[D_{wj}]_{mn}$ are for water saturated sand; and $[D_{oi}]_{mn}$ and $[D_{oj}]_{mn}$ are for LNAPL saturated sand. Equation (2) is essentially an averaging of the optical density of the three fluids (water, LNAPL, and air), each weighted by its saturation.

3. Results and Discussion

3.1. Determination of Capillary Height

The capillary height (h_c) was determined by adjusting the water table from the water reservoir to a specific desired height. After the sand was placed in the column, the drainage valve was opened instantly. The water capillarity monitoring was left for 24 h. After 24 h, the capillary height value observed was around 28 cm. Because of the small void spaces between sand particles, the rising value of h_c means that water capillarity in sand moves upwards at a faster rate. According to Liu et al. [24], several soil parameters such as effective grain size, void ratio value, and pore size distribution are the key considerations in determining the maximum height of capillary rise in soils. As a result, any changes in the specified soil properties can change the capillary height. Day [25] also concluded that the interactions between pore water pressure and atmospheric pressure play an important role in determining the capillary rise in soil caused by pore spaces in soil particles. Based on Terzaghi and Peck [26], the height of capillary can be estimated from the equations below:

$$H_{c,max} = C/eD_{10} \quad (3)$$

$$e = G_s \gamma_w / \gamma_d \quad (4)$$

$$\gamma_d = W/V \quad (5)$$

where C is the grain shape constant of sand which is in the range of 10 to 50 mm^2 [26]. D_{10} is the effective grain diameter obtained from soil classification. e: is the void ratio.

Equation (4) is used to calculate the void ratio where:

G_s is specific gravity of soil
γ_w is the unit weight of water in the soil
γ_d is unit weight of dry soil
W is the mass of the dry soil
and V is the volume of the dry soil.

Equation (3) is used to calculate the dry unit weight of soil. The total dry sand used in the column was 1100 g, thus, the dry unit weight calculated according to Equation (5) was 15.7 kN/m^3. The void ratio obtained using Equation (2) was 0.69. By substituting Equations (3) and (4) into Equation (5), the capillary height, h_c obtained was between 280 mm and 314 mm according to the value of C. The difference of h_c value is due to some parameters provided from both equations are different. According to Liu et al. [24] in determining maximum capillary rise height in a short period of time, the parameters which is contact angle, dry density, specific gravity, pore space, and hydraulic properties can be applied.

3.2. Effect of Diesel Spill Volume

Due to many factors, including entrapment of LNAPL at residual saturations and viscosity rate of LNAPLs, determining the exact amount of LNAPL spills under a leaking underground storage tank (LUST) is difficult. When LNAPL is removed, water invades the void space, causing LNAPL flow paths to become narrower, reducing the porous media's ability to allow LNAPL migration. As the LNAPL becomes immobile due to saturation, the reduced size of flow paths in porous media causes the large LNAPL spill volume to become more complicated. As a result, as the saturation of LNAPL increases, the recovery rate process becomes more difficult. Using Equation (2) and substituting with the values of the unit weight of water, γ_w, maximum dry density, γ_d, and specific gravity, G_s, the diesel spill volume needed to contaminate the sand particles can be determined based on the observation of the column test. The porosity value, n, needed to enable the diesel to fill the pore spaces can be calculated using the calculated void ratio, e, obtained from these parameters.

$$n = e/(1 + e) \tag{6}$$

The volume of void in the sand is equal to the total volume of diesel required to fill the void spaces of sand particles and contaminate it. The calculated porosity was 0.40 which resulted in a volume of diesel required to contaminate the sand particles with a value of is 0.00029 m^3. As a result, if the sand column is set up with different dimensions, the appropriate diesel spill volume would be different as well, due to the different void ratio and porosity values. The propensity of diesel to hit the ground water table was faster when enough diesel volume was released and flowed downward through the sand column. Because of the high volume and weight of the migrated diesel, it exerted a high pressure on the sand, allowing it to be polluted at a faster rate. The diesel will begin to fill the void spaces in the sand particles as it migrates downwards under the effect of gravity. Diesel displaced blue dye solution from the interior of the largest pores during this process. The ability of diesel to displace dye solution from larger pores than smaller pores has been demonstrated by selective entry of diesel into larger voids. More diesel would be displaced, resulting in a larger network of interconnected pores containing diesel. According to Simantiraki et al. [27], diesel spreads faster in the vertical direction in fine sand because fine sand particles have smaller void spaces, resulting in higher capillary pressure.

3.3. Diesel Migration Influence on Capillary Height

The volume of LNAPL spills is one of the most important factors that can influence the severity of pollution of the subsurface. Several researchers reported that even small amounts of LNAPLs spilled could seriously contaminate ground water [28]. In this study, three different volume of diesel was used as mentioned earlier. In prior to spilling the diesel, the capillary fringe height was left for 24 h. Within 24 h, saturated capillary fringe could move upwards to 28 cm from the water table and leave only 2 cm of the vadose zone. The total period to conduct this experiment was set to 2.5 h. Within this period, the volume of diesel had completely migrated downward through the sand particles. The elevation change of capillary height was also observed during the whole experiment. After 2.5 h, no change of elevation for capillary height or capillary fringe zone was observed due to diesel migration. From the visual observation, no capillary action was observed including the capillary depression in the capillary fringe zone, despite different volumes of spilled diesel used. As a result, the volume and weight of spilled diesel can be increased, putting more pressure on the diesel, and allowing it to infiltrate the capillary fringe zone.

The density of the sand in the column also plays a significant role in determining capillary rise. The porosity and void ratio of the sand particles decreased as they were compacted, resulting in smaller void spaces in the sand particles. The density of sand is constant in the study of diesel migration behavior since the weight and volume of sand used in the study are constant. Owing to gravitational force, capillary force, and viscosity factor, diesel is less dense than water, thus it migrates downward. When a large amount of diesel is spilled, the propensity for the diesel to fill up the sand void space becomes easier, compared to when a small volume of diesel is spilled, since the volume spill exerts more pressure. Figure 2 shows the migration of the 50 mL diesel volume in the sand column.

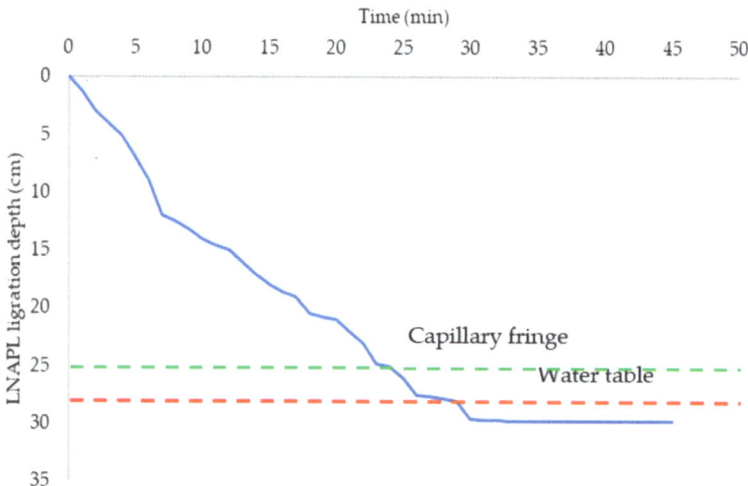

Figure 2. The behavior of 50 mL diesel volume in capillary fringe and water table.

As shown in Figure 2, the diesel migrated very fast during the first 10 min. This is due to pressure exerted by diesel in order to allow the diesel to percolate through the sand voids. After that, diesel continued its migration but with a less velocity compared to the first 10 min. After 30 min of migration, diesel velocity started to be constant and no obvious change in the velocity was observed. Diesel reached the bottom of the column after 30 min. This could be because the 50 mL of diesel spilled does not fill up all the void spaces of sand particles. Another reason is that the diesel tends to migrate and find the lowest point, as the sand may not well-compacted.

For the case of 100 mL diesel volume, diesel took longer to fill the voids and cover the sand particles in the saturated capillary fringe zone, taking more than 22 min. Furthermore, the diesel took longer to reach the ground water level. This may be explained by the fact that diesel appears to accumulate and form a plume before migrating deeper. In addition, it seems that some parts of the NAPL migrated horizontally during the start time of NAPL pouring until it filled the pores around the spillage point. After that, it started to migrate downward.

The pressure of water in the capillary fringe zone allows the diesel to accumulate. As a result, diesel attempts to conquer the water–capillary pressure and migrate lower. The increased diesel plume in the void spaces allowed for more pressure to be applied to penetrate this LNAPL. Figure 3 represents the flow of 100 mL diesel volume within the first 45 min.

For the case of 150 mL, Figure 4 shows that for the first 30 min, 150 mL of diesel spilled has a greater effect on migration depth. The large amount of diesel spilled provides higher pressure to push the migration of LNAPL in faster rate as compared to the cases of 50 mL and 100 mL diesel volume. In addition, the diesel took around 15 min to fill void spaces in the saturated capillary fringe zone. The diesel migrated deeper after 15 min, contaminating the ground water table and sand. Several factors such as soil hydraulic conductivity can also influence the LNAPL spills. This could be explained as the sand has medium degree of permeability which possess good drainage system. Good drainage paths result in a lower resistance to fluid flow. Although the volume and weight of diesel spilled is high, the permeability coefficient can be low as the diesel tends to accumulate and form contamination plume in subsurface environment.

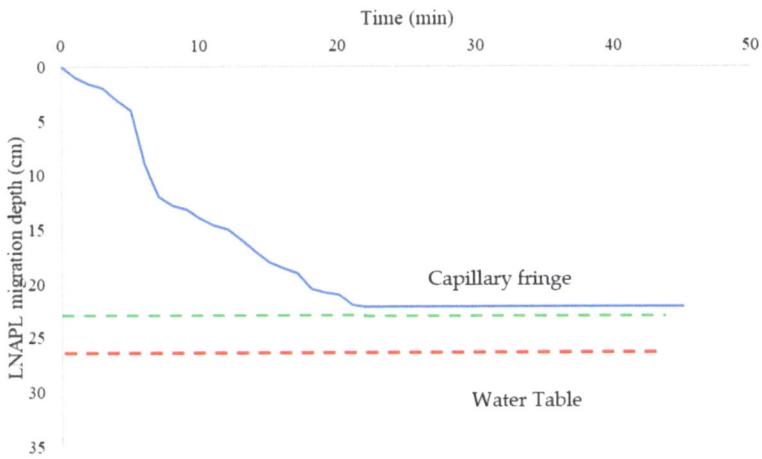

Figure 3. The behavior of 100 mL diesel volume in capillary fringe and water table.

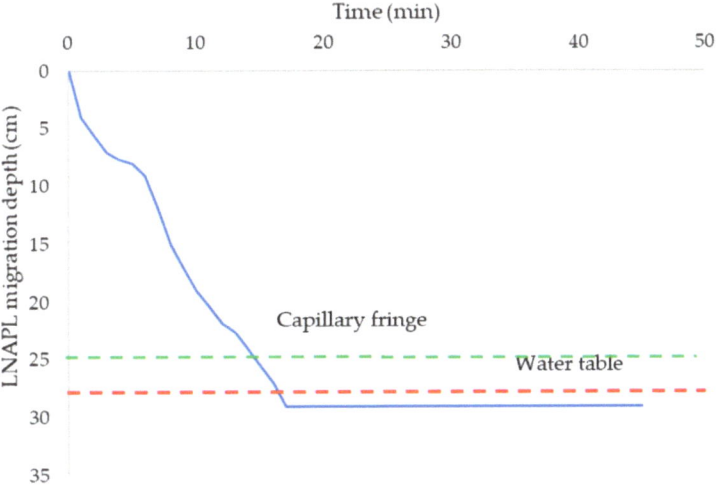

Figure 4. The behavior of 15 × 0 mL diesel volume in capillary fringe and water table.

Based on the figures provided above, it can be summarized that the greater the volume diesel spilled, the deeper the migration depth through the subsurface of sand with short time. The 150 mL of diesel portrayed great impact on the sand particles. Due to the good drainage paths of the sand, the infiltrated 150 mL diesel can migrate deeper provided with high pressure subjected to the sand due to the high-volume diesel spilled. Besides, it also required shorter time to fill up the void spaces in the saturated capillary fringe zone which is only 20 min. With high volume of spill, the migration of diesel became continuous due to constant pressure exerted. Hence, this will induce more constant rate of the 150 mL diesel infiltration as compared to 50 mL and 100 mL of diesel volumes. Figure 5 showed the images of the different volumes of diesel in the sand column. It is worth mentioning that in real life and real application, the diesel plume shape as well as velocity could be different from the results shown in this study. This is because of the column wall on the migration of fluids. The wall effect is very common in 2-D columns and results in looser packing along the column wall compared to the rest of the column parts.

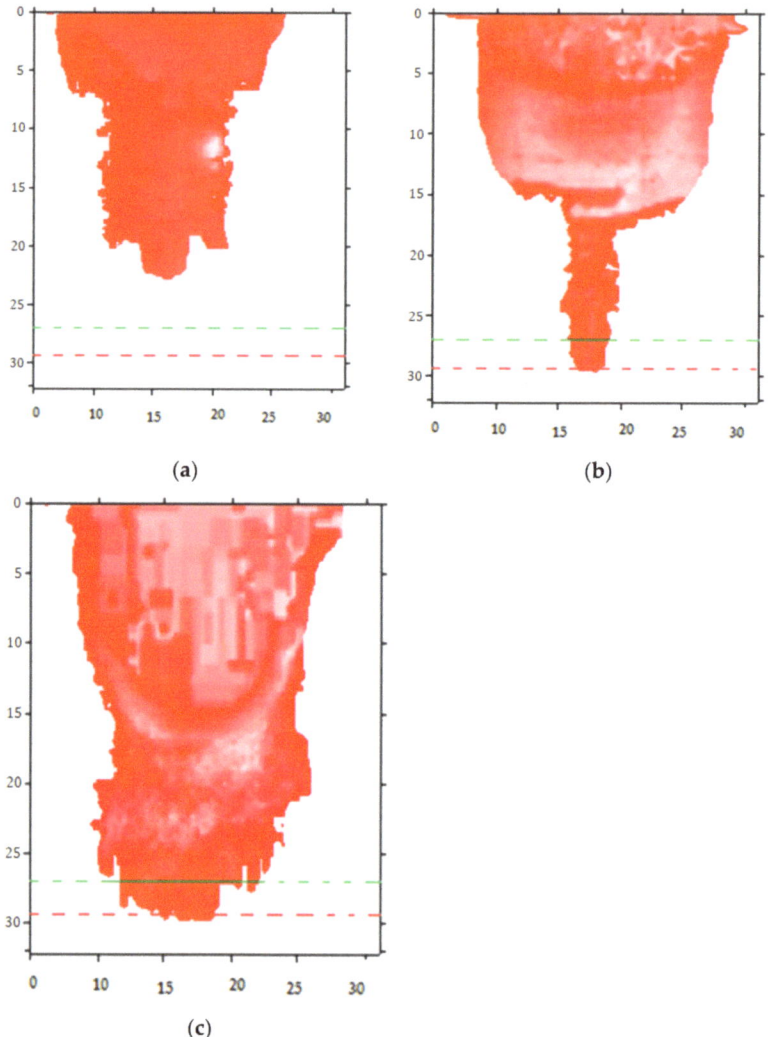

Figure 5. Samples of the image analysis showing the diesel behavior using (**a**) 50 mL, (**b**) 100 mL, and (**c**) 150 mL.

4. Conclusions

- Three 1-D laboratory experiments were carried out to investigate the effect of different LANPL volumes on its migration through sand column using simplified image analysis method.
- Diesel was used as a representative of LNAPL with three different volumes, 50 mL, 100 mL, and 150 mL.
- The water table was set at 29 cm from the bottom of the column.
- The image analysis results provided quantitative time-dependent data on the LNAPL distribution through the duration for the experiments.
- Results demonstrated that the higher diesel volume (150 mL) exhibited the faster LNAPL migration through all experiments. This observation was due to the high

- volume of diesel as compared to other cases which provides high pressure to migrate deeper in a short time.
- In all experiments, the diesel migration was fast during the first few minutes of observation, and then the velocity was decreased gradually. This is due to pressure exerted by diesel in order to allow the diesel to percolate through the sand voids.
- Overall, this study proved that the image analysis can be a good and reliable tool to monitor the LNAPL migration in porous media.
- The results of the study confirm the viability of SIAM to be used as a laboratory tool to assess the behavior of fluids in soil and subsurface as compared to other image analysis such as light transmission visualization as well as light reflection method [2].

Author Contributions: Conceptualization, M.Y.D.A.; methodology, T.A.M.; software, A.A.; validation, S.S.A.A.; formal analysis, M.F.M.A.; data curation, M.A. All authors have read and agreed to the published version of the manuscript.

Funding: The research leading to these results has received funding from Ministry of Higher Education, Research, and Innovation (MoHERI) of the Sultanate of Oman under the Block Funding Program, MoHERI Block Funding Agreement No. MoHERI/BFP/ASU/01/2020.

Institutional Review Board Statement: Not applicable.

Informed Consent Statement: Not applicable.

Data Availability Statement: The data presented in this study are available on request from the corresponding author. The data are not publicly available due to that the current project are in progress and not archived yet.

Conflicts of Interest: The authors declare no conflict of interest.

References

1. Agaoglu, B.; Copty, N.K.; Scheytt, T.; Hinkelmann, R. Interphase mass transfer between fluids in subsurface formations: A review. *Adv. Water Resour.* **2015**, *79*, 162–194. [CrossRef]
2. Alazaiza, M.Y.D.; Ngien, S.K.; Bob, M.M.; Ishak, W.M.F.; Kamaruddin, S.A. An overview of photographic methods in monitoring non-aqueous phase liquid migration in porous medium. *Spec. Top. Rev. Porous Med. Int. J.* **2015**, *6*, 367–381. [CrossRef]
3. Huling, S.G.; Weaver, J.W. *Ground Water Issue: Dense Nonaqueous Phase Liquids*; US Environmental Protection Agency: Washington, DC, USA, 1991.
4. Alazaiza, M.Y.D.; Ngien, S.K.; Bob, M.M.; Kamaruddin, S.A.; Ishak, W.M.F. Non-aqueous phase liquids distribution in three-fluid phase systems in double-porosity soil media: Experimental investigation using image analysis. *Groundw. Sustain. Dev.* **2018**, *7*, 133–142. [CrossRef]
5. Ngien, S.K.; Rahman, N.A.; Bob, M.M.; Ahmad, K.; Sa'ari, R.; Lewis, R.W. Observation of light non-aqueous phase liquid migration in aggregated soil using image analysis. *Trans. Porous Med.* **2012**, *92*, 83–100. [CrossRef]
6. Amr, S.S.A.; Alazaiza, M.Y.D.; Bashir, M.J.; Alkarkhi, A.F.; Aziz, S.Q. The performance of $S_2O_8^{2-}/Zn^{2+}$ oxidation system in landfill leachate treatment. *Phys. Chem. Earth* **2020**, *120*, 102944. [CrossRef]
7. Alazaiza, M.Y.D.; Ngien, S.K.; Bob, M.M.; Kamaruddin, S.A.; Ishak, W.M.F. Quantification of dense nonaqueous phase liquid saturation in double-porosity soil media using a light transmission visualization technique. *J. Porous Med.* **2017**, *20*, 591–606. [CrossRef]
8. Pankow, J.F.; Cherry, J.A. *Dense Chlorinated Solvents and Other DNAPLs in Groundwater: History, Behavior, and Remediation*; Waterloo Press: Portland, OR, USA, 1996.
9. Sale, T.; Newell, C.J. Impacts of Source Management on Chlorinated Solvent Plumes. In *Situ Remediation of Chlorinated Solvent Plumes*; Springer: Berlin/Heidelberg, Germany, 2010.
10. Engelmann, C.; Händel, F.; Binder, M.; Yadav, P.K.; Dietrich, P.; Liedl, R.; Walther, M. The fate of DNAPL contaminants in non-consolidated subsurface systems–Discussion on the relevance of effective source zone geometries for plume propagation. *J. Hazard. Mater.* **2019**, *375*, 233–240. [CrossRef] [PubMed]
11. Aydin, G.A.; Agaoglu, B.; Kocasoy, G.; Copty, N.K. Effect of temperature on cosolvent flooding for the enhanced solubilization and mobilization of NAPLs in porous media. *J. Hazard. Mater.* **2011**, *186*, 636–644. [CrossRef] [PubMed]
12. Kokkinaki, A.; O'Carroll, D.; Werth, C.J.; Sleep, B. An evaluation of Sherwood–Gilland models for NAPL dissolution and their relationship to soil properties. *J. Contam. Hydrol.* **2013**, *155*, 87–98. [CrossRef]
13. Karaoglu, A.G.; Copty, N.K.; Akyol, N.H.; Kilavuz, S.A.; Babaei, M. Experiments and sensitivity coefficients analysis for multiphase flow model calibration of enhanced DNAPL dissolution. *J. Contam. Hydrol.* **2019**, *225*, 103515. [CrossRef]

14. Alazaiza, M.Y.D.; Copty, N.K.; Abunada, Z. Experimental investigation of cosolvent flushing of DNAPL in double-porosity soil using light transmission visualization. *J. Hydrol.* **2020**, *584*, 124659. [CrossRef]
15. Niemet, M.R.; Selker, J.S. A new method for quantification of liquid saturation in 2D translucent porous media systems using light transmission. *Adv. Water Resour.* **2001**, *24*, 651–666. [CrossRef]
16. Bob, M.M.; Brooks, M.C.; Mravik, S.C.; Wood, A.L. A modified light transmission visualization method for DNAPL saturation measurements in 2-D model. *Adv. Water Resour.* **2008**, *31*, 727–742. [CrossRef]
17. Belfort, B.; Weill, S.; Lehmann, F. Image analysis method for the measurement of water saturation in a two-dimensional experimental flow tank. *J. Hydrol.* **2017**, *550*, 343–354. [CrossRef]
18. Alazaiza, M.Y.D.; Ramli, M.H.; Copty, N.K.; Sheng, T.J.; Aburas, M.M. LNAPL saturation distribution under the influence of water table fluctuations using simplified image analysis method. *Bull. Eng. Geol. Environ.* **2020**, *79*, 1543–1554. [CrossRef]
19. Kamon, M.; Endo, K.; Katsumi, T. Measuring the k–S–p relations on DNAPLs migration. *Eng. Geol.* **2003**, *70*, 351–363. [CrossRef]
20. Tidwell, V.C.; Glass, R.J. X ray and visible light transmission for laboratory measurement of two-dimensional saturation fields in thin-slab systems. *Water Resour. Res.* **1994**, *30*, 2873–2882. [CrossRef]
21. Alazaiza, M.Y.D.; Ramli, M.H.; Copty, N.K.; Ling, M.C. Assessing the impact of water infiltration on LNAPL mobilization in sand column using simplified image analysis method. *J. Contam. Hydrol.* **2021**, *238*, 103769. [CrossRef]
22. Yilmaz, I.T.; Gumus, M.; Akçay, M. Thermal barrier coatings for diesel engines. In Proceedings of the International Scientific Conference, Gabrovo, Bulgaria, 19–20 November 2010; pp. 19–20.
23. Gallage, C.; Kodikara, J.; Uchimura, T. Laboratory measurement of hydraulic conductivity functions of two unsaturated sandy soils during drying and wetting processes. *Soils Found* **2013**, *53*, 417–430. [CrossRef]
24. Liu, Q.; Yasufuku, N.; Miao, J.; Ren, J. An approach for quick estimation of maximum height of capillary rise. *Soils Found* **2014**, *54*, 1241–1245. [CrossRef]
25. Day, R.W. Foundation Engineering Handbook. In *Design and Construction with 2006 International Building Code*; McGraw-Hill: Sydney, Australia, 2005.
26. Terzaghi, K.; Peck, R. *Soil Mechanics in Engineering Practice*; Wiley: New York, NY, USA, 1948.
27. Simantiraki, F.; Aivalioti, M.; Gidarakos, E. Implementation of an image analysis technique to determine LNAPL infiltration and distribution in unsaturated porous media. *Desalination* **2009**, *248*, 705–715. [CrossRef]
28. Alazaiza, M.Y.D.; Ngien, S.K.; Copty, N.; Bob, M.M.; Kamaruddin, S.A. Assessing the influence of infiltration on the migration of light non-aqueous phase liquid in double-porosity soil media using a light transmission visualization method. *Hydro. J.* **2019**, *27*, 581–593. [CrossRef]

Article

Modeling of the Effects of Pleat Packing Density and Cartridge Geometry on the Performance of Pleated Membrane Filters

Dave Persaud [1,†], Mikhail Smirnov [1,†], Daniel Fong [2,‡] and Pejman Sanaei [3,*]

1. Department of Electrical and Computer Engineering, New York Institute of Technology, New York, NY 10023-7692, USA; dpersa06@nyit.edu (D.P.); msmirnov@nyit.edu (M.S.)
2. Department of Mathematics and Science, U.S. Merchant Marine Academy, Kings Point, NY 11024, USA; Daniel.Fong@usmma.edu
3. Department of Mathematics, New York Institute of Technology, New York, NY 10023-7692, USA
* Correspondence: psanaei@nyit.edu
† These authors contributed equally to this work.
‡ The views expressed in this article are D. Fong's own and not those of the U.S. Merchant Marine Academy, the Maritime Administration, the Department of Transportation, or the United States government.

Abstract: Pleated membrane filters are widely used to remove undesired impurities from a fluid in many applications. A filter membrane is sandwiched between porous support layers and then pleated and packed into an annular cylindrical cartridge with a central hollow duct for outflow. Although this arrangement offers a high surface filtration area to volume ratio, the filter performance is not as efficient as those of equivalent flat filters. In this paper, we use asymptotic methods to simplify the flow throughout the cartridge to systematically investigate how the number of pleats or pleat packing density affects the performance of the pleated membrane filters. The model is used to determine an optimal number of pleats in order to achieve a particular optimum filtration performance. Our findings show that only the "just right"—neither too few nor too many—number of pleats gives optimum performance in a pleated filter cartridge.

Keywords: filtration; porous media; pleated membrane filters; mathematical modeling

1. Introduction

Filtration is the process of separating contaminants from a liquid or gas by using filter membranes. There are a multitude of different applications of membrane filters, but some of the most common examples include use in the pharmaceutical industry, biotechnology, semiconductor fabrication, water treatment, food production, mask production, vacuum cleaners and air purifiers [1–8]. Filters are generally made of porous membranes, whose internal morphologies and pore geometry affect the filtration efficiency [9–15]. In addition, the solvent and contaminants' properties may alter the filtration performance. In the case of solid particles, the most practical method of filtration is based on the geometry and shapes of the particles, dictating the best choice of membranes for the filtration process. As a fluid flows through membrane filters, separation of particles can occur either from blockage and/or adsorption, depending on the particle sizes. If a particle enters a pore, in which its narrowest part is smaller than the particle's typical size, the pore will be blocked; otherwise, the particle can stick to the pore wall. Generally speaking, fouling mechanisms impede filter performance by *adsorption* (a collection of small particles stick to the pores' walls and shrink them); *complete blocking* (pores become fully blocked by large particles); *intermediate blocking* (medium-sized particles block a portion of the pores' volume); and by *caking* (particles at the pores' inlets accumulate and then form a cake layer on the membrane surface), as described in [10–14,16–18]. These mechanisms of fouling add more resistance to the flow, which requires a higher pressure gradient (or more energy) for the fluid to overcome the fouling and pass through the membrane [19–23]. It is important to note, as also later discussed in Section 2.2, that only two mechanisms of fouling are incorporated in this paper: small particle adsorption and blocking or sieving by large particles. It should also be noted that our model does not consider swelling of the porous medium due to water absorption, as discussed in [24].

Pleated membrane filters are considered as an efficient type of filter used in a variety of industrial applications such as drinking water production, vaccine purification, and sterilization of natural gases in order to remove undesired particles from the feed solution [11,25–27]. The three crucial design attributes of a pleated filter are the pleat height, pleat packing density, and the pleat type [28]. The pleat height of a filter is the depth of a given pleat; the packing density represents the number of pleats filled within the volume of a cartridge; the pleat type is the pattern in which the fabrics of the filter are folded to construct the pleats. There are various pleated filter geometries and we focus on the cylindrical one in this paper [28]. Typically, in a pleated membrane filter cartridge, the membrane is placed between two supporting layers and the resulting three layers are folded (pleated) and placed into an annular cylindrical cartridge.

The design and construction of a pleated filter is an important process, since it provides a fundamental understanding of the geometry and unique structure of the pleated filter, which in turn affects the filtration performance. Pleated filter cartridges are constructed by heating the membrane to a specific temperature such that the pleats in the membrane become permanent. While this process is taking place, the pleat height and the pleat type are also configured by the manufacturer. To create a cylindrical shape: (i) the ends of the pleats are sealed togetherand (ii) the lengths are adjusted accordingly in order to fit the dimensions of the cartridge. The pleats can then be packed into the cylindrical housing with both inlet and outlet units to complete the pleated membrane cartridge design [28,29].

Pleated membrane filters are more compact in comparison to equal-area nonpleated (flat) membrane filters. Therefore, they have a greater surface area to volume ratio [11,30,31]; however, this can also introduce various disadvantages. As shown in previous studies [11,31], the performance of the pleated membrane filters are inferior compared to the flat (unpleated) filters with equivalent membrane surface areas. This stems from several factors, such as the additional resistance to the overall system due to the pleat packing density (PPD), as well as the support layers, the complex fluid dynamics within the pleated membrane, and possible damage of the membrane occurring during the pleating process. Multiple studies have demonstrated that membrane permeability is not the only factor that affects the overall filtration performance. Dense pleating can also cause a significant decrease in the filtration efficiency of over 50% in comparison to a flat sheet medium with similar permeability and an equivalent surface area [32–34]. While pleated membrane filters have a larger surface area than the flat membrane filters, a greater amount of pleats (or higher PPD) worsens the filter performance by increasing the overall system resistance. Therefore, this presents the question on how to optimize the filter cartridge geometry and the pleat packing density in such a way that the filter yields maximum performance.

Recent works [11,31] only attempted to answer part of this question (the effects of pleat geometry on the filter performance) using highly simplified models that focused only on a 2D flow through the membrane, meaning that the models did not account for any variation along the filter cartridge axial direction. However, our new quasi-3D model (the schematic shown in Figure 1a) addresses the mentioned extension by including two additional regions to make the model more realistic: (i) the empty area, where the flow enters the cartridge and then passes across the pleated membrane; (ii) the hollow region, which is a central hollow duct for outflow. In total, the empty area, the pleated membrane and the hollow region are all incorporated in our model. In this paper, we focus on the pleat packing density and the filter cartridge's geometry as the main characteristics of a pleated membrane filter to achieve the optimal filtration performance.

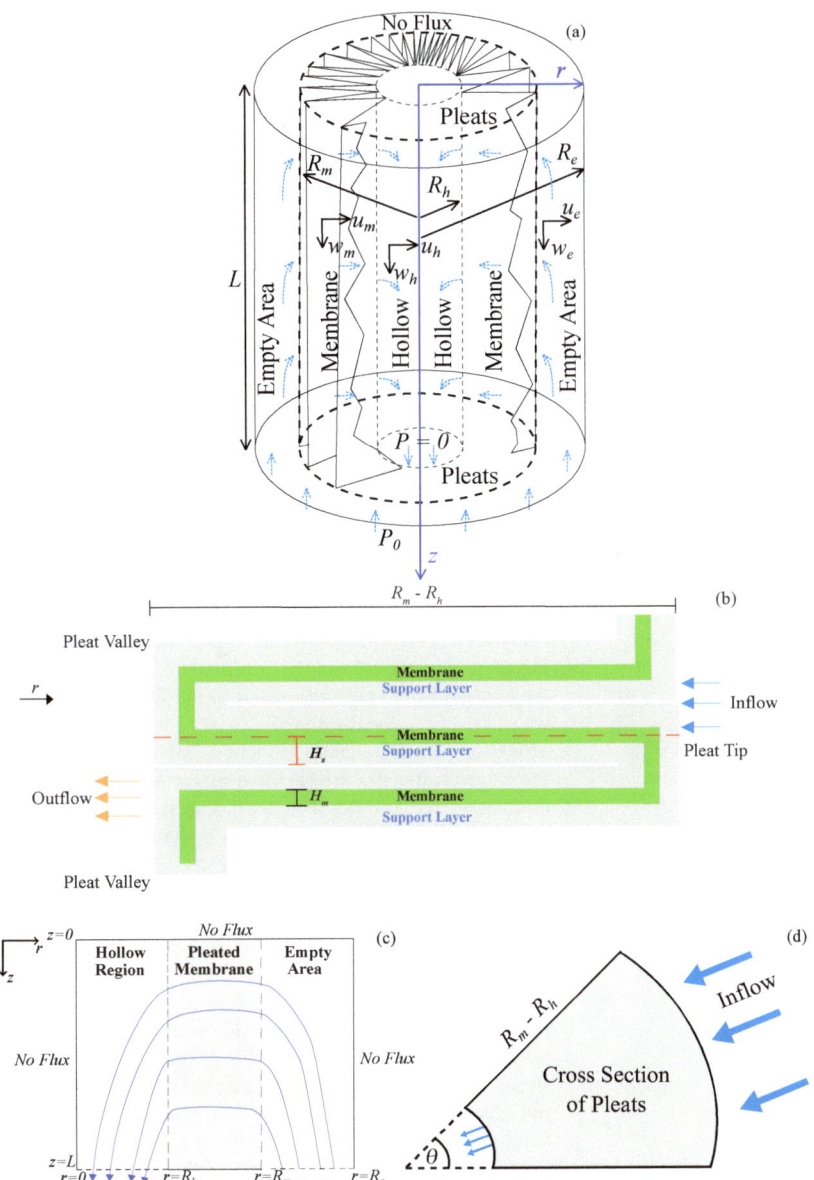

Figure 1. (**a**) Schematic of the pleated filter cartridge with height L, containing (i) a hollow duct with radius R_h; (ii) a pleated membrane with the pleat length of $R_m - R_h$; and (iii) an empty area with an outer radius R_e. (**b**) A close-up of the pleats with the membrane layer and two support layers with thicknesses of H_m and H_s, respectively. (**c**) The inlet specified pressure drop is P_0 at $z = L$ and $R_m < r < R_e$, while the outlet specified drop is 0 at $z = L$ and $0 < r < R_h$. The flow velocity is (u_i, w_i), where $i \in \{h, m, e\}$ are for the hollow region, membrane and empty area, respectively. A no-flux condition is applied at the top of the cartridge ($z = 0$), as well as its center and walls ($r = 0$ and $r = R_e$). (**d**) Cross section of the filter cartridge with pleat length of $R_m - R_h$, where θ represents the angle that sweeps across the given pleat. The filtrate enters from the empty area and travels through the pleat, exiting to the hollow region.

The paper is first laid out in Section 2 by describing the flow through the hollow region and the empty area with the incompressible axisymmetric Stokes equations for low Reynolds numbers, while the flow in the membrane area is modeled by Darcy's law. Each of these equations is given their respective boundary conditions based on the physical geometry. Next, we nondimensionalize the model in Section 2.1.1. In Section 2.1.2, we exploit the small aspect ratios of the filter cartridge and pleats and use asymptotic methods to simplify the system of equations. Then, we arrive at coupled boundary value problems for pressure and velocities in each of the three regions, as well as the permeability of the membrane and support layers, evolving with space and quasi-statically with time. These equations are solved numerically and presented in Section 3 to ultimately demonstrate the effects of pleat packing density and cartridge geometry on the filter performance. Finally, the discussion of the model and results are concluded in Section 4 in the context of industrial pleated membrane filters and of future modeling directions.

2. Mathematical Description

We consider the filtration flow problem through a pleated membrane filter cartridge consisting of a central duct surrounded by an annular pleated membrane (see Figure 1a), resulting in three different regions of the model: hollow duct, pleated membrane, and empty area. We consider the cartridge to have high pleat packing density, while assuming each pleat to be identical and part of a periodic array along the arc of the cartridge. To make the problem amenable to analytical techniques while still capturing key features of the PPD and the cartridge geometry on the filtration efficiency, we introduce an axisymmetric coordinate system, (r, z). We denote the radius of the hollow region as R_h, the length of the pleat (from the inner to outer cartridge boundary) as $R_m - R_h$, and the total radius of the pleated filter cartridge given as R_e. The height of the pleated filter cartridge is denoted by L, and we assume that $R_h/L \ll 1$ and $R_m/L \ll 1$. The porous membrane consists of a membrane layer sandwiched between two support layers, as shown in Figure 1b. The membrane thickness H_m and the support layer thickness H_s are both assumed to be much smaller than the pleat length, $R_m - R_h$. In addition, the support layer is much thicker than the membrane, which is typical in pleated membrane filters [28,29], specifically, $H_m \ll H_s \ll R_m - R_h$ (see Figure 1b).

With specified pressure P_0, the solvent enters from the cartridge inlet at the bottom of the empty area, $R_m < r < R_e, z = L$, to flow upward, travels through the membrane $R_h < r < R_m, 0 < z < L$, and flows down the hollow region towards the outlet, $0 < r < R_h$, $z = L$, where the pressure is set to 0 (see Figure 1c). It is important to note that the pressure in each of the three regions is continuous at each interface. The top of the cartridge at $z = 0$, the sides of the empty area $r = R_e$, and the center of the cartridge, $r = 0$ are all set to have no-flux boundary conditions. Since the geometry has symmetry along the centerline, we show only the right half of the pleated membrane filter cartridge in Figure 1c. While we do not account for angular dependency θ in our model, we introduce this angle, which sweeps across a pleat in order to derive an equation for the average of the support layers and membrane permeabilities (details of the derivation are included in Appendix A). Figure 1d shows an arbitrary cross section of the pleated membrane filter with angle θ and pleat length $R_m - R_h$.

2.1. Fluid Transport

We assume the feed is a dilute suspension of particles, which we treat as an incompressible Newtonian fluid. We denote the fluid velocity and pressure by $\mathbf{u} = (u, w)$ and p, with subscripts h, m and e denoting the hollow region, membrane and empty area, respectively. There are two time scales in our model: the first one is related to the filtration flow velocity and the second one is originated from the rate of change in pore radius due to particles deposition. Since the time scale of membrane morphology change due to particle deposition is much longer compared to that of flow velocity, we employ a quasi-static model for the flow here. We describe fluid flow in the hollow region ($0 < r < R_h$, $0 < z < L$) by the Stokes equations for an incompressible Newtonian fluid with a low Reynolds number. In axisymmetric cylindrical coordinates, the Stokes equations yield is

$$\frac{\partial p_h}{\partial r} = \mu \left[\frac{1}{r}\frac{\partial}{\partial r}\left(r\frac{\partial u_h}{\partial r}\right) + \frac{\partial^2 u_h}{\partial z^2} - \frac{u_h}{r^2} \right], \tag{1}$$

$$\frac{\partial p_h}{\partial z} = \mu \left[\frac{1}{r}\frac{\partial}{\partial r}\left(r\frac{\partial w_h}{\partial r}\right) + \frac{\partial^2 w_h}{\partial z^2} \right], \tag{2}$$

where μ is the viscosity of the fluid. For incompressible axisymmetric flow, it follows that the continuity equation becomes

$$\frac{1}{r}\frac{\partial}{\partial r}\left(ru_h\right) + \frac{\partial w_h}{\partial z} = 0. \tag{3}$$

The pleated membrane occupies $R_h < r < R_m$, $0 < z < L$. We model the fluid transport using Darcy's law for the flow in porous media,

$$\mathbf{u}_m = (u_m, w_m) = -\frac{k}{\mu}\nabla p_m, \quad \nabla = (\partial_r, \partial_z), \tag{4}$$

where $k(r,z)$ is the average of the support layers and membrane permeabilities. Incompressibility of the flow requires

$$\frac{1}{r}\frac{\partial}{\partial r}\left(ru_m\right) + \frac{\partial w_m}{\partial z} = 0. \tag{5}$$

In the empty area ($R_m < r < R_e$, $0 < z < L$), the fluid is also modeled using the incompressible axisymmetric Stokes equations

$$\frac{\partial p_e}{\partial r} = \mu \left[\frac{1}{r}\frac{\partial}{\partial r}\left(r\frac{\partial u_e}{\partial r}\right) + \frac{\partial^2 u_e}{\partial z^2} - \frac{u_e}{r^2} \right], \tag{6}$$

$$\frac{\partial p_e}{\partial z} = \mu \left[\frac{1}{r}\frac{\partial}{\partial r}\left(r\frac{\partial w_e}{\partial r}\right) + \frac{\partial^2 w_e}{\partial z^2} \right], \tag{7}$$

$$\frac{1}{r}\frac{\partial}{\partial r}\left(ru_e\right) + \frac{\partial w_e}{\partial z} = 0. \tag{8}$$

We assume the flow is driven by an imposed pressure difference, P_0, at the inlet $z = L$ in the empty area ($R_m < r < R_e$) and exits the filter through the outlet at $z = L$ in the hollow region ($0 < r < R_h$),

$$p_e = P_0, \quad p_h = 0 \quad \text{at} \quad z = L, \tag{9}$$

with the symmetry conditions in the hollow region

$$u_h = 0, \quad \frac{\partial w_h}{\partial r} = 0 \quad \text{at} \quad r = 0. \tag{10}$$

The housing boundaries $z = 0$ and $r = R_e$ are solid boundaries on which we impose no flux; therefore,

$$u_h = w_h = u_m = w_m = u_e = w_e = 0 \quad \text{at} \quad z = 0,$$
$$u_e = w_e = 0 \quad \text{at} \quad r = R_e. \tag{11}$$

At the interface between the hollow region and membrane, we impose continuity of pressure and velocity,

$$p_h = p_m, \quad u_h = \phi u_m \quad \text{at} \quad r = R_h, \tag{12}$$

where ϕ is the local pleated membrane porosity [35] (assumed constant). Similarly, at the interface between the membrane and empty area, we impose continuity of pressure and velocity,

$$p_e = p_m, \quad u_e = \phi u_m \quad \text{at} \quad r = R_m. \tag{13}$$

The last boundary condition represents a nonvanishing tangential slip velocity [35,36]

$$\begin{aligned}\frac{\partial w_h}{\partial r} &= -\frac{\alpha}{\sqrt{k(r,z)}}(w_h - w_m) \quad \text{at} \quad r = R_h, \\ \frac{\partial w_e}{\partial r} &= -\frac{\alpha}{\sqrt{k(r,z)}}(w_e - w_m) \quad \text{at} \quad r = R_m,\end{aligned} \tag{14}$$

where α is a dimensionless slip constant that depends on the surface properties. The flux q through the filter boundary at $z = L$ is defined by

$$q = \int_0^{R_h} 2\pi r ||\mathbf{u}_h||\Big|_{z=L} dr = \int_0^{R_h} 2\pi r \sqrt{u_h^2 + w_h^2}\Big|_{z=L} dr. \tag{15}$$

2.1.1. Nondimensionalization

Motivated by the small aspect ratio of the pleated filter cartridge, i.e., $\epsilon = R_m/L \ll 1$, we introduce dimensionless variables as follows, using hats to denote dimensionless variables:

$$\begin{aligned}(r,z) &= L(\epsilon \hat{r}, \hat{z}), \quad (p_h, p_m, p_e) = P_0(\hat{p}_h, \hat{p}_m, \hat{p}_e), \\ (u_h, w_h) &= \frac{LP_0\epsilon^2}{\mu}(\epsilon \hat{u}_h, \hat{w}_h), \quad (u_m, w_m) = \frac{KP_0}{R_m\mu}(\hat{u}_m, \hat{w}_m), \\ \phi &= \frac{L^2\epsilon^4}{K}\hat{\phi}, \quad (u_e, w_e) = \frac{LP_0\epsilon^2}{\mu}(\epsilon \hat{u}_e, \hat{w}_e), \\ k(r,z) &= K\hat{k}(\hat{r}, \hat{z}), \quad q = \frac{2\pi P_0 R_m^4}{L\mu}\hat{q},\end{aligned} \tag{16}$$

where K is the average of the support layers and the membrane permeabilities. We define the parameter $l = R_h/R_m$, which characterizes the dimensionless ratio of the inner to the outer radii of the pleated membrane, and we scale the radius of the external housing $R_e = R_m(1 + \epsilon^*)$, where $\epsilon^* \ll 1$. Dropping hats, (1)–(3) give the dimensionless system of equations in the hollow region ($0 < r < l$, $0 < z < 1$)

$$\frac{\partial p_h}{\partial r} = \epsilon^2 \frac{1}{r}\frac{\partial}{\partial r}\left(r\frac{\partial u_h}{\partial r}\right) + \epsilon^4 \frac{\partial^2 u_h}{\partial z^2} - \epsilon^2 \frac{u_h}{r^2}, \tag{17}$$

$$\frac{\partial p_h}{\partial z} = \frac{1}{r}\frac{\partial}{\partial r}\left(r\frac{\partial w_h}{\partial r}\right) + \epsilon^2 \frac{\partial^2 w_h}{\partial z^2}, \tag{18}$$

$$\frac{1}{r}\frac{\partial}{\partial r}(ru_h) + \frac{\partial w_h}{\partial z} = 0. \tag{19}$$

Similarly, the porous membrane Equations (4) and (5) become ($l < r < 1$, $0 < z < 1$)

$$u_m = -k(r,z)\frac{\partial p_m}{\partial r}, \tag{20}$$

$$w_m = -\epsilon k(r,z)\frac{\partial p_m}{\partial z}, \tag{21}$$

$$\frac{1}{r}\frac{\partial}{\partial r}(ru_m) + \epsilon \frac{\partial w_m}{\partial z} = 0. \tag{22}$$

In the empty area ($1 < r < 1 + \epsilon^*$, $0 < z < 1$), the dimensionless equations,

$$\frac{\partial p_e}{\partial r} = \epsilon^2 \frac{1}{r}\frac{\partial}{\partial r}\left(r\frac{\partial u_e}{\partial r}\right) + \epsilon^4 \frac{\partial^2 u_e}{\partial z^2} - \epsilon^2 \frac{u_e}{r^2}, \tag{23}$$

$$\frac{\partial p_e}{\partial z} = \frac{1}{r}\frac{\partial}{\partial r}\left(r\frac{\partial w_e}{\partial r}\right) + \epsilon^2 \frac{\partial^2 w_e}{\partial z^2}, \tag{24}$$

$$\frac{1}{r}\frac{\partial}{\partial r}(ru_e) + \frac{\partial w_e}{\partial z} = 0, \tag{25}$$

are obtained from (6)–(8). The boundary conditions (9)–(14) give the pressure drop boundary conditions that drive the flow

$$p_e = 1, \quad p_h = 0 \quad \text{at} \quad z = 1; \tag{26}$$

the boundary conditions on the hollow region centerline

$$u_h = 0, \quad \frac{\partial w_h}{\partial r} = 0 \quad \text{at} \quad r = 0; \tag{27}$$

on the housing boundaries

$$\begin{aligned} u_h = w_h = u_m = w_m = u_e = w_e = 0 \quad \text{at} \quad z = 0, \\ u_e = w_e = 0 \quad \text{at} \quad r = 1 + \epsilon^*; \end{aligned} \tag{28}$$

at the interface between the hollow region and membrane

$$p_h = p_m, \quad u_h = \phi u_m \quad \text{at} \quad r = l; \tag{29}$$

at the interface between the membrane and the empty area

$$p_e = p_m, \quad u_e = \phi u_m, \quad w_h = 0 \quad \text{at} \quad r = 1; \tag{30}$$

and finally

$$\begin{aligned} \frac{\partial w_h}{\partial r} = -\frac{\epsilon \alpha_0}{\sqrt{k}}\left(w_h - \frac{\epsilon}{\phi_0}w_m\right) \quad \text{at} \quad r = l, \\ \frac{\partial w_e}{\partial r} = -\frac{\epsilon \alpha_0}{\sqrt{k}}\left(w_e - \frac{\epsilon}{\phi_0}w_m\right) \quad \text{at} \quad r = 1, \end{aligned} \tag{31}$$

where $\alpha_0 = L\alpha/\sqrt{K}$ and $\phi_0 = (L^2\epsilon^4)/K$. Note that we assume $\alpha_0 = \mathcal{O}(1/\epsilon^2)$ to be able to simplify the boundary conditions (31). The total dimensionless flux q is given by

$$q = \int_0^l r\sqrt{\epsilon^2 u_h^2 + w_h^2}\Big|_{z=1} dr. \tag{32}$$

2.1.2. Asymptotic Analysis for the Flow

We use asymptotic analysis to derive a reduced system of governing equations at the leading order. We begin by expanding each dependent variable as an asymptotic series in power of ϵ as follows. For example, we let $p_h = p_h^0 + \epsilon p_h^1 + \epsilon^2 p_h^2 + \cdots$ and similarly expand the pressures p_m, p_e; the velocities $u_h, w_h, u_m, w_m, u_e, w_e$; the membrane porosity ϕ and the membrane permeability k. At the leading order (omitting the superscript 0 for brevity), the Equations (17)–(31) for the hollow region, membrane and the empty area are reduced to

$$\begin{aligned} \frac{\partial p_h}{\partial r} = 0, \quad \frac{\partial p_h}{\partial z} = \frac{1}{r}\frac{\partial}{\partial r}\left(r\frac{\partial w_h}{\partial r}\right), \\ \frac{1}{r}\frac{\partial}{\partial r}(ru_h) + \frac{\partial w_h}{\partial z} = 0, \quad 0 < r < l, \ 0 < z < 1, \end{aligned} \tag{33}$$

$$u_m = -k\frac{\partial p_m}{\partial r}, \quad w_m = 0,$$
$$\frac{\partial}{\partial r}(ru_m) = 0, \quad l < r < 1, \quad 0 < z < 1, \tag{34}$$

$$\frac{\partial p_e}{\partial r} = 0, \quad \frac{\partial p_e}{\partial z} = \frac{1}{r}\frac{\partial}{\partial r}\left(r\frac{\partial w_e}{\partial r}\right),$$
$$\frac{1}{r}\frac{\partial}{\partial r}(ru_e) + \frac{\partial w_e}{\partial z} = 0, \tag{35}$$
$$1 < r < 1+\epsilon^*, \quad 0 < z < 1,$$

$$p_e = 1, \quad p_h = 0 \quad \text{at} \quad z = 1, \tag{36}$$

$$u_h = 0, \quad \frac{\partial w_h}{\partial r} = 0 \quad \text{at} \quad r = 0, \tag{37}$$

$$u_h = w_h = u_m = w_m = u_e = w_e = 0 \quad \text{at} \quad z = 0, \tag{38}$$

$$u_e = w_e = 0 \quad \text{at} \quad r = 1+\epsilon^*, \tag{39}$$

$$p_h = p_m, \quad u_h = \phi u_m, \quad w_h = 0 \quad \text{at} \quad r = l, \tag{40}$$

$$p_e = p_m, \quad u_e = \phi u_m, \quad w_e = 0 \quad \text{at} \quad r = 1, \tag{41}$$

where the hollow region pressure $p_h(z)$ and empty area pressure $p_e(z)$ only vary axially, obtained from (33) and (35), respectively. Note that we use the assumption $\alpha_0 = \mathcal{O}(1/\epsilon^2)$ to obtain the boundary conditions in (40) and (41) for the axial velocity in the hollow region and the empty area w_h and w_e, respectively. Integrating (33) and using boundary conditions (37) and (40) give the hollow region velocities u_h and w_h in terms of the hollow region pressure gradient as

$$u_h(r,z) = \frac{r}{16}(2l^2 - r^2)\frac{d^2 p_h}{dz^2},$$
$$w_h(r,z) = -\frac{1}{4}(l^2 - r^2)\frac{dp_h}{dz}. \tag{42}$$

Similarly, by integrating (35) and applying boundary conditions (39) and (41), we obtain the following equations for the empty area velocities u_e and w_e in terms of the empty area pressure gradient

$$u_e(r,z) = -\frac{1}{16}M(r)\frac{d^2 p_e}{dz^2},$$
$$w_e(r,z) = \frac{1}{4}\left(r^2 - 1 + \gamma \ln r\right)\frac{dp_e}{dz}, \tag{43}$$

where

$$\gamma = \frac{1-(1+\epsilon^*)^2}{\ln(1+\epsilon^*)}, \quad M(r) = r\left(r^2 - 2 + \gamma(2\ln r - 1)\right) -$$
$$\frac{(1+\epsilon^*)^2}{r}\left((1+\epsilon^*)^2 - 2 + \gamma(2\ln(1+\epsilon^*) - 1)\right). \tag{44}$$

Furthermore, using the boundary conditions (40) and (41) along with integrating of (34) yield the velocity u_m and the p_m in the membrane:

$$u_m(r,z) = \frac{\tilde{u}_m(z)}{r},$$
$$p_m(r,z) = p_e(z) + \tilde{u}_m(z)\int_r^1 \frac{dr'}{r' k(r',z)}, \tag{45}$$

where
$$\tilde{u}_m(z) = (p_h(z) - p_e(z))\left(\int_l^1 \frac{dr'}{r'k(r',z)}\right)^{-1}, \qquad (46)$$

is determined as part of the solution to the reduced model. At this stage, we now emphasize each quantity that depends on space and/or time. Combining (40), (41), (42), (43), (45) and (46) gives the resulting simplified model for $p_h(z,t)$ as

$$\frac{l^4}{16}\left(\int_l^1 \frac{dr'}{r'k(r',z,t)}\right)\frac{\partial^2 p_h(z,t)}{\partial z^2} - \phi p_h(z,t) = -\phi p_e(z,t), \qquad (47)$$

$$\frac{M(1)}{16}\left(\int_l^1 \frac{dr'}{r'k(r',z,t)}\right)\frac{\partial^2 p_e(z,t)}{\partial z^2} - \phi p_e(z,t) = -\phi p_h(z,t), \qquad (48)$$

subject to boundary conditions

$$\frac{\partial p_h}{\partial z}(0,t) = 0, \quad p_h(1,t) = 0. \qquad (49)$$

$$\frac{\partial p_e}{\partial z}(0,t) = 0, \quad p_e(1,t) = 1, \qquad (50)$$

obtained by (36), (38), (42) and (43). By adding (47) and (48) together, then integrating and using boundary conditions (49) and (50), the coupled system can be simplified further

$$p_e(z,t) = 1 - \frac{l^4}{M(1)} p_h, \qquad (51)$$

$$\frac{l^4}{16}\left(\int_l^1 \frac{dr'}{r'k(r',z,t)}\right)\frac{\partial^2 p_h(z,t)}{\partial z^2} - \left(1 + \frac{l^4}{M(1)}\right)\phi p_h(z,t) = -\phi,$$
$$\frac{\partial p_h}{\partial z}(0,t) = p_h(1,t) = 0. \qquad (52)$$

Note that, after solving (52), we can obtain the velocities and pressures in all three regions using (42)–(46). In addition, the total dimensionless flux of fluid flowing across the filter outlet, introduced in (32), becomes

$$q = \int_0^l rw_h|_{z=1}dr, \qquad (53)$$

where w_h can be obtained from (42) and (52).

2.1.3. Fluid Velocity and Streamfunction

For purpose of illustration, we show the visualization of the fluid flow through the membrane and the hollow region. Recall that the flow in our model is two-dimensional (axisymmetric assumption) and quasi-static; therefore, we employ streamfunction ψ_m and ψ_h in the membrane and the hollow region, respectively,

$$u_i(r,z) = -\frac{1}{r}\frac{\partial \psi_i}{\partial z}, \quad w_i(r,z) = \frac{1}{r}\frac{\partial \psi_i}{\partial r}, \quad i \in \{m,h\}. \qquad (54)$$

Using the flow velocities for the membrane and the hollow region, (42) and (45), respectively, along with (54), give us the streamfunctions in these two regions as

$$\psi_h(r,z) = \frac{1}{4}\left(l^2\frac{r^2}{2} - \frac{r^4}{4}\right)\frac{dp_h}{dz},$$
$$\psi_m(z) = -\int_0^z \frac{p_h(z') - p_e(z')}{\int_l^1 \frac{dr'}{r'k(r',z)}} dz'. \qquad (55)$$

2.2. Membrane Fouling

In general, the average of the support layer and membrane permeabilities $k(r,z,t)$ varies in both space and time as membrane fouling occurs. For a given average permeability $k(r,z,t)$, the boundary value problem (52) is solved to determine the velocities in both the hollow region and the membrane, as explained earlier. To close the model, we need an equation for the time evolution of the average permeability due to membrane fouling. Here, we present just the average permeability estimate (the derivation is provided in Appendix A),

$$k(r,z,t) \sim \frac{\mathcal{N} k_m(r,z,t) k_s(r,z)}{\mathcal{N}^2 k_m(r,z,t) + k_s(r,z)}, \quad \mathcal{N} = \frac{\pi(R_m + R_h)}{H_s}, \tag{56}$$

where $k_s(r,z)$ is the average permeability in the support layers, $k_m(r,z,t)$ is the permeability in the membrane and $\mathcal{N}$ is a dimensionless measure of packing density, which we call the packing density factor. Note that we assume the average permeability of the support layers $k_s(r,z)$ is not a function of time, since fouling does not happen in the support layers.

To derive an asymptotic model for membrane fouling with variations along the z direction, we modify the model described in Sanaei et al. [31] by introducing nonzero z-dependent pressure at the end the membrane. The membrane permeability k_m varies quasi-statically and is expressed as a function of the pore radius $a(z,t)$ and the number of unblocked pores per unit area $n(r,z,t)$. The model assumes that membrane pores are long, thin cylindrical tubes, spanning the membrane, which initially all have the same radius. A more sophisticated model would allow for nonuniform shrinkage of pores due to the adsorption, but in our simple model we assume uniform adsorption so that the pore radius does not vary spatially. The time evolution of the membrane permeability induces time variation for the pressure in the support layer p_s. Here, we just present the final reduced model with modifications to account for the axial-dependence pressure at the interface between the membrane and the hollow region, and refer the readers to Sanaei et al. [31] for more details. The full dimensionless system is

$$a(z,t) = 1 - \beta t,$$

$$k_m(r,z,t) = a(z,t)^4 \left(n(r,z,t) + \frac{1 - n(r,z,t)}{1 + \rho_b a(z,t)^4} \right), \tag{57}$$

$$\frac{\partial n(r,z,t)}{\partial t} = -n(r,z,t) a(z,t)^4 e^{-ba(z,t)} \left(2 p_s(r,z,t) + c_1(z,t) \int_r^1 \frac{dr'}{k_s(r',z)} + c_2(z,t) \right), \tag{58}$$

$$\frac{\partial}{\partial r}\left(k_s(r,z) \frac{\partial p_s(r,z,t)}{\partial r} \right) - 2\Gamma k_m(r,z,t) p_s(r,z,t) = \Gamma k_m(r,z,t) \left(c_1(z,t) \int_r^1 \frac{dr'}{k_s(r',z)} + c_2(z,t) \right), \tag{59}$$

subject to conditions

$$n(r,z,0) = 1, \tag{60}$$

$$p_s(1,z,t) = p_e(z,t), \quad \frac{\partial p_s}{\partial r}(1,z,t) = \frac{c_1(z,t)}{k_s(1,z)}, \tag{61}$$

$$p_s(l,z,t) = -\left(c_1(z,t) \int_0^1 \frac{dr'}{k_s(z,r')} + c_2(z,t) + p_h(z,t) \right),$$
$$\frac{\partial p_s}{\partial r}(l,z,t) = 0, \tag{62}$$

where p_s is the pressure in the support layer, $\beta = (8\mu E H_m)/(\pi a_0^4 P_0 g_\infty)$ is the dimensionless pore shrinkage rate representing the time scale in which pores close due to adsorption, relative to that in which particles block individual pores from upstream, g_∞ is the total large particle concentration and E is the rate of pore radius shrinkage. $\Gamma = K_{m0}(R_m - R_h)^2/(K_s H_m H_s)$ is a dimensionless measure of the relative importance of the resistance of the packing material to that of the membrane, b is the ratio of initial pore size to characteristic particle size and ρ_b is the dimensionless parameter that characterizes blocking strength.

3. Results

The model consists of various dimensional and dimensionless parameters shown in Tables 1 and 2, respectively, with typical ranges of values in practice. The fixed parameter values used in the simulation are shown in the caption of each figure (Figures 2–7). Our numerical scheme is based on second-order accurate finite difference spatial discretization of Equations (52) and (59), with a simple first-order implicit time step in the blocking Equation (58), and trapezoidal quadrature to find the integrals in (52), (53), (55), (58), (59) and (62). The solution scheme for this system is straightforward:

1. At $t = 0$, assign $k_m(r,z,0) = k_{m0} = 1$, then find $k(r,z,0)$ from (56) by using $k_s(r,z) = 1$ for a given $\mathcal{N}$.
2. Solve (52) for the pressure in the hollow region p_h, and then use the result in (51) to find the pressure in the empty area p_e.
3. Solve (59) along with the boundary conditions (61) and (62) in order to find p_s, c_1, and c_2.
4. Use these resulting values to update the number of unblocked pores per unit area n, in (58).
5. Use (57) to update the pore radius a, and the membrane permeability, k_m, respectively.
6. Return to step (1) and repeat until the membrane pore radius $a \to 0$.

Table 1. Dimensional parameter values [11,28,31,37–39].

Parameter	Description	Typical Value
L	Length of the pleats	0.2–0.5 m
R_h	Radius of the hollow region	1–1.5 cm
R_m	Radius of the membrane area	2–3 cm
R_e	Radius of the empty area	3–4.5 cm
H_s	Support layer thickness	1 mm
H_m	Membrane thickness	300 µm
P_0	Pressure drop	10–100 K Pa
K_s	Average support layer permeability	10^{-11} m^2
K_{m0}	Clean membrane permeability	5×10^{-13} m^2
K	Average of the support layer and clean membrane permeabilities	10^{-13} m^2

Table 2. Dimensionless parameters and approximate values [11,31,40].

Parameter	Formula	Typical Value
ϵ	R_m/L	0.04–0.15
l	R_h/R_m	0.1–1
ϕ	Pleated membrane porosity	0.01–0.9
Γ	$K_{m0}(R_m - R_h)^2/(K_s H_m H_s)$	1–100
$\mathcal{N}$	$\pi(R_m + R_h)/H_s$ Packing density factor	0.4–5
ρ_b	Blocking strength	0.25–10, 2 used here
b	Ratio of initial pore size to particle size	0.2–10, 0.5 used here
β	$(8\mu E H_m)/(\pi a_0^4 P_0 g_\infty)$	0.001–0.1, 0.02 used here
ϵ^*	$R_e/R_m - 1$	0.03–1.25

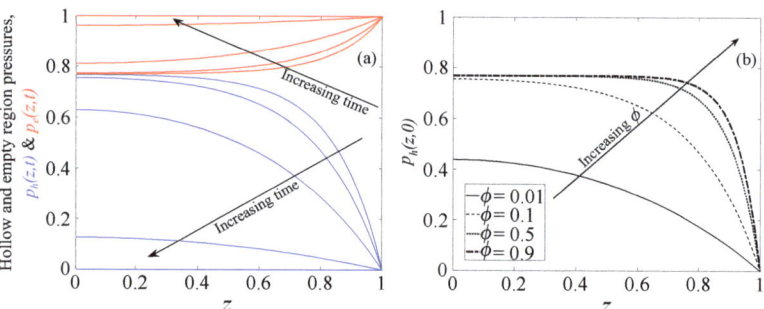

Figure 2. (a) Pressure in the hollow region $p_h(z,t)$ (blue) and empty area $p_e(z,t)$ (red), for $\phi = 0.5$ at several different times $t = 0$, $0.2t_f$, $0.4t_f$, $0.6t_f$, $0.8t_f$, and t_f (the final filtration time) with the membrane porosity $\phi = 0.5$. (b) The initial pressure in the hollow region $p_h(z,0)$ for several different values of local porosity, ϕ. In both figures, we use the parameters $l = 0.5$, $\Gamma = 10$, $\mathcal{N} = 1$, $\beta = 0.02$, $\rho_b = 2$, $b = 0.5$, and $\epsilon^* = 0.5$.

Figure 2a shows the hollow region and empty area pressures, p_h and p_e, respectively, as functions of z for several different values of time t, up to the final filtration time t_f, at which the filtration process stops. Our results demonstrate that the hollow region pressure attains its maximum value at the top of the cartridge ($z = 0$) and monotonically decreases along the positive z direction, which creates a pressure gradient that allows for downward flow for the entire duration. As time progresses, the fouling in the pleated membrane increases the system resistance, which in turn lowers the hollow region pressure until it eventually reaches a uniform value of 0 at the final time, t_f. The maximum pressure of the empty area instead happens at the bottom of the cartridge ($z = 1$) and reaches the minimum at the top of the cartridge ($z = 0$), which creates a pressure gradient that allows fluid to flow from the bottom of the cartridge to the top of the empty area, as the model intended. At a later stage of filtration, the empty area pressure becomes nearly constant along the entire length of the cartridge due to the complete fouling and blocking of the membrane. In Figure 2b, we investigated the effect of the local porosity ϕ on the initial pressure in the hollow region $p_h(z,0)$, along the axial direction of the pleated filter z. As expected, we observed that the higher pressure drop occurs with higher porosity. It is important to note that for every variation of ϕ, as z increases, the initial pressure in the hollow region tends to zero, which is consistent with the boundary condition in (52). For

$\phi = 0.9$, $p_h(z,0)$ is the largest and also witnesses the steepest drop in comparison to the other values of the local porosity used here.

We plot the streamlines of the hollow and membrane regions side by side in Figure 3 for $t = 0.4t_f$ and $0.8t_f$. It shows the fluid enters the membrane region (Figure 3b,d) at $r = 1$ from the empty area and flows into the hollow region (Figure 3a,c) and then moves out of the cartridge at $z = 1$. The flow in the membrane region is dominated by the radial direction at the leading order; as a result, the fluid travels horizontally through the membrane. In addition, as shown by the contour values, more of the fluid travels near the bottom of the cartridge due to the higher pressure difference across the membrane, which is consistent with the results of Figure 2a. On the other hand, the flow in the hollow region moves in both the radial and axial directions towards the filter outlet ($z = 1$). Our results also show that the streamlines in both regions have become more uniformly spread out along the axial direction as time evolves. This uniformity stems from the membrane fouling, which in turn causes the empty area and the hollow region pressures approach constant values of 1 and 0, respectively, as demonstrated in Figure 2a. Furthermore, the additional resistance due to the membrane fouling reduces the flow greatly in both the hollow and membrane regions at $t = 0.8t_f$ compared with $t = 0.4t_f$, as indicated by the color bars of Figure 3.

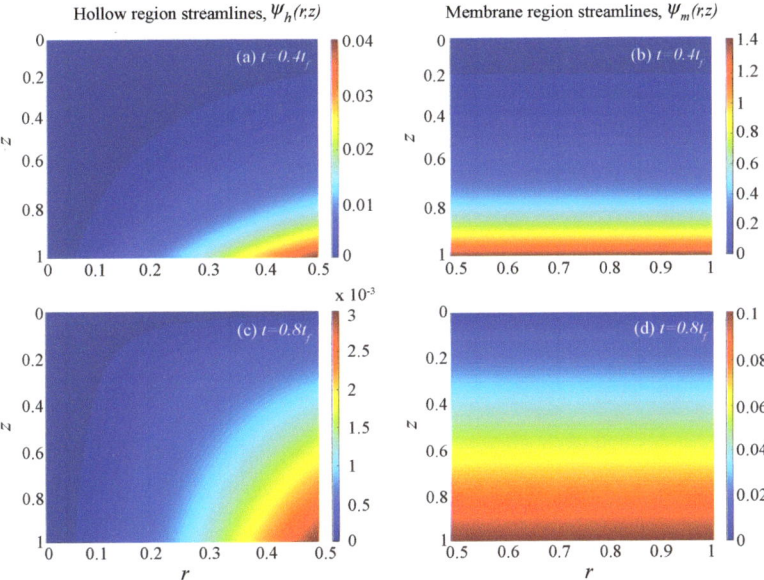

Figure 3. Contour plots showing the streamlines at time $t = 0.4t_f$ in (**a**) the hollow region and (**b**) in the membrane region. Plots (**c**,**d**) show the streamlines in the hollow region and membrane region, respectively, at time $t = 0.8t_f$. Parameter values are $\phi = 0.5$, $l = 0.5$, $\Gamma = 10$, $\mathcal{N} = 1$, $\beta = 0.02$, $\rho_b = 2$, $b = 0.5$, and $\epsilon^* = 0.5$.

Figure 4a,b show the pressure contour plot in the membrane region $p_m(r,z)$, at $t = 0.4t_f$ and $0.8t_f$, respectively. We observe that the pressure difference across the membrane has a strong axial dependency at early times (such as $t = 0.4t_f$) and becomes nearly independent of z as time evolves (e.g., $t = 0.8t_f$). This suggests more fluid travels through the membrane (or higher flow rate) near the bottom initially and gradually turns into uniform flow through the entire membrane, which is consistent with the results shown in Figure 3b,d.

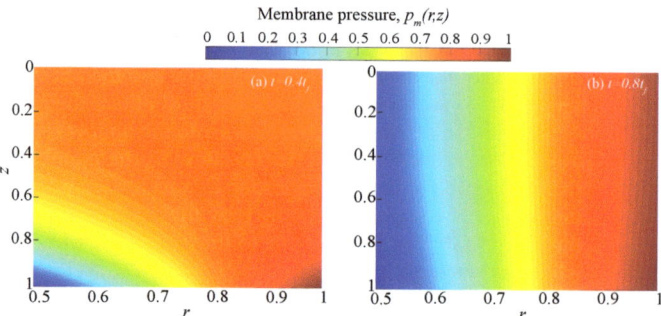

Figure 4. Contour plots showing pressure in the membrane region $p_m(r,z)$ at (**a**) $t = 0.4t_f$; and (**b**) $t = 0.8t_f$. Parameter values are $\phi = 0.5$, $l = 0.5$, $\Gamma = 10$, $\mathcal{N} = 1$, $\beta = 0.02$, $\rho_b = 2$, $b = 0.5$, and $\epsilon^* = 0.5$.

We now examine how the membrane permeability evolves as the membrane fouling occurs. This can provide us with more explanations on how the pressures and fluid streamlines change due to membrane fouling. We first plotted the membrane permeability at a very early time $t = 0.01t_f$, in Figure 5a. This figure indicates that the membrane permeability achieves its maximum at the very bottom of the cartridge $z = 1$, while the membrane becomes clogged more at the top $z = 0$, presumably due to the high fluid velocity caused by the higher pressure drop near the bottom, which results in less clogging. This figure also reveals that the membrane permeability contour curves upwards near the bottom of the membrane, suggesting that fouling occurs preferentially at the middle rather than the membrane pleat valleys and tips. This observation is the exact opposite of an earlier 2D simple model of the filter cartridge [31]. On the other hand, it is consistent with [11] in terms of membrane permeability symmetry about the midway of the membrane.

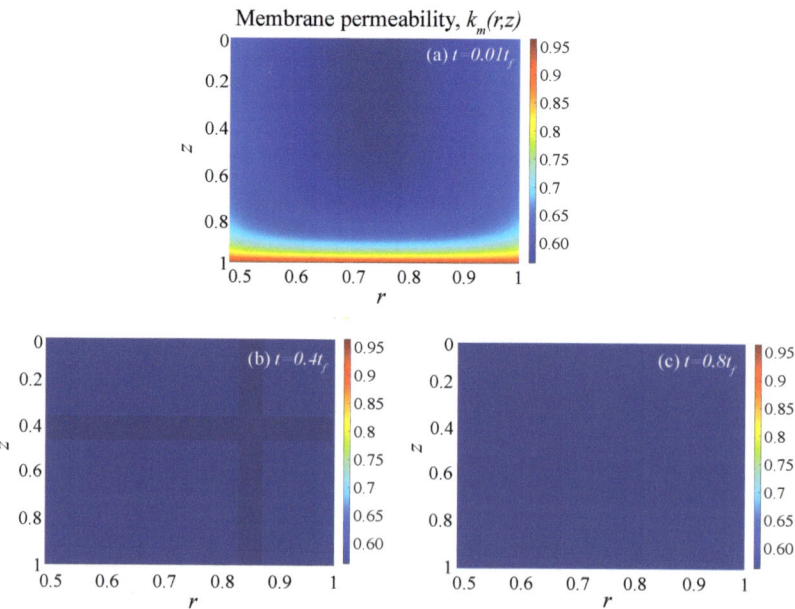

Figure 5. Contour plot showing membrane permeability $k_m(r,z)$ at times (**a**) $t = 0.01t_f$, (**b**) $t = 0.4t_f$, and (**c**) $t = 0.8t_f$. Parameter values are $\phi = 0.5$, $l = 0.5$, $\Gamma = 10$, $\mathcal{N} = 1$, $\beta = 0.02$, $\rho_b = 2$, $b = 0.5$, and $\epsilon^* = 0.5$.

Figure 5b,c show the membrane permeability at $t = 0.4t_f$ and $0.8t_f$, respectively (at the same time that the streamlines and membrane pressure were plotted earlier; see Figures 3 and 4). Our results show that after a moderate amount of fouling ($t = 0.4t_f$) as well as almost the end of the filter's life span ($t = 0.8t_f$), the membrane permeability becomes uniform over most of the (r, z) plane.

There are several ways to characterize the performance of a filter and, among them, making a flux-throughput graphs is one of the most common experimental approaches. The dimensionless flux of fluid flowing across the filter outlet $q(t)$ is given in (53) and the throughput is defined as the amount of filtrate fluid up to a certain time, i.e., $v(t) = \int_0^t q(t')dt'$. Note that it is important for a filter to have high flux, while it attains the maximize total throughput $v(t_f)$. In other words, the filter with the longer life span t_f and higher total throughput is the most desirable. Note that high flux in a filter can be achieved at the expense of having a low total throughput (e.g., the filter possesses a lower life span) or a filter may have a high throughput but in return very low flux (overall slower filtration). Both scenarios are costly in different ways, and usually in practice some compromise between these two is found [31,41]. Since the flux and throughput are both functions of time, they depend on the amount of fouling that occurs in the filter, where more fouling causes less flux and as a consequence less total throughput. As the resistance increases with time due to fouling, more energy is required to be put into the system to conserve the same initial efficiency of the filter. Therefore, it is more desirable in industry to have the flux remain as high as possible through the filter life cycle so that the minimal amount of energy will be used to overcome fouling [42,43].

Figure 6 shows flux $q(t)$ versus throughput $v(t)$, for several different values of membrane porosity ϕ. Our results demonstrate two drops for each curve: a very rapid drop at early times and a much more gradual decrease for the remaining time. The initial rapid drop may be interpreted as modeling the scenario where sieving (pore blockage from large particles) is dominant, while the latter gradual drop may be explained as fouling by adsorption (small particles stick to the pores and walls and shrink them) [4,15,17,18]. The simulated results show that the pleated filter's performance increases for larger values of ϕ, providing less resistance to the flow in the membrane.

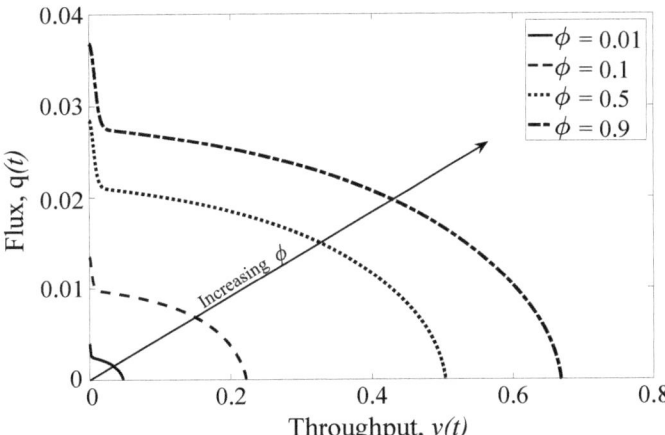

Figure 6. Flux $q(t)$, as a function of throughput $v(t) = \int_0^t q(t')dt'$, for several different values of porosity ϕ, with $l = 0.5$, $\Gamma = 10$, $\mathcal{N} = 1$, $\beta = 0.02$, $\rho_b = 2$, $b = 0.5$, and $\epsilon^* = 0.5$.

In order to examine the effects of the PPD and the filter cartridge geometry on the filtration performance, the packing density factor $\mathcal{N} = \pi(R_m + R_h)/H_s$ (defined in (56) previously) was varied in two different ways. We let the radius of the membrane area R_m vary and kept the support layer thickness H_s constant and vice versa in Figure 7a,b, respectively. Note that increasing R_m (equivalently $\mathcal{N}$) corresponds to a higher membrane surface area. These figures represent the total throughput $v(t_f)$ as a function of $\mathcal{N}$. Note that the dimensionless measure of the relative importance of the resistance of the packing material to that of the membrane ($\Gamma = K_{m0}(R_m - R_h)^2 / (K_s H_m H_s)$, see Table 2), also changed as we varied either R_m or H_s. Our results here demonstrate that the total throughput increases with R_m, as shown in Figure 7a. The physical significance of our results is that increasing the length of the pleats (as R_m increases), while keeping the support layer thickness constant, will allow more of the membrane's surface to be packed into the filter cartridge and, as a consequence, more throughput was achieved. Figure 7b investigates how the support layer thickness H_s, affects the total throughput. By decreasing H_s, the total throughput initially increases rapidly and peaks at $\mathcal{N} \approx 1.36$, but then decreases at a slower rate. Note that, initially, as H_s decreases, $\mathcal{N}$ increases, meaning that more pleats will be in the cartridge and therefore the throughput increases. Continuing to decrease H_s beyond a critical threshold will further increase the overall system resistance from the support layers. Even though there are more pleats in the filter cartridge (as $\mathcal{N}$ increases), the excessive resistance causes the total throughput to decay.

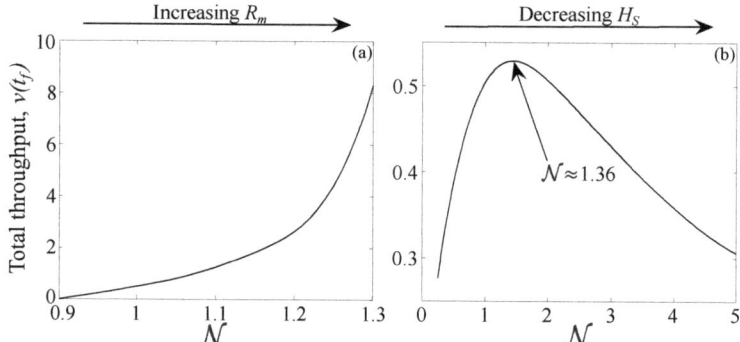

Figure 7. Total throughput $v(t_f)$ as a function of the packing density factor $\mathcal{N}$, with parameter values $\phi = 0.5, \beta = 0.02, \rho_b = 2, b = 0.5$ when: (**a**) the radius of the membrane area R_m and (**b**) the support layer thickness H_s, varies. In (**b**), $l = 0.5, \epsilon^* = 0.5$.

4. Discussion and Conclusions

In this work, we have developed a mathematical model for the fluid flow in a pleated membrane filter cartridge composed of three regions: empty area, pleated membrane, and central hollow duct (see Figure 1). We considered the effect of pressure variations along the axis of the pleated filter cartridge. Fluid was pumped into the empty area via the inlet, it passed through the pleated membrane and then flowed out of the exit into the hollow region. The fluid flow in the empty area and hollow region are governed by the axisymmetric Stokes equations and the pleated membrane region was modeled by the work of Sanaei et al. [31] with modifications to account for the axial pressure variations. The governing equations have been reduced through careful asymptotic analysis to exploit the separation of length scales inherent in the filtration flow problem through a pleated membrane filter cartridge. The resulting leading-order system of equations has been solved numerically.

Our model is able to describe the flow through the whole pleated filter cartridge. Figure 2a shows that the pressure gradients in the empty area and hollow region are positive and negative, respectively, which creates a flow driven through the membrane from the inlet to the outlet. In addition, our model demonstrates that the pressure difference across the membrane increases with the axial direction as well as time (see Figures 2 and 4). These were not investigated in the simple 2D models in the previous studies [11,31]. Our model also reveals the permeability variations in both the axial and radial directions within the membrane region in Figure 5. We examined the change in membrane permeability k_m

in the (r, z) plane at various times and found that there is a strong axial dependence at early times, which fades away as time evolves.

The main focus of this work is to understand the importance of the pleat packing density on the performance of pleated membrane filters. When determining the effectiveness of any membrane filter, there are multiple metrics which can be used. Here, we focus on the total throughput, which is the most common experimental characteristic in industry. Specifically speaking, we defined the packing density factor as $\mathcal{N} = \pi(R_m + R_h)/H_s$ in (56) and plotted the total throughput versus $\mathcal{N}$ in Figure 7. Our results show that the optimum filtration performance was achieved under two conditions: (i) when the length of the pleat was relatively long (compared with the cartridge housing radius), in order to increase the total membrane surface area, and (ii) when the optimal number of pleats were packed into the filter cartridge. Taken together, our findings show that only the "just right"—neither too few nor too many—number of pleats, ensures optimum performance in a pleated filter cartridge.

Our model is a valuable step towards a qualitative understanding of the effects of pleat packing density and axial-pressure variations on the filtration performance of pleated membrane filters, which can be used to inform future design. The results of our work in this paper are a novel extension to the earlier effort of Sanaei et al. [31]. Solving the full problem numerically (in the whole filtration region, including the inlet and outlet) would give great insight into the validity of the model reduction presented here. However, this would be very difficult numerically and computationally intensive, and is beyond the scope of this paper. Moreover, we focused on the geometry of the pleated membrane filter rather than particle capture efficiency and more complex fouling mechanisms. It would be an interesting extension of the model to incorporate the transport of particles, which can lead to multiple modes of fouling.

Author Contributions: D.P., M.S., D.F. and P.S. composited the manuscript, implemented and carried out the simulations. P.S. designed the study. All authors have read and agreed to the published version of the manuscript.

Funding: D.P., M.S. and P.S. are supported by an Institutional Support of Research and Creativity (ISRC) grant provided by New York Institute of Technology. D.P. and M.S. acknowledge the Edward Guiliano Global Fellowship from New York Institute of Technology.

Institutional Review Board Statement: Not applicable.

Informed Consent Statement: Not applicable.

Data Availability Statement: Data that support the findings of this study are available within the article. Further information is available from the corresponding author upon reasonable request.

Conflicts of Interest: The authors declare no conflict of interest.

Appendix A. Derivation of $k(r, z)$

In order to derive the expression for the average of the support layers and membrane permeabilities $k(r, z)$, given in (56), we need to balance the resistance across a pleat, as shown in Figure 1d. Note that the resistance of the membrane, for the top and bottom support layers, are in series with each other; therefore, the summation of these two resistances is equal to the average resistance of the support layers and membrane:

$$\frac{1}{k(r,z)\pi(R_m^2 - R_h^2)\theta} = \frac{2}{K_s(R_m - R_h)H_s} + \frac{1}{K_m(R_m - R_h)H_m}, \quad \text{(A1)}$$

where K_s is the average support layer permeability, K_m is a typical initial clean membrane permeability, and θ is the angle that sweeps a pleat in the cartridge, which is $\mathcal{O}(2 \times 10^{-2})$. Note that the resistance is the reciprocal of permeability times area, which was used in deriving (A1). By using the scaling for the membrane and support layers permeabilities given in [31], as well as hats to denote dimensionless variables, we obtain

$$k_s(r,z) = K_s \hat{k}_s(\hat{r}, \hat{z}), \quad k_m(r,z,t) = K_m \hat{k}_m(\hat{r}, \hat{z}, \hat{t}), \qquad (A2)$$

along with the scalings in (16). Dropping hats, (A1) simplifies to

$$\frac{1}{k} = \frac{2K\theta}{K_s} \frac{\mathcal{N}}{k_s} + \frac{\pi^2 K\theta (R_m + R_h)^2}{K_m H_m H_s} \frac{1}{\mathcal{N} k_m}. \qquad (A3)$$

Assuming that $\theta \sim \mathcal{O}(2 \times 10^{-2})$ and using the dimensional parameter values given in Table 2, we come to the conclusion that $\frac{2K\theta}{K_s}$ and $\frac{\pi^2 K\theta (R_m + R_h)^2}{K_m H_s H_m}$ both are $\mathcal{O}(1)$; therefore, (A3) gives

$$k(r,z,t) \sim \frac{\mathcal{N} k_m(r,z,t) k_s(r,z)}{\mathcal{N}^2 k_m(r,z,t) + k_s(r,z)}, \quad \mathcal{N} = \frac{\pi (R_m + R_h)}{H_s}, \qquad (A4)$$

as in (56).

References

1. Noble, R.D.; Stern, S.A. *Membrane Separations Technology: Principles and Applications*; Elsevier: Amsterdam, The Netherlands, 1995.
2. Bowen, W.R.; Jenner, F. Theoretical descriptions of membrane filtration of colloids and fine particles: An assessment and review. *Adv. Colloid Interface Sci.* **1995**, *56*, 141–200. [CrossRef]
3. Lonsdale, H. The growth of membrane technology. *J. Membr. Sci.* **1982**, *10*, 81–181. [CrossRef]
4. Sanaei, P.; Cummings, L.J. Flow and fouling in membrane filters: Effects of membrane morphology. *J. Fluid Mech.* **2017**, *818*, 744–771. [CrossRef]
5. Printsypar, G.; Bruna, M.; Griffiths, I.M. The influence of porous media microstructure on filtration. *J. Fluid Mech.* **2018**, *861*, 484–516. [CrossRef]
6. O'Dowd, K.; Nair, K.M.; Forouzandeh, P.; Mathew, S.; Grant, J.; Moran, R.; Bartlett, J.; Bird, J.; Pillai, S.C. Face masks and respirators in the fight against the COVID-19 pandemic: A review of current materials, advances and future perspectives. *Materials* **2020**, *13*, 3363. [CrossRef]
7. Gu, B.; Renaud, D.; Sanaei, P.; Kondic, L.; Cummings, L. On the influence of pore connectivity on performance of membrane filters. *J. Fluid Mech.* **2020**, *902*, A5. [CrossRef]
8. Griffiths, I.; Mitevski, I.; Vujkovac, I.; Illingworth, M.; Stewart, P. The role of tortuosity in filtration efficiency: A general network model for filtration. *J. Membr. Sci.* **2020**, *598*, 117664. [CrossRef]
9. Sanaei, P.; Cummings, L.J. Membrane filtration with complex branching pore morphology. *Phys. Rev. Fluids* **2018**, *3*, 094305. [CrossRef]
10. Griffiths, I.; Kumar, A.; Stewart, P. A combined network model for membrane fouling. *J. Colloid Interface Sci.* **2014**, *432*, 10–18. [CrossRef]
11. Sun, Y.; Sanaei, P.; Kondic, L.; Cummings, L.J. Modeling and design optimization for pleated membrane filters. *Phys. Rev. Fluids* **2020**, *5*, 044306. [CrossRef]
12. Sanaei, P.; Cummings, L.J. Membrane filtration with multiple fouling mechanisms. *Phys. Rev. Fluids* **2019**, *4*, 124301. [CrossRef]
13. Dalwadi, M.P.; Griffiths, I.M.; Bruna, M. Understanding how porosity gradients can make a better filter using homogenization theory. *Proc. R. Soc. Math. Phys. Eng. Sci.* **2015**, *471*, 20150464. [CrossRef]
14. Dalwadi, M.P.; Bruna, M.; Griffiths, I.M. A multiscale method to calculate filter blockage. *J. Fluid Mech.* **2016**, *809*, 264–289. [CrossRef]
15. Griffiths, I.; Kumar, A.; Stewart, P. Designing asymmetric multilayered membrane filters with improved performance. *J. Membr. Sci.* **2016**, *511*, 108–118. [CrossRef]
16. Miller, D.J.; Kasemset, S.; Paul, D.R.; Freeman, B.D. Comparison of membrane fouling at constant flux and constant transmembrane pressure conditions. *J. Membr. Sci.* **2014**, *454*, 505–515. [CrossRef]
17. Fong, D.; Cummings, L.; Chapman, S.; Sanaei, P. On the performance of multilayered membrane filters. *J. Eng. Math.* **2021**, *127*, 1–25. [CrossRef]
18. Liu, S.Y.; Chen, Z.; Sanaei, P. Effects of Particles Diffusion on Membrane Filters Performance. *Fluids* **2020**, *5*, 121. [CrossRef]
19. Jaffrin, M.Y. Hydrodynamic techniques to enhance membrane filtration. *Annu. Rev. Fluid Mech.* **2012**, *44*, 77–96. [CrossRef]
20. Yang, Z.; Peng, X.; Chen, M.Y.; Lee, D.J.; Lai, J. Intra-layer flow in fouling layer on membranes. *J. Membr. Sci.* **2007**, *287*, 280–286. [CrossRef]
21. Bessiere, Y.; Abidine, N.; Bacchin, P. Low fouling conditions in dead-end filtration: Evidence for a critical filtered volume and interpretation using critical osmotic pressure. *J. Membr. Sci.* **2005**, *264*, 37–47. [CrossRef]

22. Herterich, J.G.; Xu, Q.; Field, R.W.; Vella, D.; Griffiths, I.M. Optimizing the operation of a direct-flow filtration device. *J. Eng. Math.* **2017**, *104*, 195–211. [CrossRef]
23. Mondal, S.; Griffiths, I.M.; Ramon, G.Z. Forefronts in structure–performance models of separation membranes. *J. Membr. Sci.* **2019**, *588*, 117166. [CrossRef]
24. Matias, A.F.; Coelho, R.C.; Andrade, J.S., Jr.; Araújo, N.A. Flow through time–evolving porous media: Swelling and erosion. *J. Comput. Sci.* **2021**, *53*, 101360. [CrossRef]
25. Emami, P.; Motevalian, S.P.; Pepin, E.; Zydney, A.L. Impact of module geometry on the ultrafiltration behavior of capsular polysaccharides for vaccines. *J. Membr. Sci.* **2018**, *561*, 19–25. [CrossRef]
26. Yogarathinam, L.T.; Gangasalam, A.; Ismail, A.F.; Arumugam, S.; Narayanan, A. Concentration of whey protein from cheese whey effluent using ultrafiltration by combination of hydrophilic metal oxides and hydrophobic polymer. *J. Chem. Technol. Biotechnol.* **2018**, *93*, 2576–2591. [CrossRef]
27. Liu, Z.; Ji, Z.; Shang, J.; Chen, H.; Liu, Y.; Wang, R. Improved design of two-stage filter cartridges for high sulfur natural gas purification. *Sep. Purif. Technol.* **2018**, *198*, 155–162. [CrossRef]
28. Brown, A.I. An Ultra Scale-down Approach to the Rapid Evaluation of Pleated Membrane Cartridge Filter Performance. Ph.D. Thesis, UCL (University College London), London, UK, 2011.
29. Jornitz, M.W. Filter construction and design. In *Sterile Filtration*; Springer: Berlin/Heisenberg, Germany, 2006; pp. 105–123.
30. Chen, D.R.; Pui, D.Y.; Liu, B.Y. Optimization of pleated filter designs using a finite-element numerical model. *Aerosol Sci. Technol.* **1995**, *23*, 579–590. [CrossRef]
31. Sanaei, P.; Richardson, G.; Witelski, T.; Cummings, L. Flow and fouling in a pleated membrane filter. *J. Fluid Mech.* **2016**, *795*, 36–59. [CrossRef]
32. Velali, E.; Dippel, J.; Stute, B.; Handt, S.; Loewe, T.; von Lieres, E. Model-Based Performance Analysis of Pleated Filters with Non-Woven Layers. In *Separation and Purification Technology*; Elsevier: Amsterdam, The Netherlands, 2020; p. 117006.
33. Brown, A.; Levison, P.; Titchener-Hooker, N.; Lye, G. Membrane pleating effects in 0.2 μm rated microfiltration cartridges. *J. Membr. Sci.* **2009**, *341*, 76–83. [CrossRef]
34. Lutz, H. Rationally defined safety factors for filter sizing. *J. Membr. Sci.* **2009**, *341*, 268–278. [CrossRef]
35. Griffiths, I.; Howell, P.; Shipley, R. Control and optimization of solute transport in a thin porous tube. *Phys. Fluids* **2013**, *25*, 033101. [CrossRef]
36. Beavers, G.S.; Joseph, D.D. Boundary conditions at a naturally permeable wall. *J. Fluid Mech.* **1967**, *30*, 197–207. [CrossRef]
37. Lo, L.M.; Hu, S.C.; Chen, D.R.; Pui, D.Y. Numerical study of pleated fabric cartridges during pulse-jet cleaning. *Powder Technol.* **2010**, *198*, 75–81. [CrossRef]
38. Waghode, A.; Hanspal, N.; Wakeman, R.; Nassehi, V. Numerical analysis of medium compression and losses in filtration area in pleated membrane cartridge filters. *Chem. Eng. Commun.* **2007**, *194*, 1053–1064. [CrossRef]
39. Wakeman, R.; Hanspal, N.; Waghode, A.; Nassehi, V. Analysis of pleat crowding and medium compression in pleated cartridge filters. *Chem. Eng. Res. Des.* **2005**, *83*, 1246–1255. [CrossRef]
40. Fotovati, S.; Hosseini, S.; Tafreshi, H.V.; Pourdeyhimi, B. Modeling instantaneous pressure drop of pleated thin filter media during dust loading. *Chem. Eng. Sci.* **2011**, *66*, 4036–4046. [CrossRef]
41. Van der Sman, R.; Vollebregt, H.; Mepschen, A.; Noordman, T. Review of hypotheses for fouling during beer clarification using membranes. *J. Membr. Sci.* **2012**, *396*, 22–31. [CrossRef]
42. Meng, F.; Chae, S.R.; Drews, A.; Kraume, M.; Shin, H.S.; Yang, F. Recent advances in membrane bioreactors (MBRs): Membrane fouling and membrane material. *Water Res.* **2009**, *43*, 1489–1512. [CrossRef] [PubMed]
43. Zhang, W.; Zhu, Z.; Jaffrin, M.Y.; Ding, L. Effects of hydraulic conditions on effluent quality, flux behavior, and energy consumption in a shear-enhanced membrane filtration using box-behnken response surface methodology. *Ind. Eng. Chem. Res.* **2014**, *53*, 7176–7185. [CrossRef]

Article

Oscillatory Reversible Osmotic Growth of Sessile Saline Droplets on a Floating Polydimethylsiloxane Membrane

Pritam Kumar Roy [1], Shraga Shoval [2], Leonid A. Dombrovsky [3,4] and Edward Bormashenko [1,*]

1. Chemical Engineering Department, Faculty of Engineering, Ariel University, P.O. Box 3, Ariel 407000, Israel; pritamr256@gmail.com
2. Department of Industrial Engineering and Management, Faculty of Engineering, Ariel University, P.O. Box 3, Ariel 407000, Israel; shraga@ariel.ac.il
3. X-BIO Institute, University of Tyumen, 6 Volodarskogo St, 625003 Tyumen, Russia; ldombr@yandex.ru
4. Heat Transfer Department, Joint Institute for High Temperatures, 17A Krasnokazarmennaya St, 111116 Moscow, Russia
* Correspondence: edward@ariel.ac.il

Abstract: We report a cyclic growth/retraction phenomena observed for saline droplets placed on a cured poly (dimethylsiloxane) (PDMS) membrane with a thickness of 7.8 ± 0.1 μm floating on a pure water surface. Osmotic mass transport across the micro-scaled floating PDMS membrane provided the growth of the sessile saline droplets followed by evaporation of the droplets. NaCl crystals were observed in the vicinity of the triple line at the evaporation stage. The observed growth/retraction cycle was reversible. A model of the osmotic mass transfer across the cured PDMS membrane is suggested and verified. The first stage of the osmotic growth of saline droplets is well-approximated by the universal linear relationship, whose slope is independent of the initial radius of the droplet. The suggested physical model qualitatively explains the time evolution of the droplet size. The reported process demonstrates a potential for use in industrial desalination.

Keywords: osmotic membrane; polydimethylsiloxane; saline droplet; mass transport; evaporation; reversible cycle

Citation: Roy, P.K.; Shoval, S.; Dombrovsky, L.A.; Bormashenko, E. Oscillatory Reversible Osmotic Growth of Sessile Saline Droplets on a Floating Polydimethylsiloxane Membrane. *Fluids* **2021**, *6*, 232. https://doi.org/10.3390/fluids6070232

Academic Editors: Manfredo Guilizzoni and Mehrdad Massoudi

Received: 8 May 2021
Accepted: 15 June 2021
Published: 22 June 2021

Publisher's Note: MDPI stays neutral with regard to jurisdictional claims in published maps and institutional affiliations.

Copyright: © 2021 by the authors. Licensee MDPI, Basel, Switzerland. This article is an open access article distributed under the terms and conditions of the Creative Commons Attribution (CC BY) license (https://creativecommons.org/licenses/by/4.0/).

1. Introduction

Polydimethylsiloxane (PDMS) membranes are broadly used for the manufacturing of microfluidic devices [1], separation of organics from water [2,3], gas separation [4–9], and removing aldehydes from the reactants [10]. Osmotic mass transport across PDMS-based liquid layers has already been implemented for the controlled crystallization of proteins [11,12]. In our recent research, we demonstrated osmotic mass transport between the saline water encapsulated within a composite liquid marble coated with liquid PDMS and the supporting layer of water [13]. Our present research is devoted to osmotic mass transport across cross-linked (cured), floating PDMS membranes. PDMS membranes may be manufactured by dip- or spin-coating [14] or by 3D printing [15]. Micro-scale thickness cured PDMS osmotic membranes may be manufactured by drop-casting on the water/vapor interface [16]. We demonstrate that cured floating osmotic PDMS membranes enable completely reversible growth/retraction oscillations of sessile saline droplets deposited on the membranes. The observed oscillations of the droplet size are explained by the osmotic mass transport and evaporation cycles as described below in detail. The investigated process demonstrates the potential of cross-linked PDMS membrane use for desalination.

2. Materials and Methods

2.1. Materials

The following materials were used in the recent experiment: polystyrene Petri dish (55 mm × 16 mm); poly (dimethylsiloxane) (PDMS) Sylgard 184, supplied by Dow Corning,

USA (with the following characteristics: molecular weight 207.4 g/mol, viscosity 5.5 Pa × s, surface tension 20.4 mN/m); deionized water (DI) from Millipore SAS (France) (with the following characteristics: specific resistivity $\hat{\rho} = 18.2$ MΩ ×cm at 25 °C, surface tension $\gamma = 72.9$ mN/m; viscosity $\eta = 8.9 \times 10^{-4}$ Pa × s); sodium chloride (NaCl) was supplied by Melach Haaretz Ltd., Kfar Monash, Israel. Droplets of 5 mL of the saturated aqueous NaCl solution (25.9% w/w) were used in the experiment. The thickness of the PDMS membrane was established by weighting as 7.8 ± 0.1 μm.

2.2. Methods

The floating PDMS membrane was prepared as depicted in Figure 1 by pouring a mixture containing liquid PDMS and a curing agent on the distilled water/vapor interface (see also [16]). Afterwards, a 5 μL saline droplet was placed on the PDMS membrane, floating within the closed vessel (chamber), as shown in Figure 2.

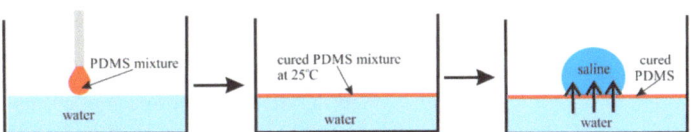

Figure 1. Schematic representation of the PDMS membrane preparation method.

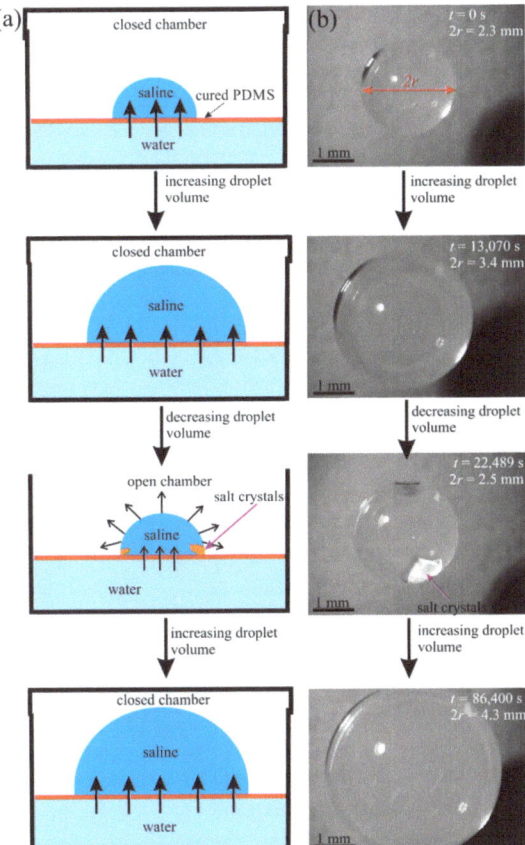

Figure 2. Growth and decay of a 5 μL saline droplet on a floating PDMS membrane due to osmosis and evaporation; (**a**) schematic and (**b**) images.

A mixture of PDMS and the crosslinker (Sylgard 184 silicone elastomer curing agent) was prepared with a weight ratio of 10:1. After that, a polystyrene Petri dish was taken and half-filled with DI water. Then, 10 µL of the PDMS mixture was gently deposited on top of the surface and kept at room temperature for curing. After complete curing of the PDMS mixture at room temperature (the curing time was 48 h), a 5 µL saline droplet (saturated solution) was deposited, and simultaneously, a video was captured to observe the changes in the droplet diameter over time. The preparation method is schematically shown in Figure 1. The apparent contact angle of the saline droplet on the PDMS membrane is $\theta = 87° \pm 2°$. Uniformity of the PDMS membrane was controlled with an optical digital microscope BW1008-500X (New Taipei City, Taiwan)

A BW1008-500X digital microscope and Ramé-Hart advanced goniometer model 500-F1 (Succasunna, NJ, USA) were used to capture images and movies of the saline droplet. The experiments were carried out at an ambient air temperature of $t = 25\,°C$. The relative humidity of air was equal to RH = 44 ± 2%.

3. Results and Discussion

The floating PDMS membrane was prepared as depicted in Figure 1 by pouring a mixture containing liquid PDMS and a curing agent on the distilled water/vapor interface as described in detail in the Materials and Methods Section (see also [16]). Afterwards, 5–15 µL saline droplets were placed on the PDMS membrane floating within the closed vessel (chamber), as shown in Figure 2. Osmotic mass transport across the PDMS membrane gave rise to the increase in the volume of the droplet, accompanied by advancing motion of the triple (three-phase) line, as depicted in Figure 2 (diffusion of water through thin oil layers was reported recently in [17]). We performed two series of experiments with (i) long-range and (ii) short-range cycles of the osmotic growth/evaporation of droplets as described below. Time evolution of the droplet contact radius $r(t)$ and the apparent contact angle $\theta_{app}(t)$ shown in Figure 4 were registered with the goniometer.

1. In the long-time experiments, the stage of growth continued for $\tau_{gr} = 13070 \pm 0.2$ s. During this time, the droplet volume increased from 5 µL to 30 µL. Afterwards, the chamber was opened, as shown in Figure 2 and Video S1 and the droplet was evaporated during $\tau_r = 11419 \pm 0.2$ s. Increasing the evaporation time scale gave rise to the formation of the NaCl crystals in the vicinity of the triple line as shown in Figure 2. At this stage, the volume of the droplet was decreased, and the triple line retracted. We performed $n = 2$ cycles of the long-time osmotic growth/evaporation of a droplet and observed that the process is reversible. Statistical scattering of the contact radius and volume of the droplet within growth/evaporation cycles were established as ±0.05 mm and ±0.3 µL, respectively.

2. Short-time growth/evaporation (retraction) experiments are illustrated in Figure 3 and Video S2, depicting the cyclic change in the volume V and contact diameter D of the droplet. In these experiments, the time scales were $\tau_{gr} = \tau_r = 3600 \pm 0.2$ s; $n = 5$, and reversible growth/evaporation cycles were performed.

The initial volume of the droplets in these experiments was confined within the range of 5 µL ≤ V ≤ 15 µL. The final volume of these droplets was in the range of 7 µL ≤ V ≤ 19 µL. The changes in both the volume and the contact area diameter of a droplet were accompanied by a change in the apparent contact angle, illustrated in Figure 5. The range of the apparent contact angles registered during the osmotic growth/retraction cycles was established as 65° ± 2° < θ_{app} < 87° ± 2°. This change in the apparent contact angle is reasonably attributed to the phenomenon of the contact angle hysteresis [18–24]. In our experiments, this hysteresis is strengthened by the pinning of the triple line arising from the coffee-stain effect inevitable under evaporation of saline droplets [24–30]. The coffee-stain effect is evidenced by the formation of NaCl crystals close to the triple line, as depicted in Figure 2.

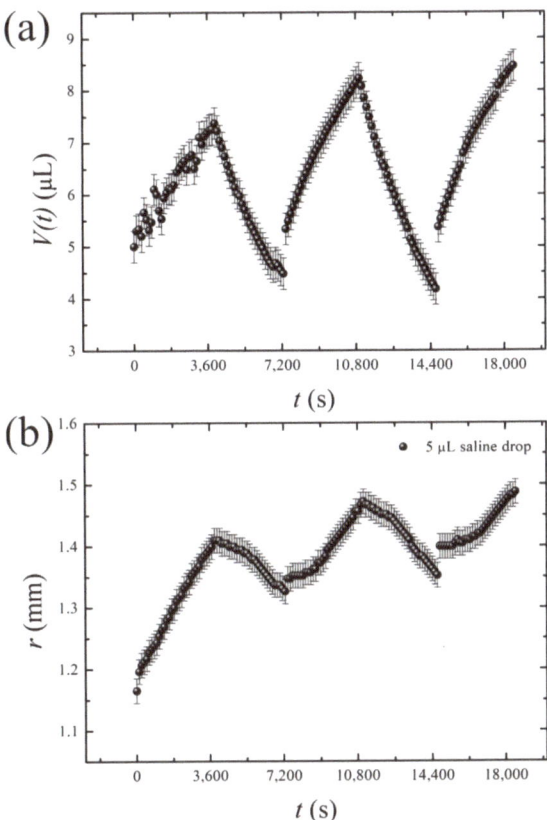

Figure 3. Oscillatory behavior of the droplets exposed to the osmotic mass transport/evaporation cycles is depicted. (**a**) Time evolution of the drop volume $V(t)$; (**b**) the time dependence of the contact radius r.

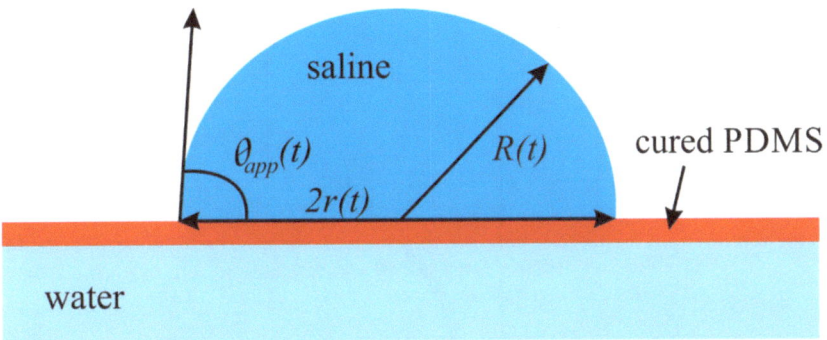

Figure 4. Geometrical parameters of saline droplet placed on the floating PDMS membrane are depicted. $R(t)$ is the current radius of the droplet; $r(t)$ is the current radius of the contact area and $\theta_{app}(t)$ is the apparent contact angle of the droplet.

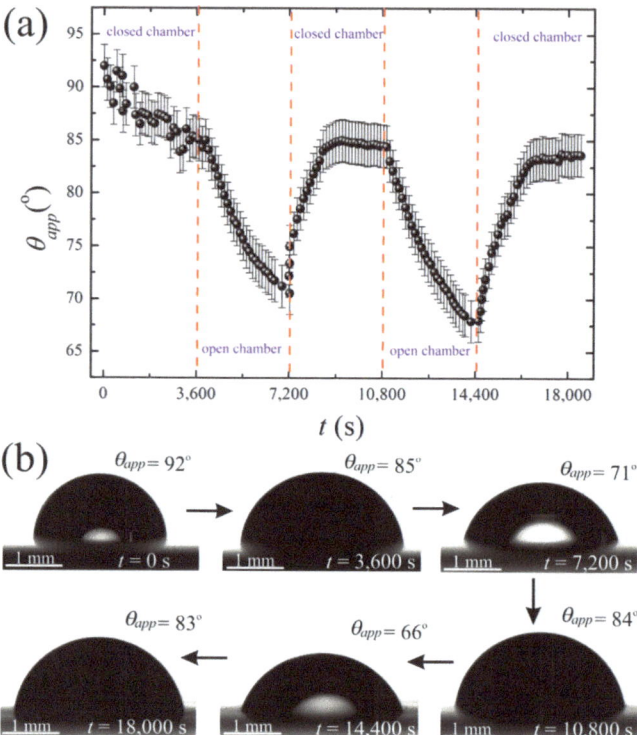

Figure 5. (**a**) Time evolution of the apparent contact angle $\theta(t)$ is depicted. (**b**) Sequence of images illustrating the side view of the time evolution of the droplet placed on the PDMS membrane is shown.

We now address the stage of the osmotic growth of the droplets in more detail as illustrated in Figure 6. Consider an approximate model of the osmotic growth of the water droplet observed in recent experiments.

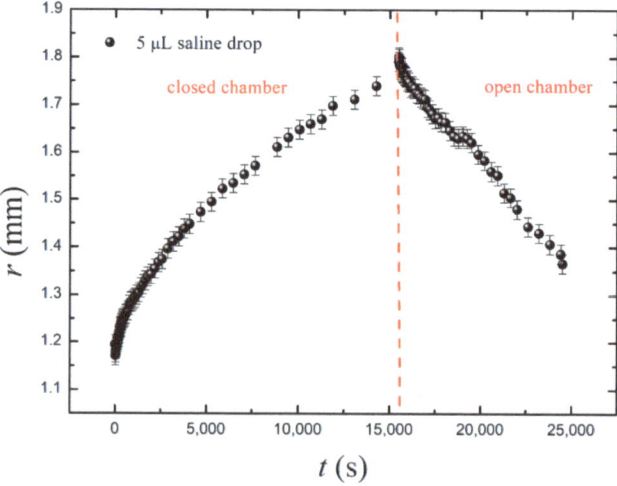

Figure 6. Growth and decay of a 5 µL saline drop on a floating PDMS membrane due to osmosis and evaporation.

In contrast to spherical small liquid marbles, considered recently in [13], the water droplet is almost hemispherical. The current density of salt water in the droplet can be expressed as follows:

$$\rho(t) = \rho_0 - \left(1 - 1/\overline{R}^3\right)(\rho_0 - \rho_w) \qquad (1)$$

where $\overline{R}(t) = R/R_0$, $R(t) \geq R_0$ is the current radius of the droplet, R_0 and ρ_0 are the initial radius of the droplet and the initial density of salt water, respectively, and ρ_w is the density of pure water under the membrane. According to Equation (1), $\rho(0) = \rho_0$ and the value of ρ decreases with time due to the osmotic growth of the droplet.

The balance equation for the volume of the growing hemispherical droplet appears as follows:

$$2\pi R^2 \dot{R} = \left(\dot{m} \times \pi R^2\right)/\rho_w \qquad (2)$$

where $\dot{m}$ is the constant osmotic flow rate of pure water measured in kg/(m²·s). It is natural to assume that $\dot{m}$ is directly proportional to the following difference in densities:

$$\rho - \rho_w = (\rho_0 - \rho_w)/\overline{R}^3 \qquad (3)$$

This proportionality can be written as:

$$\dot{m} = \psi_{osm}/\overline{R}^3 \qquad (4)$$

where ψ_{osm} is the unknown phenomenological osmotic parameter of the membrane [13]. It is convenient to introduce the characteristic time of the process:

$$\tau_{osm} = 2\rho_w R_0/\psi_{osm} \qquad (5)$$

and rewrite Equation (2) as follows:

$$\tau_{osm} \overline{R}^3 \dot{\overline{R}} = 1 \qquad (6)$$

The obvious initial condition for the droplet radius is:

$$\overline{R}(0) = 1 \qquad (7)$$

The analytical solution to the Cauchy problem, defined by Equations (6) and (7) is simply:

$$\overline{R} = \sqrt[4]{1 + \overline{t}}, \qquad (8)$$

where $\overline{t} = t/\tau_{osm}$ is the dimensionless time of the osmotic evolution of the droplet. At the beginning of the process (at $\overline{t} \ll 1$), Equation (8) is approximated as:

$$\overline{R} = 1 + 0.25\overline{t} \qquad (9)$$

It is interesting that Equations (8) and (9) do not contain any dependence on the initial radius of the droplet. Thus, the experimental date obtained with different initial radii of the droplets should be fitted with the universal straight line. This prediction is confirmed by the measurements carried out for the saline water droplets of different initial volumes (5 µL, 10 µL, and 15 µL). One recognizes from Figure 7 that the initial time dependences of the growing droplet radius are almost linear, and the slope of all the curves is very close to that predicted by Equation (9). This enables us to estimate the characteristic time of the process, which appears to be in the range of

$$1.35 \pm 0.25 \text{ h} < \tau_{osm} < 1.40 \pm 0.25 \text{ h} \qquad (10)$$

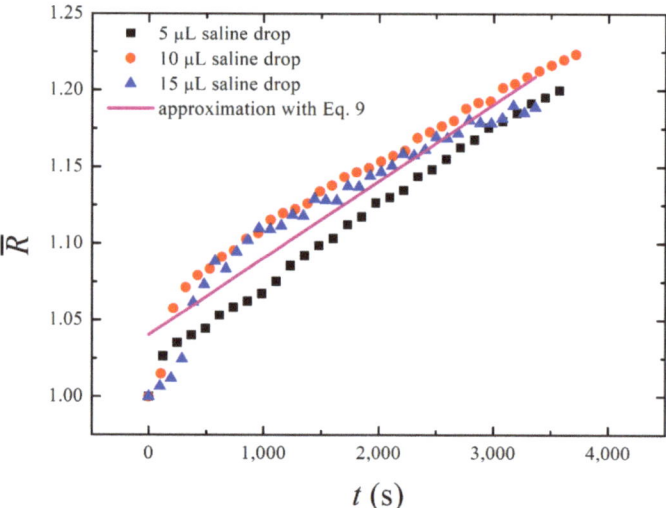

Figure 7. Experimental curves describing the osmotic evolution of a droplet of different initial volumes V_0 during the "short-time" growth are presented (black squares—$V_0 = 5$ µL, red circles—$V_0 = 10$ µL; blue triangles—$V_0 = 15$ µL); the solid line demonstrates the best linear fitting of the experimental data for $V_0 = 15$ µL. The osmotic growth is satisfactorily approximated by the universal linear time dependence, reproduced by the solid line, whose slope is independent of the initial volume of a droplet.

Radii of the studied droplets were smaller than the capillary length, which is $l_{ca} = 2.71$ mm for water droplets [24], and the apparent contact angles were close to $\frac{\pi}{2}$; thus, the shape of the droplets is close to hemispherical.

Note that a more comprehensive model of the osmotic growth of droplets should take into account the non-perfectly spherical shape of the droplets and inevitably uneven distribution of salt over the droplet volume. Most likely, the latter effect is responsible for the relatively fast increase in the droplet size at the very beginning of its osmotic growth (see Figure 7).

The "long-time" non-linear osmotic growth of droplets may be described by the phenomenological equation, suggested recently in [13].

$$\overline{R}(\bar{t}) = \sqrt[4]{\alpha - \beta exp(-\gamma \bar{t})} \qquad (11)$$

where the triad α, β, and γ were taken as parameters. The best possible fit established with the least square method, depicted in Figure 8, was obtained at $\alpha = 7.23$, $\beta = 6.22$, and $\gamma = 0.71$.

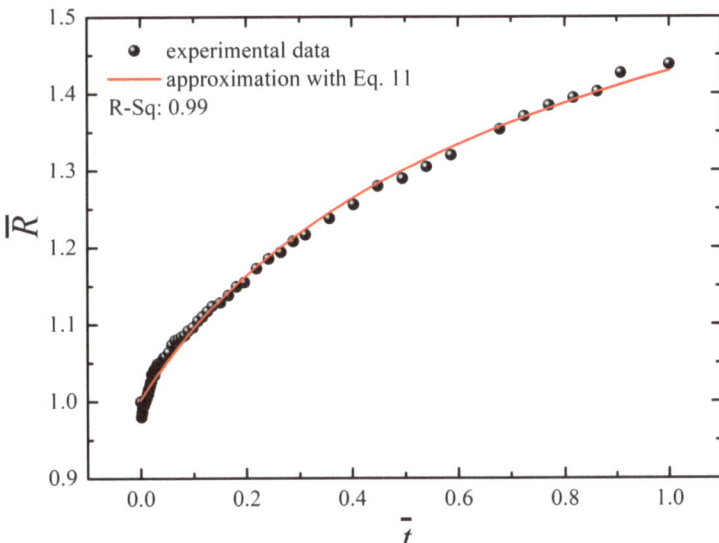

Figure 8. Time evolution of the dimensionless radius of the droplet (black circles) and its fit with Equation (11) at R_0 = 1.2 mm and τ_{osm} = 4875 s.

4. Conclusions

We conclude that water diffusion across floating micro-scale thickness cured PDMS film gives rise to the osmotic growth of a saline water droplet placed on the film. The phenomenological model of the osmotic mass transfer to the droplet is suggested. The dimensionless equation describing the osmotic growth of a droplet is shaped: $\overline{R} = \sqrt[4]{1+\overline{t}}$. The initial stage of osmotic growth of saline droplets is satisfactorily approximated by the universal linear time dependence, whose slope is independent of the initial radius of a droplet. The calculations using this physical model are in good agreement with the experimental data for droplets of different initial volume. The osmotic growth of the droplet followed by the evaporation of the droplet yields reversible growth/retraction cycles. NaCl crystals were observed in the vicinity of the triple line at the evaporation stage due to the pinning of the triple line and the coffee-stain effect [21–26]. The reversibility of the reported growth/retraction cycles should be emphasized. The "long-time" osmotic growth of droplets is described by the phenomenological equation: $\overline{R}(\overline{t}) = \sqrt[4]{\alpha - \beta exp(-\gamma \overline{t})}$. The characteristic time scale of the osmotic mass transport is established as 1.35 ± 0.25 h < τ_{osm} < 1.40 ± 0.25 h. The reported cured PDMS membranes and the process have potential for desalination [31] and development of separators for batteries [32].

Supplementary Materials: The following are available online at https://www.mdpi.com/article/10.3390/fluids6070232/s1, Video S1: Long-time growth/evaporation, Video S2: Short-time growth/evaporation.

Author Contributions: Conceptualization, P.K.R., S.S., and E.B.; methodology, P.K.R. and L.A.D.; software, P.K.R.; validation, P.K.R. and E.B.; physical modeling, L.A.D.; investigation, P.K.R., S.S., E.B., and L.A.D.; resources, S.S.; data curation, P.K.R.; writing—original draft preparation, E.B.; writing—review and editing, S.S., E.B., and L.A.D.; visualization, P.K.R.; supervision, S.S. and E.B.; project administration, S.S. All authors have read and agreed to the published version of the manuscript.

Funding: This research received no external funding.

Institutional Review Board Statement: Not applicable.

Informed Consent Statement: Not applicable.

Data Availability Statement: The data presented in this study are available on request from the corresponding author.

Conflicts of Interest: The authors declare no conflict of interests.

References

1. Mitrovski, S.M.; Elliott, L.C.C.; Nuzzo, R.G. Microfluidic devices for energy conversion: Planar integration and performance of a passive, fully immersed H_2-O_2 fuel cell. *Langmuir* **2004**, *20*, 6974–6976. [CrossRef] [PubMed]
2. Vankelecom, I.F.J.; de Kinderen, J.; Dewitte, B.M.; Uytterhoeven, J.B. Incorporation of hydrophobic porous fillers in PDMS membranes for use in pervaporation. *J. Phys. Chem.* **1997**, *101*, 5182–5185. [CrossRef]
3. Li, L.; Xiao, Z.; Tan, S.; Pu, L.; Zhang, Z. Composite PDMS membrane with high flux for the separation of organics from water by pervaporation. *J. Membr. Sci.* **2004**, *243*, 177–187. [CrossRef]
4. Fu, Y.-J.; Qui, H.-Z.; Liao, K.-S.; Lue, S.J.; Hu, C.-C.; Lee, K.-R.; Lai, J.-Y. Effect of UV-ozone treatment on poly(dimethylsiloxane) membranes: Surface characterization and gas separation performance. *Langmuir* **2010**, *26*, 4392–4399. [CrossRef]
5. Firpo, G.; Angeli, E.; Repetto, L.; Valbusa, U. Permeability thickness dependence of polydimethylsiloxane (PDMS) membranes. *J. Membr. Sci.* **2015**, *481*, 1–8. [CrossRef]
6. Nour, M.; Berean, K.; Balendhran, S.; Ou, J.Z.; Du Plessis, J.; McSweeney, C.; Bhaskaran, M.; Sriram, S.; Kalantar-zadeh, K. CNT/PDMS composite membranes for H_2 and CH_4 gas separation. *Int. J. Hydrog. Energy* **2013**, *38*, 10494–10501. [CrossRef]
7. Nour, M.; Berean, K.; Griffin, M.J.; Matthews, G.I.; Bhaskaran, M.; Sriram, S.; Kalantar-zadeh, K. Nanocomposite carbon-PDMS membranes for gas separation. *Sens. Actuator B Chem.* **2012**, *161*, 982–988. [CrossRef]
8. Sanchis-Perucho, P.; Robles, Á.; Durán, F.; Ferrer, J.; Seco, A. PDMS membranes for feasible recovery of dissolved methane from AnMBR effluents. *J. Membr. Sci.* **2020**, *604*, 118070. [CrossRef]
9. Ataeivarjovi, E.; Tang, Z.; Chen, J. Study on CO_2 Desorption Behavior of a PDMS–SiO_2 Hybrid Membrane Applied in a Novel CO_2 Capture Process. *ACS Appl. Mater. Interfaces* **2018**, *10*, 28992–29002. [CrossRef]
10. Logemann, M.; Alders, M.; Wist, M.; Pyankova, V.; Krakau, D.; Gottschalk, D.; Wessling, M. Can PDMS membranes separate aldehydes and alkenes at high temperatures? *J. Membr. Sci.* **2020**, *615*, 118334. [CrossRef]
11. Zheng, B.; Tice, J.D.; Roach, L.S.; Ismagilov, R.F. A droplet-based, composite PDMS/glass capillary microfluidic system for evaluating protein crystallization conditions by microbatch and vapor-diffusion methods with on-chip X-ray diffraction. *Angew. Chem. Int.* **2004**, *43*, 2508–2511. [CrossRef]
12. Zheng, B.; Gerdts, C.J.; Ismagilov, R.F. Using nanoliter plugs in microfluidics to facilitate and understand protein crystallization. *Curr. Opin. Struct. Biol.* **2005**, *15*, 548–555. [CrossRef]
13. Roy, P.K.; Legchenkova, I.; Shoval, S.; Dombrovsky, L.A.; Bormashenko, E. Osmotic evolution of composite liquid marbles. *J. Colloid Interface Sci.* **2021**, *592*, 167–173.
14. Thangawng, A.L.; Ruoff, R.S.; Swartz, M.A.; Glucksberg, M.R. An ultra-thin PDMS membrane as a bio/micro–nano interface: Fabrication and characterization. *Biomed. Microdevices* **2007**, *9*, 587–595. [CrossRef]
15. Koo, J.W.; Ho, J.S.; An, J.; Zhang, Y.; Chua, C.K.; Chong, T.H. A review on spacers and membranes: Conventional or hybrid additive manufacturing? *Water Res.* **2020**, *188*, 116497. [CrossRef]
16. Kim, D.; Kim, S.-H.; Park, Y. Floating-on-water fabrication method for thin Polydimethylsiloxane membranes. *Polymers* **2019**, *11*, 1264. [CrossRef]
17. Zhang, R.; Liao, W.; Wang, Y.; Wang, Y.; Wilson, D.I.; Clarke, S.M.; Yang, Z. The growth and shrinkage of water droplets at the oil-solid interface. *J. Colloid Interface Sci.* **2021**, *584*, 738–748. [CrossRef]
18. Erbil, H.Y. *Surface Chemistry of Solid and Liquid Interfaces*; Blackwell: Oxford, UK, 2006.
19. Tadmor, R. Line energy and the relation between advancing, receding, and young contact angles. *Langmuir* **2004**, *20*, 7659–7664. [CrossRef]
20. Tadmor, R. Approaches in wetting phenomena. *Soft Matter* **2011**, *7*, 1577–1580. [CrossRef]
21. Tadmor, R.; Yadav, P.S. As-placed contact angles for sessile drops. *J. Colloid Interface Sci.* **2008**, *317*, 241–246. [CrossRef]
22. Liu, J.; Xia, R.; Zhou, X. A new look on wetting models: Continuum analysis. *Sci. China Phys. Mech. Astron.* **2012**, *55*, 2158–2166. [CrossRef]
23. Liu, J.; Mei, Y.; Xia, R. A new wetting mechanism based upon triple contact line pinning. *Langmuir* **2011**, *27*, 196–200. [CrossRef] [PubMed]
24. Bormashenko, E. *Physics of Wetting Phenomena and Applications of Fluids on Surfaces*; Walter de Gruyter: Berlin, Germany, 2017.
25. Deegan, R.D.; Bakajin, O.; Dupont, T.F.; Huber, G.; Nagel, S.R.; Witten, T.A. Capillary flow as the cause of ring stains from dried liquid drops. *Nature* **1997**, *389*, 827–829. [CrossRef]
26. Deegan, R.D.; Bakajin, O.; Dupont, T.F.; Huber, G.; Nagel, S.R.; Witten, T.A. Contact line deposits in an evaporating drop. *Phys. Rev. E* **2000**, *62*, 756–765. [CrossRef]
27. Hu, H.; Larson, R.G. Analysis of the effects of Marangoni stresses on the microflow in an evaporating sessile droplet. *Langmuir* **2005**, *21*, 3972–3980. [CrossRef]
28. Zhong, X.; Ren, J.; Duan, F.J. Wettability effect on evaporation dynamics and crystalline patterns of sessile saline droplets. *J. Phys. Chem. B* **2017**, *121*, 7924–7933. [CrossRef]

29. Shao, X.; Duan, F.; Hou, Y.; Zhong, X. Role of surfactant in controlling the deposition pattern of a particle-laden droplet: Fundamentals and strategies. *Adv. Colloid Interface Sci.* **2020**, *275*, 102049. [CrossRef]
30. Shao, Z.; Hou, Y.; Zhong, X. Modulation of evaporation-affected crystal motion in a drying droplet by saline and surfactant concentrations. *Colloids Surfaces A* **2021**, *623*, 126701. [CrossRef]
31. Yao, M.; Tijing, L.D.; Naidu, D.; Kim, S.-H.; Matsuyama, H.; Fane, A.G.; Shon, K. A review of membrane wettability for the treatment of saline water deploying membrane distillation. *Desalination* **2020**, *479*, 114312. [CrossRef]
32. Angulakshmi, N.; Stephan, A.M. Efficient electrolytes for lithium–sulfur batteries. *Front. Energy Res.* **2015**, *3*, 1–8. [CrossRef]

Article

Turbulent Flow through and over a Compact Three-Dimensional Model Porous Medium: An Experimental Study

James Kofi Arthur

Department of Mechanical Engineering, Bucknell University, Lewisburg, PA 17837, USA; james.arthur@bucknell.edu

Abstract: There are several natural and industrial applications where turbulent flows over compact porous media are relevant. However, the study of such flows is rare. In this paper, an experimental investigation of turbulent flow through and over a compact model porous medium is presented to fill this gap in the literature. The objectives of this work were to measure the development of the flow over the porous boundary, the penetration of the turbulent flow into the porous domain, the attendant three-dimensional effects, and Reynolds number effects. These objectives were achieved by conducting particle image velocimetry measurements in a test section with turbulent flow through and over a compact model porous medium of porosity 85%, and filling fraction 21%. The bulk Reynolds numbers were 14,338 and 24,510. The results showed a large-scale anisotropic turbulent flow region over and within the porous medium. The overlying turbulent flow had a boundary layer that thickened along the stream by about 90% and infiltrated into the porous medium to a depth of about 7% of the porous medium rod diameter. The results presented here provide useful physical insight suited for the design and analyses of turbulent flows over compact porous media arrangements.

Keywords: porous medium; turbulent flow; particle image velocimetry; boundary layer

Citation: Arthur, J.K. Turbulent Flow through and over a Compact Three-Dimensional Model Porous Medium: An Experimental Study. *Fluids* **2021**, *6*, 337. https://doi.org/10.3390/fluids6100337

Academic Editor: Manfredo Guilizzoni

Received: 31 August 2021
Accepted: 15 September 2021
Published: 23 September 2021

Publisher's Note: MDPI stays neutral with regard to jurisdictional claims in published maps and institutional affiliations.

Copyright: © 2021 by the author. Licensee MDPI, Basel, Switzerland. This article is an open access article distributed under the terms and conditions of the Creative Commons Attribution (CC BY) license (https:// creativecommons.org/licenses/by/ 4.0/).

1. Introduction

The flow of fluids in composite porous-clear domains is a prevalent phenomenon in many natural and industrial applications. Consequently, such has been the focus of several research studies, covering both laminar and turbulent flow considerations [1–11].

In this work, attention is focused on the study of turbulent flow over a porous medium, and the concomitant effects on the flow through the coupled porous medium. Even so, the impact of such flows is far-reaching. Turbulent flow mechanisms are, for example, known to influence the hyporheic exchange of pollutants to regions below and adjacent to streams and rivers [12]. In canopies of submerged aquatic vegetation, biogeochemical processes are regulated through unsteady inertial and turbulent exchanges of mass and momentum between the canopy (acting as a permeable medium) and the ambient water [13]. Turbulent transport of wind within and above forest canopies also plays an integral role in the atmospheric exchange of carbon dioxide (therefore forcing climate change [14]). They are also important in making site assessments for architectural structures and wind turbine performance [15]. In industrial settings, on the other hand, porous assemblies in the form of pin fin arrays have been used to increase the turbulence and unsteadiness of coolant flow to guarantee efficient cooling [16]. In the casting of alloys, turbulent flows generated due to electromagnetic stirring has been found to decrease channel formation in the mushy zone and associated patterns of segregation [17].

While the range of studies varies considerably in the application, analyses of turbulent flows over porous media are facilitated by distinguishing the flow domain into a channel (or free or open) flow region, and a porous flow region [18]. These two regions are separated by an interfacial zone that is sometimes discussed as a separate region [6]. Concerning the open flow region, a great deal of what is presently known about the pertinent turbulent flow has been shaped by canopy flows [13,19–21]. From this perspective, the flow features are characterized by wall-normal variations of the mean streamwise velocities that inflect at the interfacial zone. Structurally, the flow in that region is conceived as similar to a mixing layer. However, the inflectional profile renders the flow prone to the development of vortices originating from an instability of the Kelvin–Helmholtz type. Indeed, these vortices grow into other secondary vortical structures that often propagate a coherent wave-like oscillatory phenomenon. Nevertheless, the flow in this channel flow region is often considered to be dominated by large coherent structures. It is noteworthy that other related studies associated with permeable bed models (such as cubes, spheres, nets, and foams) have given insight into the dynamics of flow in the open region. Accordingly, compared to solid walls, friction factors, wall shear stress, momentum flux, and Reynolds stress were found to be enhanced at porous walls due to porous matrix factors such as porosity and permeability [22–26]. By using a double-averaging method [5,27], the global effects of the porous medium factors were evaluated and found to be different from the roughness effects [25]. In this regard, Suga et al. [7] conducted extensive particle image velocimetry (PIV) measurements of the boundary layer above a porous boundary. Models of porous media of variant permeability were tested over a wide range of Reynolds numbers. The researchers demonstrated that the streamwise mean velocity did fit a log-law form after accounting for a displacement height and equivalent roughness height as well as a von Kármán coefficient (different from the 0.38–0.41 range conventionally applied to flows associated with smooth and rough walls). In various PIV experiments, the extent of the logarithmic layer has also been assessed, along with correlations of the log-law parameters with characteristic permeability Reynolds numbers [7,28,29]. While these results are concrete, it must also be conceded that they were notably obtained from fully developed flows over porous media. Thus, they do not account for cases of compact porous media arrangements where, at least, the streamwise velocity variations are dependent of the streamwise location.

The interfacial zone is the transitional layer between the open and the porous regions where effects of the turbulent motion of the surface flow are persistent [18]. It is therefore the region where momentum, energy, and scalers are exchanged [6]. Despite the importance of the interfacial zone in the general flow dynamics, the information provided in the literature has been relatively sparse. The simulations of Breugem et al. [25] indicate that there is a link between the sign of the flow across the interfacial zone with the transporting fluid; and that this is specified by the fluctuating velocity component. This was later confirmed by Kim et al. [6] using PIV to provide velocities of a refractive-index-matched system of turbulent flow over a cubic-packed arrangement of uniform spheres. Before this confirmation, however, Manes et al. [18] conducted velocity measurements in a turbulent open channel flow over cubically packed spheres. Instead of a gradual monotonic decay in streamwise mean velocity just below the interface between the open and porous regions, rather, they observed the velocity dampen to a minimum and then rise toward a constant value within the porous medium. In a subsequent study, Pokrajac et al. [30] explained this to be the product of enhanced turbulence in the pores closer to the interface. They noted that this resulted in the viscous drag extracting momentum from the flow, thus allowing higher mean velocities to form in the lower pores. In another surprising discovery, the canopy flow studies of Florens et al. [5] indicated that contrary to popular assumption, the dispersive stress near the top of the porous medium is not negligible. Such unexpected observations point to significant gaps in the current understanding of the flow at the interfacial zone.

For the flow within the porous region, the available data are still rare, and the information has been mixed. The numerical computations of Breugem et al. [25] as well as Sharma and García-Mayoral [31] showed an exponential decay in the mean velocity profile, turbulence intensities, and pressure fluctuations inside the porous medium. This was in part, corroboration of the experimental findings of Ruff and Gelhar [22] and Vollmer et al. [24]. Florens et al. [5], on the other hand, reported linear variations of mean wall-normal velocity, mean spanwise velocity, and total stress in the wall-normal direction. Kim et al. [6] also observed channeling of flow along the trough regions of their cubic-packed arrangements. They also noted that the crest region was the region of secondary motion resulting from neighboring trough channels. There are several studies on pin-fin heat transfer that have provided direct detailed measurements of the flow field within the arrays. These studies have provided valuable information on flow dynamics and structures such as vortex shedding strength with geometric changes [32], near wake flow development [33], and the nature of horseshoe vortex systems [34]. However, such studies usually presume turbulent flow within the arrays and do not consider the case of the effects of turbulent flow over the arrays.

In summary, it is important to note that while several studies have been conducted on turbulent flow over porous media, most of these works have been focused on the open flow region [7,28,29]. Experimental work on flow at the interfacial and porous regions, however, is scarce. Thus, while several conclusive observations have been made on the flow phenomena about changes in porosity, permeability, and Reynolds number, the same cannot be said about the interfacial and porous regions. Furthermore, it is worth noting that most of the studies on turbulent flows associated with porous boundaries appear to be focused on acquiring information from a fully developed section of the turbulent flow [6,7,18,28,30]. Consequently, some of the definite conclusions in the literature regarding the variation of mean velocity, momentum, and turbulent statistics may not necessarily apply to cases associated with compact porous media. It must be emphasized that such a lack is not trivial, because some pertinent natural scenarios may only be suited for analyses when considered as turbulent flows over compact porous media. These include terrestrial or aquatic canopy flows over regions of land with small/confined arrangements of sparse vegetation or building structures. Additionally, the design and evaluation of engineering systems such as pin-fin arrays of rods for heat transfer enhancement may not be sufficiently helped by considering what is currently in the literature when there is a turbulent flow over the pin-fin array. Specifically, fundamental questions remain regarding the nature of boundary layer development, the spatial variations of the mean turbulence, and other statistics due to three-dimensional effects, the mode, and development of interactions between the open region and the porous region at the interfacial zone, and the porous flow phenomenon itself. To provide some answers to these questions, an experimental research program was designed to conduct velocity measurements of a turbulent flow over and penetrating through a compact model porous medium. In this particular work, the objectives were to study the turbulent boundary layer development over the porous medium and its penetration into the porous flow, three-dimensional effects, and Reynolds number effects of the flow. The physical system was achieved by using a square array of rods to model the three-dimensional porous medium and testing it in a flow channel. The porous medium was arranged to cover a single porosity value of 85%, and to fill 21% of the depth of the flow channel. The uniqueness of this work lies in the fact that the porous medium model used was designed to be compact, having only twelve columns and nine rows of rods. Furthermore, a planar PIV system was used to conduct detailed velocity measurements in several streamwise-wall-normal planes through and over the porous medium in a turbulent flow over a range of Reynolds numbers. The mean flow and turbulent statistics were assessed to characterize the flow through and over the compact porous medium.

2. Measurement System and Method

2.1. Test System and Measurement Procedure

The experiments were conducted in a model test channel that was placed into an open flow transport channel supplied by *TecQuipment* Ltd. (Portland, OR, USA). The transport channel was designed to be operated in a closed water circuit. For such an arrangement, water from a supply tank is pumped to the channel inlet through a precision control valve and a flow conditioning unit. Uniform flow of relatively low background turbulence from the conditioning unit is then directed through the test section and then returned into the supply tank through a filter. The test section of the flow channel had internal dimensions of 2500 mm (length) × 80 mm (width) × 250 mm (depth). To facilitate optical access, the side walls of the test section were constructed from a transparent acrylic material.

The model test channel used in this work was built from 6 mm thick transparent acrylic plates glued together in a closed channel with the internal dimensions of 2500 mm (length) × 69 mm (width) × 43 mm (depth, H). The channel consisted of a flow entry section and a downstream section with a compact porous model. For the flow entry section, fourteen equally spaced out 3.18 mm square rods were glued on the first 90 mm portion of both upper and lower walls of the entrance of the channel. These served as trips to assure the rapid development of the turbulent boundary layer. The compact porous model in the test section was assembled by mounting transparent acrylic rods into a square array of holes drilled into the lower wall. The rods were of nominal diameter d = 3.18 mm and average height h = 9.06 mm. With 12 rows and nine columns of such an array of rods spaced out equally at distance l = 7.19 mm between adjacent rod centers, a porous model of porosity 85%, and of filling fraction h/H = 21% was achieved. This model is much more compact than previous related studies [7,25,31]. Based on the density and submergence classification parameter provided in the review by Nepf [35], the current arrangement had a submergence ratio H/h of 4.78 and frontal area density dh/l^2 of 0.56, which may be classified as a (transitionally) dense canopy of shallow submergence. The salient geometric features of the porous model are summarized in Table 1. A schematic diagram of the test channel highlighting the porous model is also shown in Figure 1. In this figure, the Cartesian coordinate system used in this work is also identified. It should be noted that the origin of the streamwise axis x (i.e., x = 0) was fixed at the center of the most upstream column of rods. Thus, a 1200 mm length of flow entry was allowed before reaching the model porous medium (Figure 1a). The origins of the wall-normal direction y (i.e., y = 0) and the spanwise direction (i.e., z = 0) were also respectively located at the lower wall and the middle of the channel span. The model test channel was assembled into the transport channel so that the midspans of both channels were fixed and coincident at z = 0.

Table 1. Summary of geometric parameters of test model, boundary layer parameters of entry flow, and friction parameters of entry flow.

Geometric Parameters of Test Model					
Porous Medium Depth h (mm)	Channel Depth H (mm)	Filling Fraction h/H	Rod Diameter of Porous Medium d (mm)	Distance between Adjacent Rod Centers l (mm)	Porosity of Porous Medium ε
9.06	43	0.21	3.18	7.19	0.85
Boundary Layer Parameters of Entry Flow					
Test Label	Bulk Velocity U_b (m/s)	Maximum Velocity U_e (m/s)	Boundary Layer Thickness δ (mm)	Displacement Thickness δ^* (mm)	Momentum Thickness θ^* (mm)	Shape Factor H
A	0.3347	0.3927	12.24	1.87	1.17	1.60
B	0.5721	0.6453	10.78	1.38	0.94	1.48
Friction Parameters of Entry Flow					
Test Label	Friction Velocity U_τ (m/s)		Skin Friction Coefficient C_f	Bulk Reynolds Number Re_H	Momentum Thickness Reynolds Number Re_θ
A	0.0145		0.003	14,338	461
B	0.0222		0.002	24,510	604

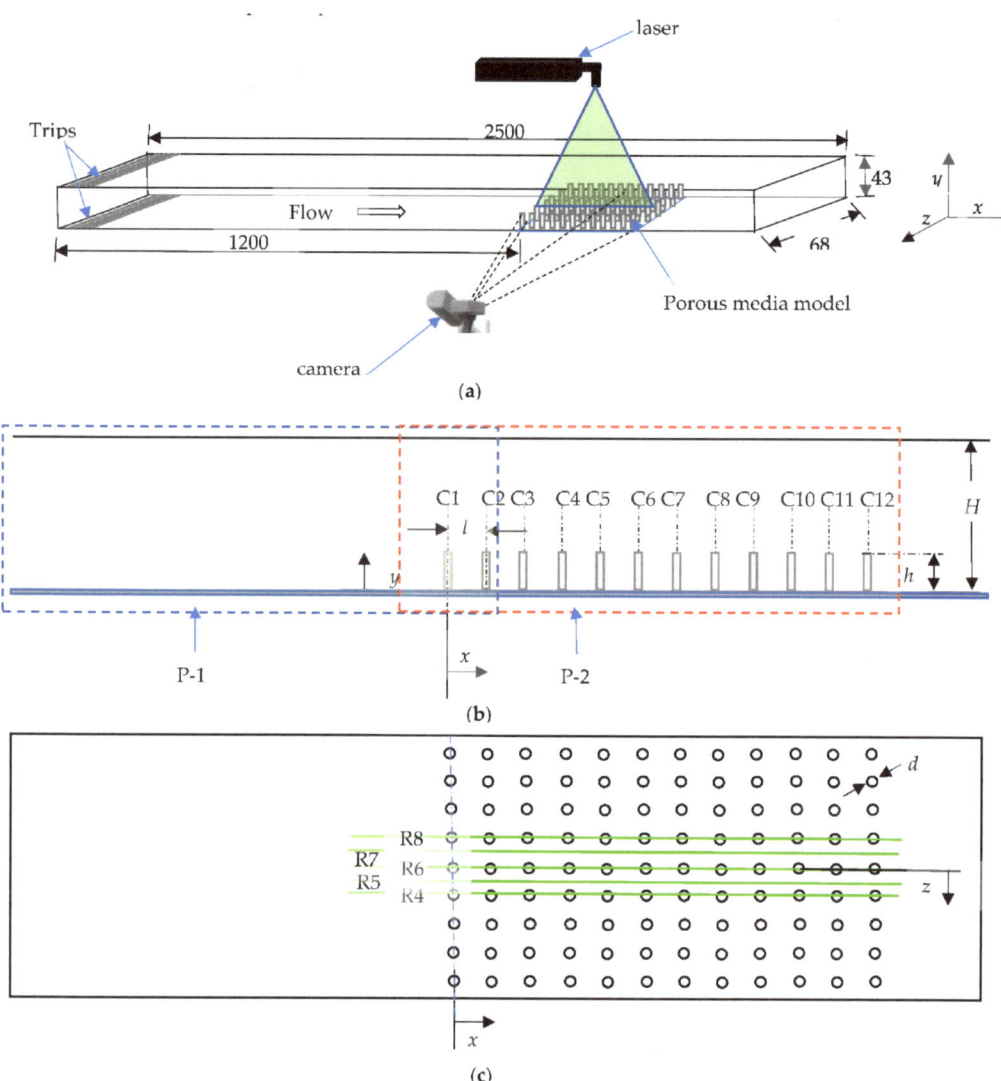

Figure 1. Schematics showing (**a**) the model test channel and PIV system; (**b**) front view showing details of porous arrangements in the test channel as well as streamwise planar coverage of PIV measurements (shown using blue and boxes with dashed line boundaries); and (**c**) top view of the porous model arrangement in the model test channel with green lines indicating lateral locations of measurement with the laser sheet. All numeric dimensions are in millimeters.

The measurement of flow velocity was accomplished using a planar particle image velocimetry (PIV) system supplied by *LaVision* Inc. (Ypsilanti, MI, USA). In using the PIV technique, water (the working fluid) was seeded with silver-coated hollow glass spheres of mean diameter 10 μm and specific gravity 1.4. The flow field was illuminated by a thin sheet of light generated by a *Quantel Evergreen* (Edinburgh, UK) Nd:YAG Dual Cavity 200 mJ/pulse laser with a wavelength of 532 nm, connected to a set of cylindrical lenses. The light scattered by the seeding particles in the flow was then captured and recorded as digital images using a 12-bit charged couple device camera with a 1608 × 1208-pixel array, and 7.4 μm pixel pitch. The camera was coupled to a 50-mm focal length *Nikon* lens. A programmable timing unit was used to synchronize the trigger rate of the laser as well as the image capturing rate of the camera. Data acquisition was enabled and controlled using PIV software (*DaVis*-10) installed on a dual processor computer with a 32-gigabyte random access memory. Instantaneous digital images were saved and processed using a multi-pass cross-correlation algorithm within the *DaVis* software.

It has been noted that for the optical measurements of flows through porous media, distortions are best eliminated by using a transparent solid matrix, and a perfect refractive index matching (RIM) of the working fluid and the solid matrix. This is particularly vital if the assessment of velocity data involves the use of images obtained from passing light simultaneously through sections of different solid models and/or fluids within the domain of interest [36,37]. However, other works have noted that a reasonable quality of velocity vectors could still be obtained from PIV measurements without RIM [4,5]. This is possible by using simple porous media test arrangements of relatively high porosity such as that used in the current work. It is also possible by employing appropriate filters and seeding particles that reduce distortions around walls or using multiple cameras arranged to capture specific flow domains subject to and within the optical path of the same refractive index within the test section. In this work, optical distortions due to a non-RIM arrangement were effectively reduced through the combined use of an appropriate porous medium model, filters, and seeding particles.

The choice of silver-coated hollow glass spheres was guided by their spherical shapes and thin reflectivity enhancing the silver coatings. These qualities make them well suited for scattering sufficient light for camera detection, even through the pores of the simulated porous medium. To assure neutral buoyancy in water, the particles were assessed using the settling velocity v_s and the response time τ_R parameters [38]: $v_s = (\rho_p - \rho_f)g d_p^2/(18\,\rho_f \nu)$ and $\tau_R = (\rho_p - \rho_f)d_p^2/(18\,\rho_f \nu)$. The symbols $\rho_p, \rho_f, g, d_p,$ and ν are the particle density, fluid density, the gravitational acceleration constant, the particle diameter, and the kinematic viscosity of the fluid, respectively. The particle settling velocity and response time were estimated to be 2.18×10^{-5} m s^{-1} and 2.22×10^{-6} s, respectively. As these values were insignificant compared to the mean velocities and sampling time used in the tests, it was predicted that within the fluid domain, the seeding particles would faithfully follow the fluid flow. The response time prediction was later verified using the viscous time scale $\tau_f = \nu/U_\tau^2$ as the characteristic time scale of the flow and the Stokes number evaluation (τ_R/τ_f). For the conditions tested, the time scales of the entry flow were 4.76×10^{-3} s and 2.03×10^{-3} s, respectively. These yielded Stokes numbers of 4.67×10^{-4} and 1.10×10^{-3}, respectively, which were less than 5×10^{-2}, as recommended by Samimy and Lele [39].

Deliberate steps were taken to focus the laser into a thin sheet of light of ~1.5 mm, keep the laser light sheet thickness in the area of interest, reduce glare on solid surfaces within the flow, and avoid reflections of solid components within the immediate environs of the field of view. Furthermore, the intensity of the laser was carefully modulated so that a sufficient level of illumination was maintained within all parts of the desired test section (including the porous medium). The laser pulse separation time was also selected through an iterative process so that the particle displacement was less than a quarter of the interrogation area [40]. Flows over porous media are often characterized by large mean velocity gradients at the interface between the porous medium and the overlying free zone flow. To keep velocity gradient bias errors due to such large gradients negligible, the displacement field variation was set at a value far less than the root mean square of the pixel size and particle image diameter. This was accounted for in the estimation of the laser pulse separation time.

To ensure that the images captured were entirely due to the light scattered within the desired plane of interest, the test section was darkened and exposed only to light from the laser. Furthermore, the camera lens was fitted with an orange filter of a band-pass wavelength of 532 nm $\pm$ 10 nm. By utilizing particle image diameters of 2 to 4 pixels, the effects of peak-locking were minimized. Using a field of view of 107 mm $\times$ 87 mm in the x and y directions, respectively, the scale factor of the measurement was 15 pixels per mm. For each measurement, 6000 instantaneous image pairs were acquired. Of this, at least 4000 were found to be more than sufficient to meet statistical convergence and to guarantee a substantial level of accuracy. The image sampling rate for each measurement was fixed at 4 Hz. During the processing stage, a substantial portion of the test section outside the flow domain was masked out. To obtain an ample number of valid vectors, the initial interrogation area was set to a size of 64 pixels × 64 pixels. Several iterations were performed, followed by a validation step to remove outliers. Ultimately, each interrogation window was also subdivided into 32 pixels × 32 pixels, so that for a sub-pixel accuracy of 0.1, the dynamic range was estimated to be 80. Furthermore, by setting the overlap between neighboring interrogation areas at 75%, additional vectors were provided so that the distances between neighboring vectors in physical units were 0.53 mm in both the x and y directions.

2.2. Velocity Data and Measurement Uncertainty

Measurements were conducted to capture six x–y (or z) planes of the test section per test condition. To facilitate such measurements, the laser and the camera were fixed on a translation stage so that they could be traversed in a composite manner in each streamwise and spanwise plane without distorting or modifying the distance between the laser and camera. Giving the precision of the translation stage, specific z plane locations could only be set to position within ± 0.5 mm. Regarding the schematic diagram in Figure 1b, the upstream conditions for each of these flow speeds were measured at x–y plane P1 for which the laser sheet of light was located at $z = 0$. Additionally, to study the spanwise variations in velocity over at least two unit cells of the porous medium array, five x–y planes were measured for the laser sheet of light located at the porous medium model section (i.e., P2). These five were conducted at $z = l$, $0.5l$, 0, $-0.5l$, and $-l$, respectively, indicated by lines R4, R5, R6, R7, and R8 in Figure 1c. This translates to a spanwise resolution measurement of ~3 mm, which is reasonable given that the thickness of the sheet of laser light was not better than 1.5 mm.

In this work, time-averaged instantaneous velocities and turbulence statistics as well as line-averaged values of the time-averaged components are reported. The components of the time-averaged velocities in the streamwise (x) and wall-normal (y) directions are respectively signified by mean velocities U and V. Line-averaged velocities and statistics of velocity data are designated by the corresponding parameters in angle brackets. Thus, for a parameter A (such as the streamwise velocity U or a turbulent statistical parameter), the line-averaged component computed between two rods in the Cnth and C$(n+1)$th column is given by:

$$<A> = \frac{1}{l} \int_{Cn}^{C(n+1)} A(x, y = b, z = R) dx \quad (1)$$

The symbols b and R in Equation (1) are given locations in the y and z directions. For further clarity, the reader is referred to column designations in Figure 1b.

An examination of errors and their propagation was undertaken as conducted by Wieneke [41]. The uncertainties of the mean velocities, turbulence intensities, Reynolds normal stresses, and Reynolds shear stresses were estimated to be no more than ±1.8%, ±2.3%, ±2.5%, and ±3.5% of their respective peak values.

2.3. Test Conditions

The experiments consisted of two test conditions in which the same porous medium model was subjected to two flow speeds. The test conditions are summarized in Table 1. The data for the entry flow shown in the table were extracted at $x = -10l$. As shown, the bulk (or global) flows measured were such that yielded a bulk Reynolds number Re_H (based on the bulk streamwise velocity U_b, and channel depth H) of 14,338 for the lower Reynolds number condition (labeled test condition 'A'), and Re_H of 24,510 for the higher Reynolds number condition (denominated as test condition 'B').

Salient boundary layer parameters such as the entry flow maximum velocity U_e; boundary layer thickness δ (wall-normal distance at $U = 0.99U_e$); displacement thickness δ^* ($\approx \int_0^\delta (1 - U/U_e) dy$); momentum thickness θ^* ($\approx \int_0^\delta U/U_e (1 - U/U_e) dy$); and shape factor H ($=\delta^*/\theta^*$) were computed for the two test conditions. The Reynolds numbers based on U_e and θ^* were shown to be 461 and 604. Each of the streamwise components of the mean flow data for the respective test conditions was plotted in the outer coordinates in Figure 2a and inner coordinates in Figure 2b. In the latter, the plot was fitted to the classical log law [42]

$$U^+ = \frac{1}{\kappa} \ln y^+ + B \qquad (2)$$

where $U^+ = U/U_\tau$ and $y^+ = y\, U_\tau/\nu$; and the von Kármán constant κ and logarithmic law constant B used were respectively 0.41 and 5 [25,43].

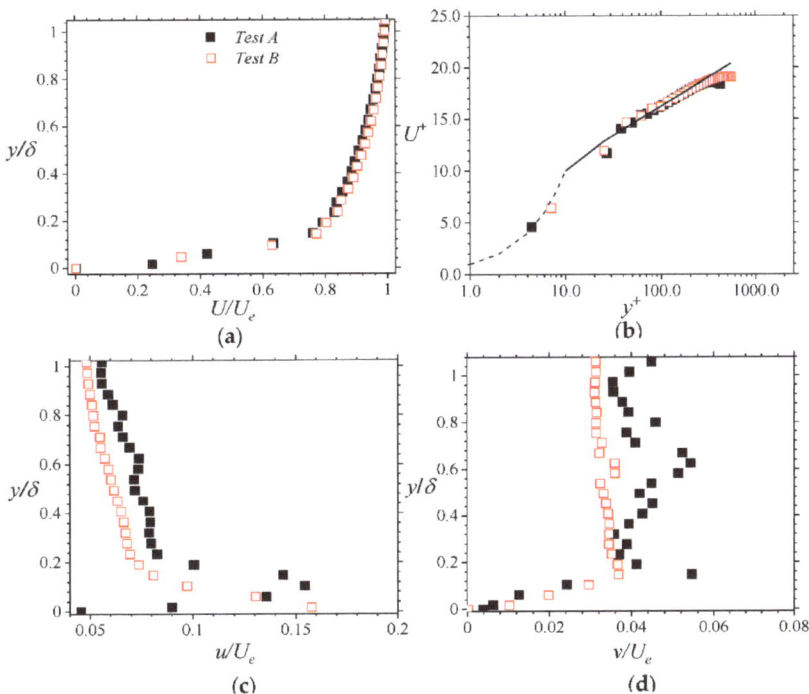

Figure 2. Boundary layer profiles of entry flow in the (**a**) outer and (**b**) inner coordinates for the two test cases. Plots in (**b**,**c**) are the turbulence intensities in the streamwise direction (*u*), and turbulence intensities in the wall-normal direction (*v*). Note that U_e is the maximum velocity of the entry flow streamwise component of the mean velocity (U); $U^+ = U/U_\tau$ where U_τ is the friction velocity, and $y^+ = y\, U_\tau/\nu$ where ν is the kinematic viscosity of the fluid. Note that the legend shown in (**a**) applies to (**b**–**d**).

Using the Clauser plot technique, [44], the (global) friction velocity U_τ for the entry flow was determined for each test condition. The skin-friction coefficient C_f was also computed as $2\,(U_\tau/U_e)^2$. The streamwise and wall-normal turbulence intensities (denoted by u and v, respectively) are also plotted in Figure 2c,d. As shown, the relative background turbulence level (u/U_e, v/U_e) measured at the edge of the boundary layer were approximately 5% and 4%, respectively in the x and y directions. This level of turbulence compares reasonably well with values of approximately 0.04 ± 0.01 reported in other zero pressure gradient turbulent boundary layer studies.

Following the procedure used in prior work [45], the Kolmogorov length scale (η) and the Taylor microscale (λ) were assessed, assuming local equilibrium. Hence, for the entry flow, η ($\approx 3\nu/U_\tau$) was found to be 25% to 39% of the vector spacing. Similarly, λ ($\approx \sqrt{15(U_e\eta^2)/\nu}$) was also found to be 86 to 122 times the vector spacing. This means that although the spatial resolution of the current tests was inadequate to resolve the Kolmogorov (smallest) length scale, the Taylor micro-scales (which contribute effectively to turbulence statistics) were sufficiently resolved.

3. Results and Discussion

In this section, the results of the tests are presented by considering the flow within the porous region, at the interfacial region, and then in the open region, respectively. The presentation and discussion are illustrated with tabulated results, one-dimensional line-averaged plots, vector plots, and contour plots. However, to maintain brevity and clarity of presentation, for one-dimensional plots, streamwise variations of the flow phenomena are largely demonstrated using multiple line-averaged data obtained from measurements made in the $z = 0$ plane. The line averages were conducted between adjacent centers of selected rods as schematized and denominated in Figure 1b. The selected pair of columns of rods were columns 1 and 2 (labeled C1–2), columns 4 and 5 (labeled C4–5), columns 6 and 7 (labeled C6–7), columns 8 and 9 (labeled C8–9), and columns 11 and 12 (labeled C11–12). Likewise, for conciseness, assessments of spanwise variations were undertaken using one-dimensional plots that were largely limited to line-averaged data from multiple spanwise planar measurements. The line averages were carried out between columns 8 and 9, and the measurement planes from which data were extracted are $z = l$, $0.5l$, 0, $-0.5l$, and $-l$. These planes are respectively indicated by R4, R5, R6, R7, and R8 in the plots as previously shown in Figure 1c. It should also be noted that for both vectors, streamlines, and contour plots, data/levels were respectively skipped to avoid congestion and to highlight prominent characteristics.

3.1. Flow within the Porous Region, Flow at Porous-Open Flow Interface

3.1.1. Mean Velocities, Vectors, Streamlines, and Vorticity

Some of the major aims of this study were to ascertain the extent to which the overlying mean turbulent flow influences the porous region, how that varies along the length and span of the compact porous medium, and how that varied with Reynolds number. To this end, line-averaged velocity plots of the mean streamwise and wall-normal velocities of the flow within the porous medium were first assessed. Notably, as indicated in Figure 3, the mean velocities were found to be three-dimensional, and the magnitudes of the flow were relatively low. The low flow within the porous region was not a surprise, given the obstruction posed by the rods. However, the magnitudes measured are worth highlighting given the high porosity of the porous medium under consideration. Even at 85% porosity, less than 1% of the mass flow was diverted through the porous region when the overlying flow was turbulent (Tables 2 and 3). Thus, the average streamwise velocity within the porous region ($<U_{b,p}>$) was of the order of just 1 mm s^{-1}, and the streamwise velocities were no more than 8% of the corresponding local line-averaged maximum velocity $<U_{max}>$.

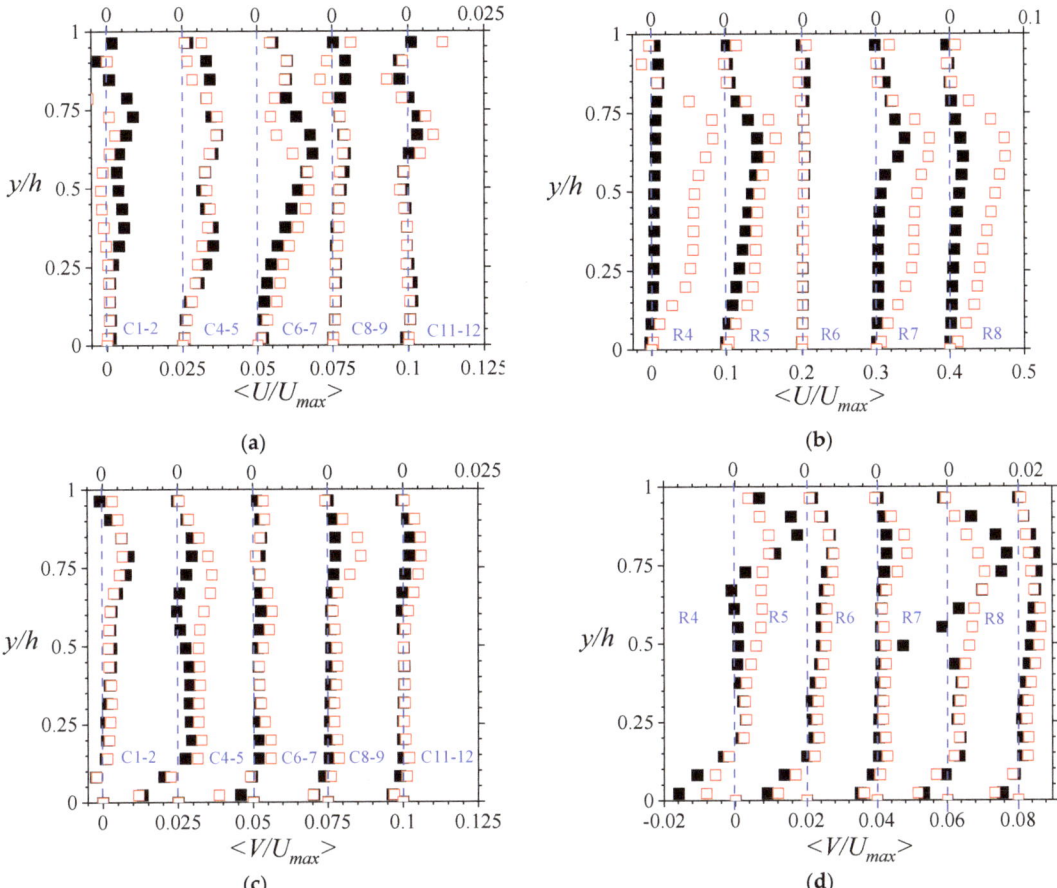

Figure 3. Parameters <U/U_{max}> and <V/U_{max}> are respectively line-averaged streamwise and wall-normal components of the mean velocity normalized by the maximum localline-averaged streamwise mean velocity in both open and porous regions. Here, the profiles for the flow within the porous medium are shown. Staggered plots (**a,c**) show the streamwise variations whereas (**b,d**) demonstrate the spanwise variations. The denominations C and R respectively stand for columns and rows, and are illustrated in Figure 1. Black-filled boxes are the results for Test A, and red-open boxes are the results for Test B.

It is also important to point out that the variation of the flow for the wall-normal coordinate is not uniformly linear. Even though the mid-plane flow is linear (as expected of flow in a plane passing through the rods), the majority of the cases indicated streamwise velocities that rose from zero at the lower wall. The velocities reached a maximum at about three-quarters of the depth of porous medium, and then dipped toward the interface. Contrary to observations by Florens et al. [5], the wall-normal mean velocities also followed a similar non-linear variation with the wall-normal coordinate (albeit muted in magnitude). Conspicuously, this variation includes negative gradients particularly close to the lower wall and in the planes outside of the midspan. These phenomena are not suggestive of flow channeling within the pores of the porous medium, nor can they imply a two-dimensional system. Rather, they are indicative of a complex flow activity within the porous medium.

Table 2. Summary results of porous media parameters showing streamwise variation in midspan plane.

Test	Parameter	Averaged Value at Location in Porous Medium				
		Column 1 and 2 (C1–2)	Column 4 and 5 (C4–5)	Column 6 and 7 (C6–7)	Column 8 and 9 (C8–9)	Column 11 and 12 (C11–12)
A	$<U_{b,p}>$ (mm/s)	1.38×10^{-3}	2.97×10^{-3}	3.99×10^{-3}	9.49×10^{-4}	8.46×10^{-6}
	$<U_s>$ (m/s)	4.32×10^{-3}	-5.60×10^{-4}	2.44×10^{-4}	-9.49×10^{-4}	2.61×10^{-3}
	Line-Averaged Shear Rate $<\Upsilon_0>$(1/s)	-1.41×10^1	4.83×10^{-1}	4.61×10^0	-1.52×10^0	-1.02×10^1
	$<U_{max}>$ (m/s)	4.19×10^{-1}	4.36×10^{-1}	4.46×10^{-1}	4.50×10^{-1}	4.54×10^{-1}
	Percentage Flow in Porous Medium (%)	1.09×10^{-1}	2.33×10^{-1}	3.17×10^{-1}	7.13×10^{-2}	6.29×10^{-4}
	$<U_s/U_{b,p}>$	3.13×10^0	-1.88×10^{-1}	6.12×10^{-2}	-1.00×10^0	3.08×10^2
	$<U_s/(\Upsilon_0 d)>$	-9.61×10^{-2}	-3.64×10^{-1}	1.67×10^{-2}	1.96×10^{-1}	-8.07×10^{-2}
	$<U_s/U_{max}>$	1.03×10^{-2}	-1.28×10^{-3}	5.47×10^{-4}	-2.11×10^{-3}	5.75×10^{-3}
B	$<U_{b,p}>$ (m/s)	8.68×10^{-4}	4.66×10^{-3}	6.57×10^{-3}	1.29×10^{-3}	8.29×10^{-4}
	$<U_s>$ (m/s)	3.87×10^{-2}	1.31×10^{-2}	4.23×10^{-4}	1.61×10^{-2}	2.17×10^{-2}
	Line-Averaged Shear Rate $<\Upsilon_0>$(1/s)	-8.65×10^1	-1.07×10^1	-1.72×10^0	-5.10×10^1	-5.93×10^1
	$<U_{max}>$ (m/s)	7.02×10^{-1}	7.22×10^{-1}	7.38×10^{-1}	7.47×10^{-1}	7.54×10^{-1}
	Percentage Flow in Porous Medium (%)	3.89×10^{-2}	2.06×10^{-1}	2.90×10^{-1}	5.51×10^{-2}	3.51×10^{-2}
	$<U_s/U_{b,p}>$	4.46×10^1	2.80×10^0	6.44×10^{-2}	1.24×10^1	2.62×10^1
	$<U_s/(\Upsilon_0 d)>$	-1.41×10^{-1}	-3.83×10^{-1}	-7.72×10^{-2}	-9.91×10^{-2}	-1.15×10^{-1}
	$<U_s/U_{max}>$	5.52×10^{-2}	1.81×10^{-2}	5.73×10^{-4}	2.15×10^{-2}	2.88×10^{-2}

Table 3. Summary results of porous media parameters showing spanwise variation for flow between columns 8 and 9.

Test	Parameter	Spatially Averaged Value at Location in Porous Medium				
		Row 4 (R4)	Row 5 (R5)	Row 6 (R6)	Row 7 (R7)	Row 8 (R8)
A	$<U_{b,p}>$ (m/s)	1.94×10^{-3}	8.61×10^{-3}	9.49×10^{-4}	4.44×10^{-3}	2.61×10^{-3}
	$<U_s>$ (m/s)	1.26×10^{-3}	1.37×10^{-3}	-9.49×10^{-4}	-9.08×10^{-4}	-1.49×10^{-3}
	Line-Averaged Shear Rate $<\Upsilon_0>$(1/s)	-1.41×10^1	4.83×10^{-1}	4.61×10^0	-1.52×10^0	-1.02×10^1
	$<U_{max}>$ (m/s)	4.43×10^{-1}	4.48×10^{-1}	4.50×10^{-1}	4.52×10^{-1}	4.52×10^{-1}
	Percentage Flow in Porous Medium (%)	1.53×10^{-3}	6.75×10^{-3}	7.53×10^{-4}	3.34×10^{-3}	1.94×10^{-3}
	$<U_s/U_{b,p}>$	6.50×10^{-1}	1.59×10^{-1}	-1.00×10^0	-2.04×10^{-1}	-5.71×10^{-1}
	$<U_s/(\Upsilon_0 d)>$	-2.80×10^{-2}	8.91×10^{-1}	-6.47×10^{-2}	1.88×10^{-1}	4.61×10^{-2}
	$<U_s/U_{max}>$	2.84×10^{-3}	3.06×10^{-3}	-2.11×10^{-3}	-2.01×10^{-3}	-3.30×10^{-3}
B	$<U_{b,p}>$ (m/s)	3.15×10^{-2}	2.50×10^{-2}	1.34×10^{-3}	3.00×10^{-2}	2.99×10^{-2}
	$<U_s>$ (m/s)	1.13×10^{-2}	2.63×10^{-2}	1.61×10^{-2}	3.44×10^{-2}	2.25×10^{-2}
	Line-Averaged Shear Rate $<\Upsilon_0>$(1/s)	-4.63×10^1	-6.36×10^1	-5.10×10^1	-7.93×10^1	-6.87×10^1
	$<U_{max}>$ (m/s)	7.28×10^{-1}	7.37×10^{-1}	7.47×10^{-1}	7.47×10^{-1}	7.51×10^{-1}
	Percentage Flow in Porous Medium (%)	1.41×10^{-2}	1.10×10^{-2}	5.94×10^{-4}	1.28×10^{-2}	1.27×10^{-2}
	$<U_s/U_{b,p}>$	3.58×10^{-1}	1.05×10^0	1.20×10^1	1.15×10^0	7.53×10^{-1}
	$<U_s/(\Upsilon_0 d)>$	-7.66×10^{-2}	-1.30×10^{-1}	-9.91×10^{-2}	-1.36×10^{-1}	-1.03×10^{-1}
	$<U_s/U_{max}>$	1.55×10^{-2}	3.57×10^{-2}	2.15×10^{-2}	4.60×10^{-2}	3.00×10^{-2}

To investigate this complicated flow activity further, vector plots, streamlines, and contours were examined. The mean vector plots for the conditions of measurement planes passing through rods of the porous medium (i.e., $z = 0, \pm l$) are shown in Figure 4. For both Reynolds numbers tested, the midspan vectors within the porous medium showed a combination of upwelling and downwelling events in no definite order. For the off-midspan planes, increasing the Reynolds number of the flow led to distinct patterns of flow circulation occurring at finite regions close to the lower wall and the interface. In Figure 5, it is also noted from the streamlines that for spanwise planes of flow between rods (i.e., $z = \pm 0.5l$), the vectors were of some structural order. Thus, vortices of different sizes and counter-rotating pairs were found in different locations of the flow in the porous region. The iso-contours of the mean spanwise vorticity ($\Omega_z = \partial V/\partial x - \partial U/\partial y$) presented in Figure 6 show that within the porous region, pockets of weak, but significant vorticity formed in both test cases, but in different spanwise locations. Of note, some of the spanwise vorticities were negative in value because the mean wall-normal velocity gradient in the streamwise coordinate ($\partial V/\partial x$) was less than the mean streamwise velocity gradient in the wall-normal direction ($\partial U/\partial y$) in those regions. Additionally, the vorticity pockets were also mostly located close to the lower wall, or at about two-thirds of the depth of the porous medium. The longitudinal extent of these pockets ranged from 10 to 30% of the depth of the porous medium. It is also important to point out that while the pockets of vorticity were longer in the upstream section of the porous medium, they shrunk and ultimately degenerated in the downstream section.

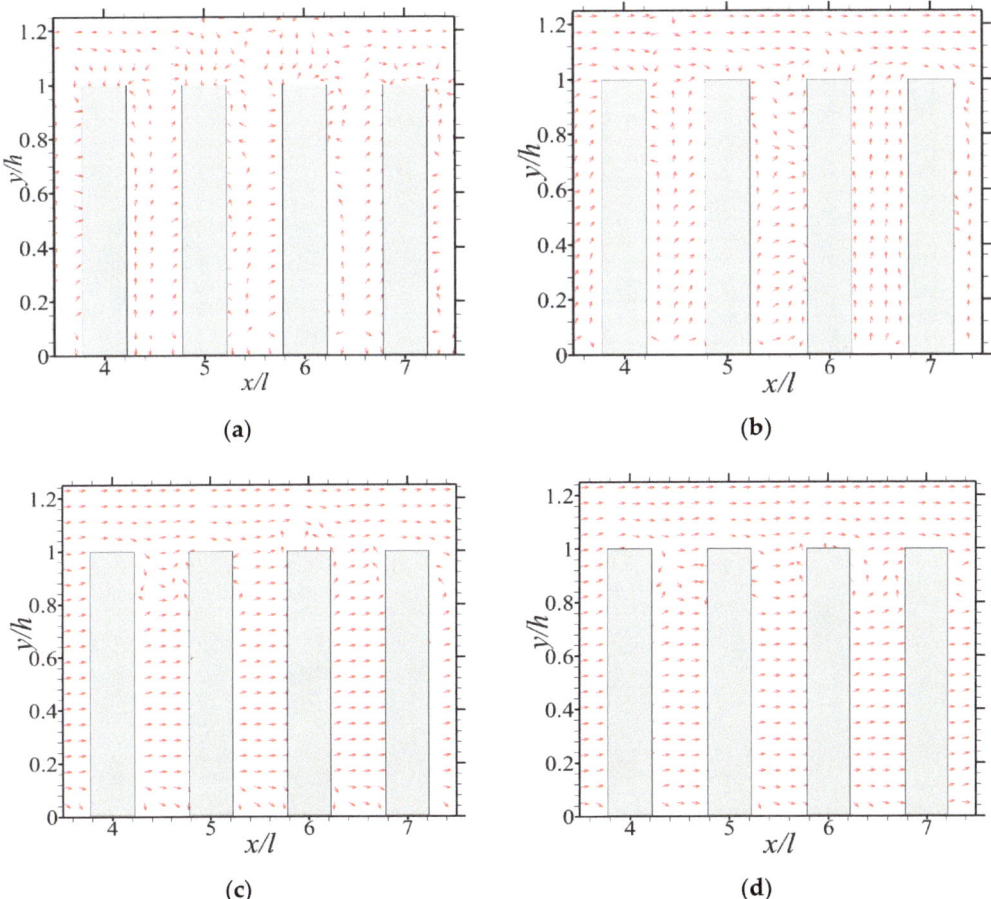

Figure 4. Vector plots for the selected rods at (**a**) $z = 0$ in Test A; (**b**) $z = 0$ in Test B; (**c**) $z = l$ in Test B; (**d**) $z = -l$ in Test B. Note that for each plot, gray-filled boxes indicate locations of porous medium rods. Vectors have also been skipped to maintain clarity.

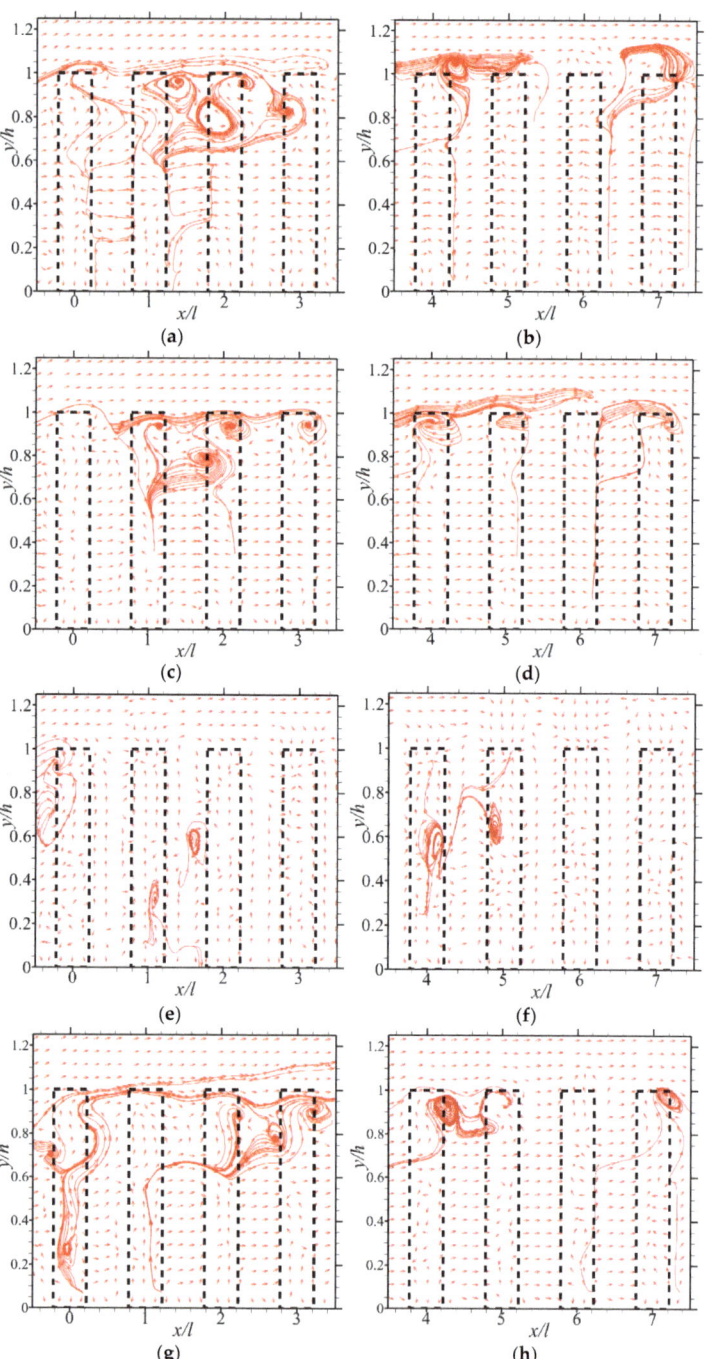

Figure 5. Vector plots with streamlines superimposed for selected rods at (**a**,**b**) $z = l/2$ in Test A; (**c**,**d**) $z = l/2$ in Test B; (**e**,**f**) $z = -l/2$ in Test A; (**g**,**h**) $z = -l/2$ in Test B. Note that for each plot, dashed lines indicate locations of porous medium rods. Vectors and streamlines have also been skipped in order to maintain clarity and to highlight features.

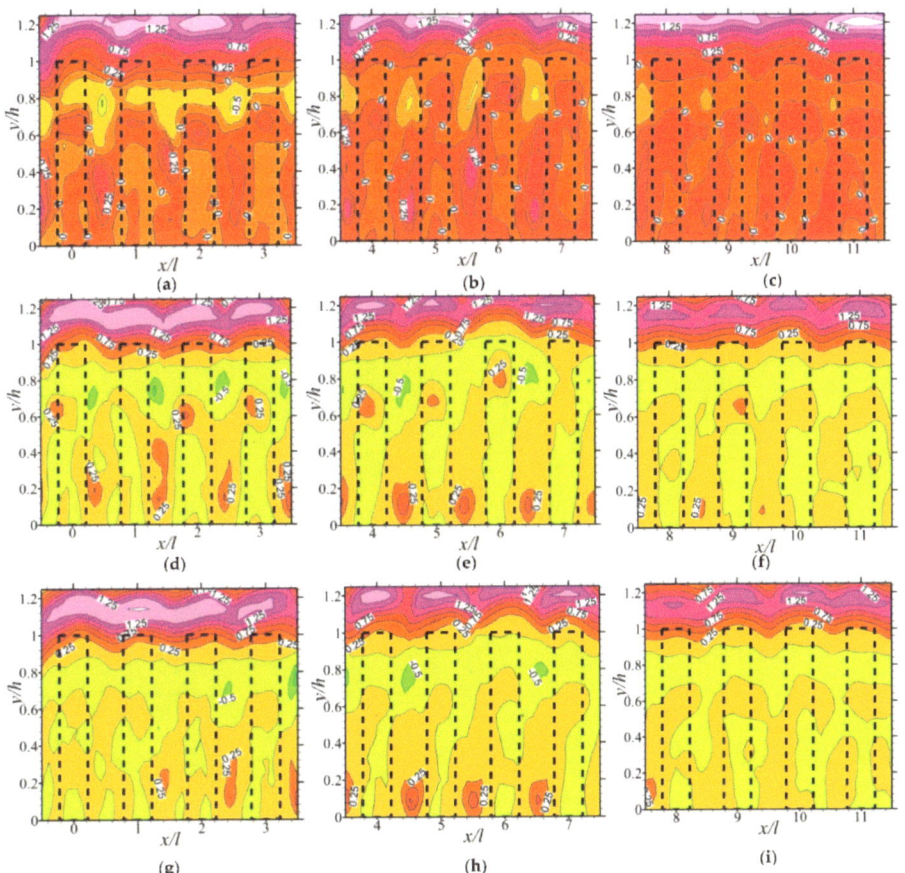

Figure 6. Iso-contours of mean spanwise vorticity $\Omega_z = \partial V/\partial x - \partial U/\partial y$ normalized by the maximum entry flow velocity and the rod diameter. Plots (**a**–**c**) are results of measurements taken at $z = l/2$ in Test A. Plots (**d**–**f**) were extracted from measurements taken at $z = l/2$ in Test B. Plots (**g**–**i**) were extracted from measurements taken at $z = -l/2$ in Test B. Note that for each plot, dashed lines indicate locations of porous medium rods.

The source of these vortical activities is open to debate. It is possible that these activities are entirely or partially characteristic of the flow behavior of the associated porous flow regime. Within the porous region, the interstitial velocities amount to a local hydraulic Reynolds number ranging from 18 to 668. Such a scope of Reynolds number translates to porous flow in a steady inertial flow regime, or in the extreme, a chaotic (turbulent) regime [46]. These regimes are respectively characterized by vortex formation or sub-pore scale turbulence along with intense vortex shedding [46–49]. Thus, it is possible that with the present porous medium flow being in the inertial flow regime (at the very least), vortices are being generated by shear close to the particle surfaces. A reasonable alternative explanation about the origins of this vorticial activity may, however, be found in the turbulent flow at the unobstructed sections close to the porous region. At such sections, there are zones of strong shear in the velocity close to the edge of the porous medium. Such shear triggers an instability of the Kelvin–Helmholtz (K–H) type, generating the canopy-scale turbulence observed in this work [13,35]. It is worth pointing out that in the particular case of a compact porous medium, prospective vorticity-generating sources at the edge of the porous medium are of two kinds. There is the leading edge of the porous region (i.e., the most upstream rods at $x = -d/2$), and the interface between the porous region at the overlying open flow region (at $y = h$). As both locations are regions of intense shear layers, either of them is capable of inducing K–H vortices. From thence, the vortices are advected elsewhere within the porous medium.

3.1.2. Turbulence Intensities and Reynolds Stresses

The nature of turbulence within the porous medium was first analyzed using plots such as those in Figure 7. In that figure, the line-averaged turbulence intensities in the streamwise direction $<u>$, and the corresponding component in the wall-normal direction $<v>$ are shown. While the pore-level turbulence intensities were less than 6% of the maximum line-averaged local streamwise mean velocity, they vary in the streamwise, wall-normal, and spanwise directions. Furthermore, they are Reynolds number dependent.

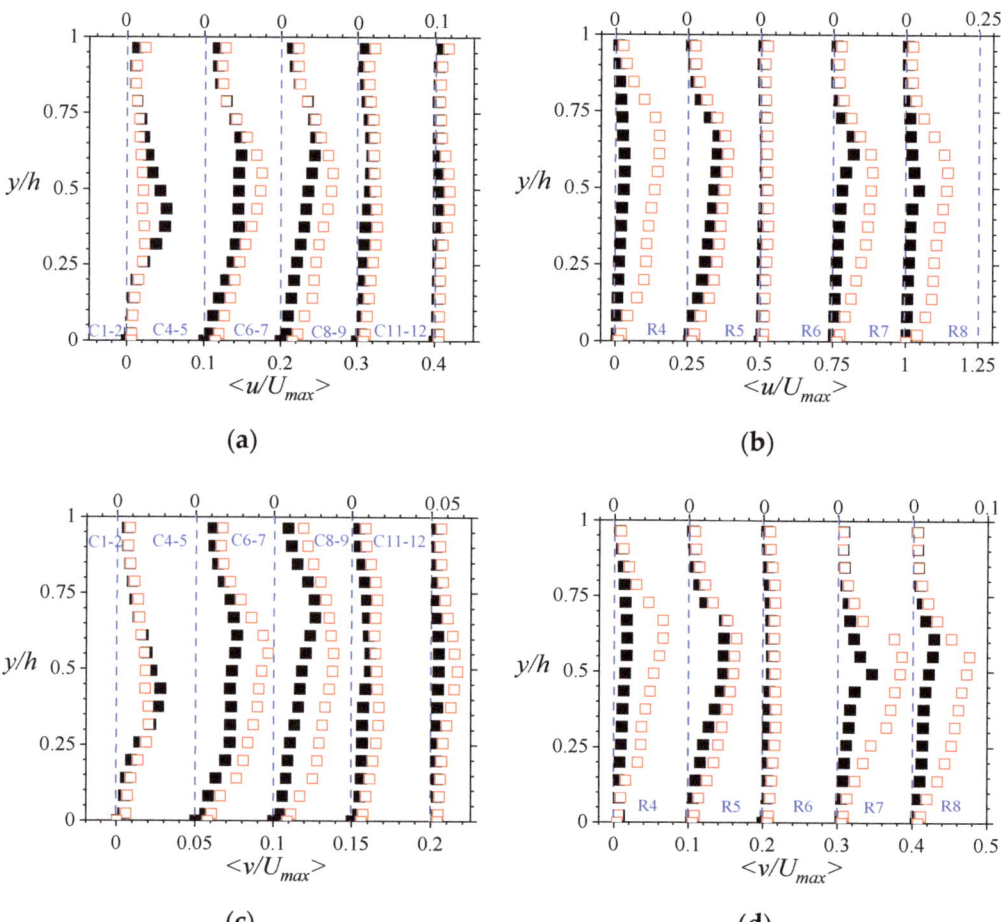

Figure 7. Parameters $<u/U_{max}>$ and $<v/U_{max}>$ are respectively line-averaged streamwise and wall-normal turbulence intensities normalized by the maximum local line-averaged streamwise mean velocity in both open and porous regions. Here, the profiles for the flow within the porous medium are shown. Staggered plots (**a**,**c**) show the streamwise variations whereas (**b**,**d**) demonstrate the spanwise variations. The denominations C and R respectively stand for columns and rows, and are illustrated in Figure 1. Black-filled boxes are for results for Test A, and red-open boxes are for results for Test B.

It is worth noting, however, that in all cases, the streamwise components of the turbulence intensities were larger than the wall-normal components. This is expected, given that the mean flow is predominantly driven in the streamwise direction, and the wall-normal components are going to be suppressed by the walls that are normal to it. Along the stream, however, the deviation of the velocity components from the mean increased over the first seven columns of rods. The profiles also showed that within the porous medium, the peak intensities were registered at about 0.5–0.6 of the depth of the porous medium. The observance of high intensities at rod mid-height and upstream rod areas can be explained by the compounding disturbance of the flow due to the rods as well as the vortical activity close to those regions. Consequently, the turbulence intensities decline as the flow moves past the eighth column of rods where vortices devolve thereon, and the flow encounters an unobstructed domain downstream. These observations suggest then that the utmost utility of a rod arrangement such as a turbulence generator is gained using a limited number of columns of rods (<8) along the stream.

The spanwise variations of the turbulence intensities also revealed low turbulence levels at the midspan, compared with other spanwise locations. Indeed, the intensities in the midspan plane were so low that their variations with depth appeared to be linear. The reason for these differences in velocities in the present study may be rationally attributed to the side wall effects due to the narrow wall channel used in the tests. The sheer proximity of the solid side walls is expected to increase the turbulence intensities close to those walls. This explains why some of the highest intensities were recorded at measurements taken at $z = \pm l$. Concerning Reynolds number variation, intensity increments are expected, as the random nature of the flow within the porous medium will only worsen as the flow speed (and consequently the Reynolds number) is increased.

The normalized Reynolds stress distributions were considered to determine if there were any further significant turbulent features within the porous medium. These are plotted in Figure 8a–d for line-averaged normal stresses in the streamwise direction $<u^2>$ and wall-normal direction $<v^2>$, and in Figure 8e,f for line-averaged shear stresses in the streamwise-wall-normal plane $<-uv>$. With a greater presence of solid boundaries along the stream, it is expected that the velocities normal to those surfaces will be suppressed more than the component parallel to the surface [50]. Hence, $<v^2>$ was found to be less than $<u^2>$, suggesting an anisotropic turbulent field. It was also observed that some of the identifiable features of the turbulence intensities were present in the Reynolds normal stresses. The peak values were at approximately the same location as that of the turbulence intensities; additionally, the upstream columns of rods also showed some significant streamwise and wall-normal variations that were Reynolds number dependent. It is therefore reasonable to assume that some of the physics dictating the trends of the turbulence intensities and the Reynolds normal stresses are similar. The Reynolds shear stresses, however, were complex and very different from those reported by Manes et al. [18]. For the upstream rods, the Reynolds shear stresses generally declined with porous medium depth to a negative value, and then resulted in a null value. While the downstream rods showed a linear profile, there were distinctive changes in the off-midspan plane profiles. As these Reynolds shear stresses were produced by the mean shear in the streamwise-wall-normal plane within the turbulent flow, they recorded significant values at regions of substantial shear such as observed at regions close to the sidewalls.

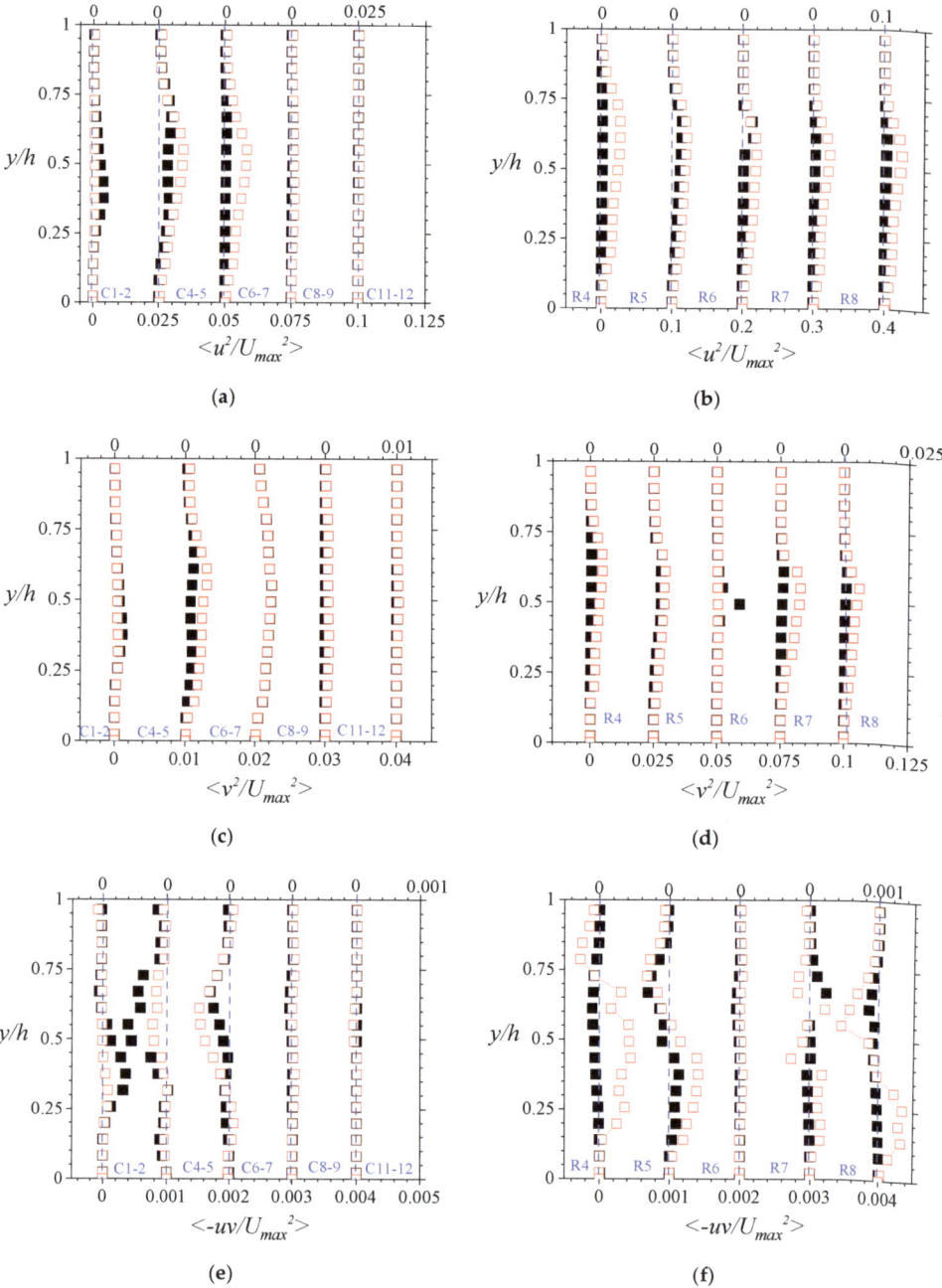

Figure 8. Turbulence quantities $<u^2/U_{max}^2>$, $<v^2/U_{max}^2>$, and $<-uv/U_{max}^2>$ are respectively line-averaged streamwise Reynolds normal stress, wall-normal Reynolds normal stress, and Reynolds shear stress in the streamwise-wall-normal plane, normalized by the maximum local line-averaged streamwise mean velocity in both open and porous regions. Here, the profiles for the flow within the porous medium are shown. Staggered plots (**a**,**c**,**e**) show the streamwise variations whereas (**b**,**d**,**f**) demonstrate the spanwise variations. The denominations C and R respectively stand for columns and rows, and are illustrated in Figure 1. Black-filled boxes are the results for Test A, and red-open boxes are the results for Test B.

3.1.3. Interfacial Flow

For a dense submerged canopy such as that under study, the drag due to the porous medium is expected to be larger than that of the lower wall. Thus, the discontinuity in drag that occurs at the interfacial edge will give rise to a region of shear that resembles a free shear layer with an inflection point near the interface [35]. The generation and propagation of K–H vortices from this shear layer suggest that the interface is a determinant of the degree of communication between the open and porous regional flows. Therefore, an understanding of the interfacial flow is integral to a proper view of the entire flow domain. The interfacial flow is known to be a function of several parameters including the specific permeability of the porous medium, average velocity within the porous medium, the velocity at that interface (i.e., the slip velocity), shear rate at the interface (i.e., $\gamma_0 = dU/dy |_{y=h}$), and the channel's local maximum velocity [51]. In this work, the interfacial flow was analyzed by focusing attention on measurements (or interpolations) of the slip velocity and its associated shear rate. The spatial variations of these select parameters were then studied with regard to changes in the bulk Reynolds numbers of the flow.

The reported slip velocity $<U_s>$ is defined as the line-averaged streamwise mean velocity at the plane tangent to the edges of the rods in most immediate contact with the open flow region (i.e., $y = h$). Particle image velocimetry measured velocities over a finite interrogation area whose center was not located exactly at this defined interface. Thus, it was not possible to provide the averaged slip velocities at the interface. Consequently, in this study, slip velocities could only be determined within interfacial location uncertainties of ± 0.27 mm (which is half the size of an interrogation window). The line-averaged interfacial mean shear rate $<\gamma_0>$ was also determined by using a curve of six or more streamwise velocity data points located at the region close to the interface. To assure some accuracy in the procedure, interpolations of the line-averaged data were first obtained. Differentiation of the curve was then performed smoothly over data points covering a distance of over 0.27 mm. The slip velocity and interfacial shear rates were then studied using three dimensionless parameters, namely $<U_s/U_{b,p}>$, $<U_s/(\gamma_0 d)>$, and $<U_s/U_{max}>$. By using $<U_s/U_{b,p}>$ and $<U_s/(\gamma_0 d)>$, the line-averaged slip velocities were respectively assessed in terms of the streamwise bulk flow within the porous medium $<U_{b,p}>$ and the local penetration by the shear. On the other hand, by using $<U_s/U_{max}>$, the relative effects of the overlying turbulent flow on the line-averaged slip velocities were determined. The results are summarized in Tables 2 and 3.

The data in both tables show that the interfacial velocities dropped from their initial values in the upstream columns of rods, and then rose again with downstream rods. The rate at which shearing was applied increased to a peak just over the columns of rods, and subsequently declined as the region of obstruction was approached. The results also indicate that there were sharp flow reversals at the interfacial region, resulting in negative slip velocities (in some cases). Furthermore, there were negative shear rates at the interfaces. The flow reversals are consistent with some of the observations made regarding vector plots in Figure 4. While the negative shear rates were more confounding, they are still reasonable given that velocities higher than the slip velocities were identified just beneath the interface in Figure 3a. On the whole, it is remarkable that relative to the bulk velocity within the porous medium, the slip velocities were dependent on the spatial location and the Reynolds number, and not a constant value as observed by Suga et al. [7]. For $<U_s/(\gamma_0 d)>$ and $<U_s/U_{max}>$, the slip velocities were also found to be greatest just after the most upstream rods. They also increased with Reynolds number. Across the span of the porous medium, however, both slip velocity and shear rates were dependent on the spatial location, and all the dimensionless parameters ($<U_s/U_{b,p}>$, $<U_s/(\gamma_0 d)>$, and $<U_s/U_{max}>$) tended to be symmetrical about the midspan at the higher Reynolds number.

The overall conclusions regarding the interfacial flow are that while the slip velocities recorded in this work were largely substantial in value compared to the bulk flow within the porous medium, they were mostly a small fraction of the maximum velocity ($\leq 6\%$). If the absolute value of $<U_s/\gamma_0>$ is a measure of the depth of penetration of the turbulent flow, then the $<U_s/(\Upsilon_0\,d)>$ parameter determined tells us that the average depth of such infiltration through the interfacial zone is small (~7% of the rod diameter). However, this penetration is effective in creating some significant measure of porous flow just beneath the interface.

3.2. Flow Development over the Porous Wall

In this section, the results of the flow above the porous wall are presented and discussed. Consideration of the flow is limited to 1.25 h above the porous wall.

3.2.1. Boundary Layer Characteristics

Having considered the flow within the porous medium, attention is now turned to the overlying turbulent flow. To characterize the boundary layers of the selected line-averaged streamwise velocities, values of the maximum streamwise component of the mean velocity $<U_{max}>$, boundary layer thickness δ, streamwise location of the edge of the boundary layer (i.e., y/h value at $0.99<U_{max}>$), displacement thickness δ^*, momentum thickness θ^*, Reynolds number based on momentum thickness and maximum velocity Re_θ, and shape parameter H were determined from the line-averaged plots. These were obtained for selected locations, according to definitions of parameters given in Section 2.3. Apart from $<U_{max}>$, (which is summarized in Tables 2 and 3), all the other relevant boundary layer values are presented in Tables 4 and 5 and plotted in Figures 9 and 10. It should be noted that for the convenience of analyses, the streamwise locations of the boundary layer data are taken as the center between the rods. Furthermore, the values of relevant entry flow boundary layer parameters are included in Figure 9 to show the effects of the porous medium on the flow.

Table 4. Boundary layer and friction parameters for flow over porous media in the midspan plane.

Test	Mean Streamwise Distance x (mm)	Boundary Layer Thickness δ (mm)	Displacement Thickness δ^* (mm)	Momentum Thickness θ^* (mm)	Shape Factor H	Momentum Thickness Reynolds Number Re_θ	Friction Velocity (m/s)
A	10.79	5.18	2.20	0.54	4.07	226	0.272
	32.36	6.99	3.18	0.67	4.74	293	0.315
	46.74	8.50	4.38	0.76	5.74	340	0.307
	61.12	8.35	3.13	0.91	3.45	409	0.257
	82.69	9.61	3.11	1.09	2.84	496	0.200
B	10.79	4.66	1.46	0.49	3.00	341	0.323
	32.36	6.58	2.51	0.73	3.46	525	0.345
	46.74	7.02	3.26	0.86	3.77	637	0.440
	61.12	7.89	2.63	0.95	2.76	711	0.359
	82.69	8.89	2.54	1.09	2.33	822	0.319

Table 5. Boundary layer and friction parameters for flow over porous media. Here, the results show spanwise variation for flow between columns 8 and 9.

Test	Mean Streamwise Distance x (mm)	Boundary Layer Thickness δ (mm)	Displacement Thickness δ^* (mm)	Momentum Thickness θ^* (mm)	Shape Factor H	Momentum Thickness Reynolds Number Re_θ	Friction Velocity (m/s)
A	7.2	8.66	3.62	0.91	3.97	404	0.266
	3.6	8.79	3.05	0.95	3.22	425	0.244
	0.0	8.35	3.13	0.91	3.45	409	0.257
	−3.6	8.56	3.43	0.88	3.88	399	0.272
	−7.2	9.25	3.31	0.89	3.73	400	0.258
B	7.2	7.92	2.52	0.90	2.81	653	0.365
	3.6	8.10	2.43	1.00	2.43	737	0.333
	0.0	7.89	2.63	0.95	2.76	711	0.359
	−3.6	8.10	2.35	0.99	2.36	743	0.335
	−7.2	8.10	2.26	0.91	2.48	686	0.348

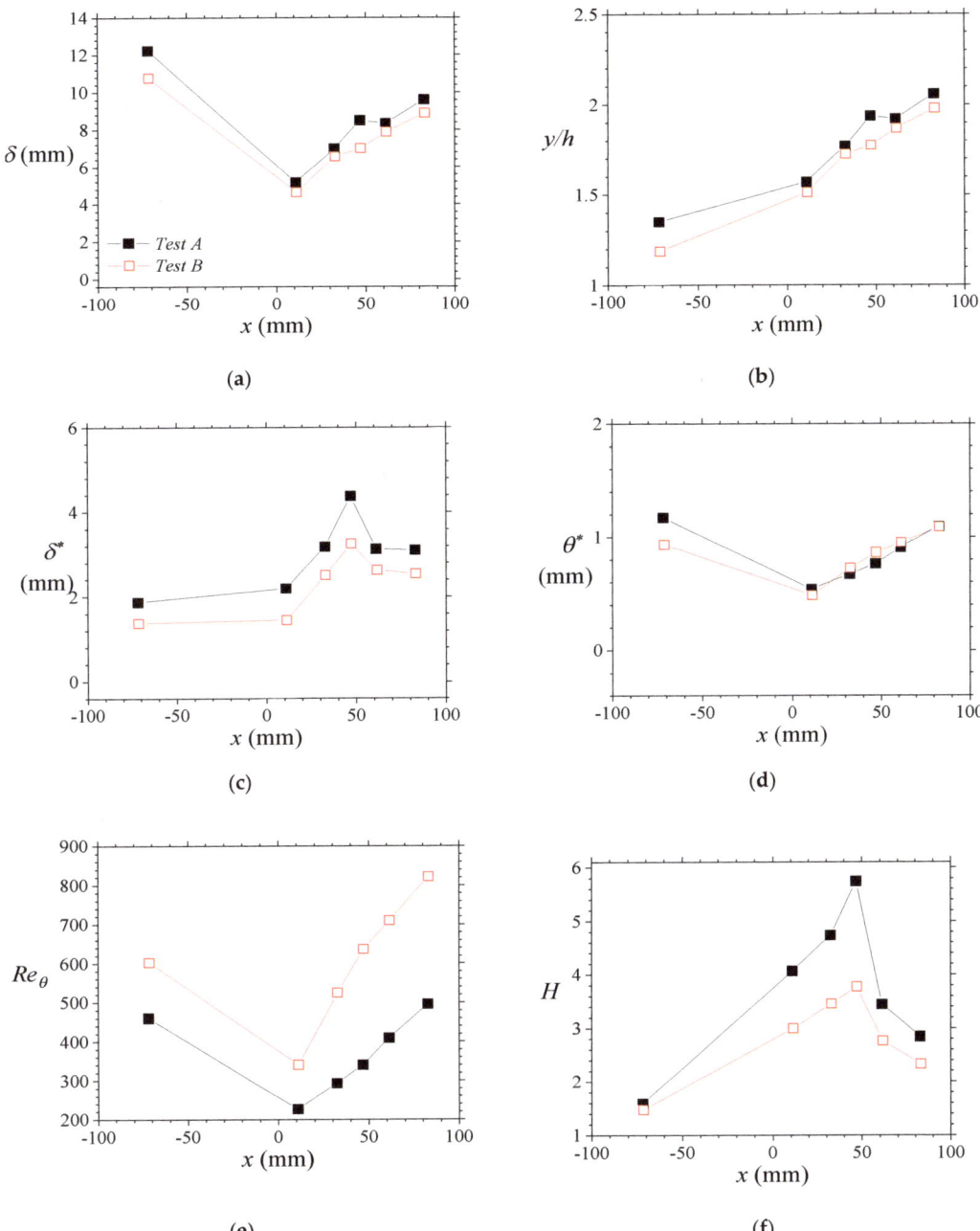

Figure 9. Streamwise variation of (**a**) boundary layer thickness; (**b**) location of the edge of the boundary layer (i.e., y/h values at $0.99<U_{max}>$); (**c**) displacement thickness; (**d**) momentum thickness; (**e**) Reynolds number based on momentum thickness and maximum velocity; and (**f**) shape parameter. Note that the legend showed in (**a**) applies to (**b**–**f**).

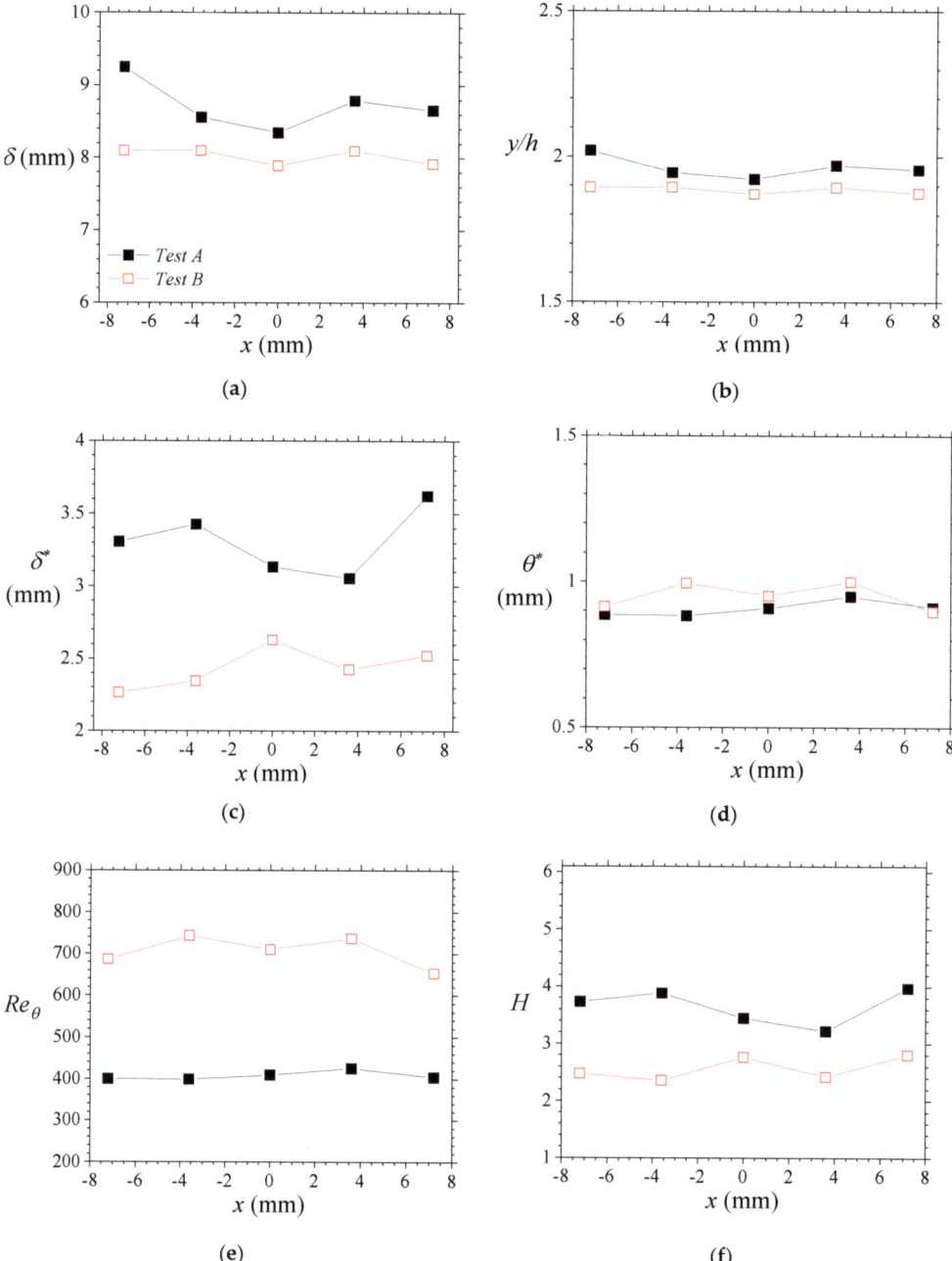

Figure 10. Spanwise variation of (**a**) boundary layer thickness; (**b**) location of the edge of the boundary layer (y/h values at $0.99<U_{max}>$); (**c**) displacement thickness; (**d**) momentum thickness; (**e**) Reynolds number based on momentum thickness and maximum velocity; and (**f**) shape parameter. Note that the legend showed in (**a**) applies to (**b**–**f**).

By introducing the porous medium, a 21% blockage was created in the flow section, resulting in an increased acceleration of flow in the open region. Some of the results of this increment were a 27% to 32% rise in maximum velocities in the open region, and an upward shift in the location of the maximum flow. The latter can particularly be seen in Figure 9b, where the edge of the boundary layer changes substantially from ~1.3 h at the entry flow to ~1.5 h at the upstream rods. There were also significant reductions in δ and θ^*, especially at the most upstream rods relative to the entry flow, following the characteristics associated with flows undergoing favorable pressure gradient changes [52,53]. However, due to the opposing effects of the porous medium (in generating significant mass flux deficit), similar reductions in δ^* were resisted.

On the other hand, the porous boundary appeared to diffuse viscous effects into the mean turbulent flow in the open region. Thus, as the boundary layer developed over the porous medium, it thickened along the stream by about 90% at the trailing rods compared to its value over the leading rods. This boundary layer development also led to a monotonic increase in θ^*, and an initial augmentation in δ^*. It must be noted that both δ^* and θ^* are respectively indicators of the mass and momentum flux deficits. For the present tests, both parameters ultimately increased in value at the trailing rods (compared to that at the leading rods of the porous medium). Thus, it may be concluded that the current porous medium significantly enhanced mass and momentum deficit (by 42–102%). Consequently, with the 30% increase in maximum streamwise mean velocity, Re_θ magnified much more significantly, and more prominently at the higher Reynolds number flow.

The relative effects of δ^* and θ^* in the flow arrangement are accounted for in the shape factor H trends indicated in Figure 9f and Table 4. For the entry flow, H compared well with previous results of turbulent flows over a smooth wall at zero pressure gradient and at similar Re_θ [54]. However, along the streamwise direction on the porous wall, the values of H were found to be high, and characterized by two distinct zonal attributes. There was the first zone upstream of the porous medium, marked by a sharp increase in mass deficit flux compared to momentum deficit flux along the stream. There was also the second zone where in anticipation of lower flux deficits in the unobstructed flow downstream, the displacement thickness dropped sharply to a constant value while the momentum deficit flux continued to increase incrementally toward the trailing edge of the porous medium. Thus, H declined in the flow over the trailing rods of the porous medium. The value of H measured between columns 11 and 12 of the porous medium rods showed that the porous medium more than doubled the shape parameter of the boundary layer relative to the entry flow. This is predictable for obstructions in flow that generate higher drag characteristics such as that expected of the porous medium in the current flow arrangement [55].

Regarding spanwise changes in boundary layer characteristics, Figure 10 and Table 5 reveal variations were less dramatic compared to the streamwise changes. However, it is noteworthy from the data that lower flow speeds tended to display three-dimensional effects. In the current arrangement, this resulted in relative deviations that were no more than 16% of the value at the mid-plane. Such maximum deviations occurred at locations closer to the sidewalls.

3.2.2. Mean Velocities, Momentum Flux, and Vorticity

The mean velocity profiles in the streamwise directions are shown in the outer co-ordinates in Figure 11. As indicated in Section 3.1.3, there were flow reversals at the interface, specifically for mid-span line-averaged measurements between columns 4 and 5, and columns 8 and 9, respectively. The line averaged shear rates were also negative in value. All of these, along with the high H values observed in Figures 9 and 10, point to complications in flow characteristics similar to those observed in adverse pressure gradient flows [55]. This behavior is possibly due to the intermittent pores on the surface of the porous medium.

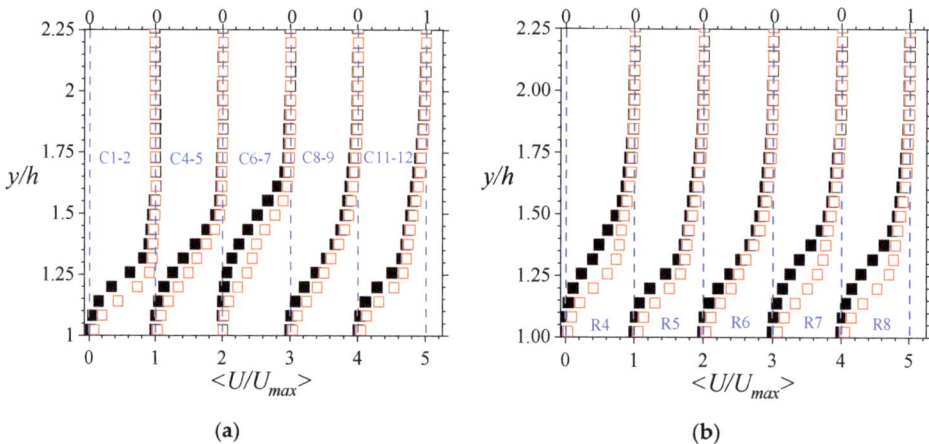

Figure 11. Turbulence quantity <U/U$_{max}$> is the line-averaged mean streamwise velocity normalized by the maximum local line-averaged streamwise mean velocity in open and porous regions. Here, the profiles for the flow in the open region are shown. Staggered plots in (**a**) show the streamwise variations whereas (**b**) demonstrate the spanwise variations. The denominations C and R respectively stand for columns and rows, and are illustrated in Figure 1. Black-filled boxes are the results for Test A, and red-open boxes are the results for Test B.

The streamwise mean velocities were also critically examined to ascertain the validity of a logarithmic region in the flow profile over the porous wall during flow development. The associated friction velocities U_τ were also determined. In doing this, a modified form of the procedure outlined in Section 2.3 for rough wall turbulent flow [6,55,56] was followed. The normalized mean plots were matched with the following modified log law [42,55].

$$U^+ = \frac{1}{\kappa} ln(y^* + y_o)^+ + B - \Delta B^+ \qquad (3)$$

where $U^+ = U/U_\tau$ and the von Kármán constant κ and logarithmic constants B were 0.41 and 5, respectively. It should be noted, however, that in this modified log law, the variable $y^+ = y\, U_\tau/\nu$ is defined as $(y^* + y_o)\, U_\tau/\nu$. This is in line with the customary practice of rough wall experiments where the 'roughness' effects are considered in the definition of y. In the current setup, y is defined as the sum of the wall-normal distance from the top plane of the rods y^*, and a virtual origin y_o. The parameter ΔB^+ is a 'roughness' function that measures the increase in local drag due to the presence of roughness elements on the surface [57]. Thus, it is seen as a downward shift of the log law profile of a comparable turbulent flow over a smooth wall. Following precedence then [6,55], the Clauser plot technique was used to simultaneously estimate U_τ and y_o. Previous DNS results of turbulent flow over two-dimensional rods showed that at $1 < x/l < 9$ (i.e., the initial developing flow region), y_o/h ranges from 0.51 to 0.52 [58]. On this basis, an initial guess within this range was used. The optimized value of $y_o = 0.515\, h$ yielded U_τ values listed in Tables 4 and 5 with an uncertainty limit of $\pm 10\%$. Selected streamwise mean velocity profiles in the inner coordinates are presented in Figure 12. Clearly, the associated roughness functions were high; indeed, they were found to range from 21 to 24 for the streamwise variation plots, and 18 to 23 for the spanwise variation plots. These values are much higher than what was indicated in other turbulent flows over rough surfaces [55,57,59], suggestive of the possibility of a different von Kármán constant than what was employed in this work.

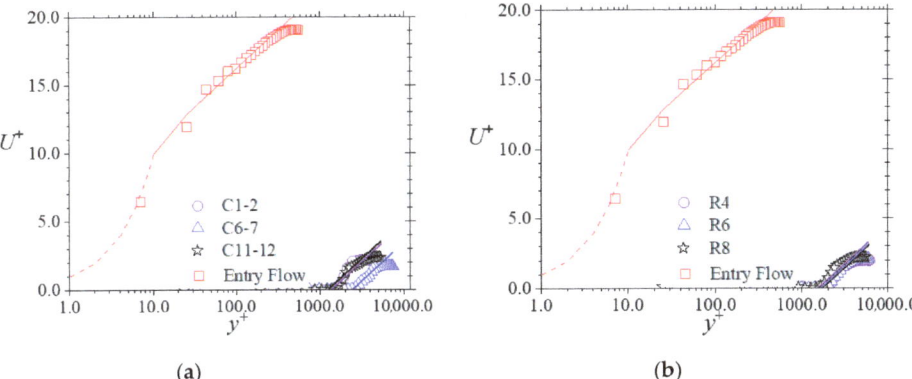

Figure 12. Mean streamwise line-averaged velocity profiles in inner coordinates for selected Test *B* conditions. Plots in (**a**) show streamwise variations compared with entry flow. Plots in (**b**) show spanwise variations compared with entry flow. The solid red line represents log law (Equation (2)). The red dashed line represents the law of the wall $U^+ = y^+$. Other solid lines represent modified log law for rough walls (Equation (3)). The denominations C and R respectively stand for columns and rows, and are illustrated in Figure 1. Black-filled boxes are the results for Test *A*, and red-open boxes are the results for Test *B*.

From these assessments, it is evident that if existent, the logarithmic layer occupies a very limited region. The friction velocities U_τ (in Tables 4 and 5) were substantially (9 to 18 times) higher over the porous surface compared to the entry flow over the smooth wall. As the flow over the porous surface developed along the stream, this friction velocity increased to a peak and declined thereafter. This is consistent with the trends seen for the displacement thickness observed in Figure 9, and mean velocity profiles in Figure 11. A similar trend in U_τ was reported in the simulations of Lee and Sung [58] for a turbulent flow over two-dimensional rods. However, in that work, their peak value was just about a quarter of the value recorded in this work. The higher values observed in the present tests stem from the increased drag due to the three-dimensional porous medium rod arrangement. The lateral deviations of U_τ, on the other hand, were insignificant considering the uncertainty limits.

Remarkably, the mean wall-normal profiles in Figure 13a,b were not negligible as often observed in turbulent flows over smooth walls of zero pressure gradient [55,58]. Indeed, in the flow over the first seven columns of rods, the maximum values were 5 to 12% of the line-averaged local maximum mean streamwise velocity. Compared with Figure 11, it is patent that the flow over those rods is three-dimensional, and that the wall-normal velocities contribute significantly to the mean momentum transport over the porous surface. Consequently, it is not surprising that mean momentum fluxes in Figure 13c,d were also significant. The high fluxes imply that they are also important factors in the production of shear stresses in those regions [57]. The dynamic roles of the mean wall-normal velocities and the momentum fluxes, however, declined downstream as the flow approached the unobstructed domain.

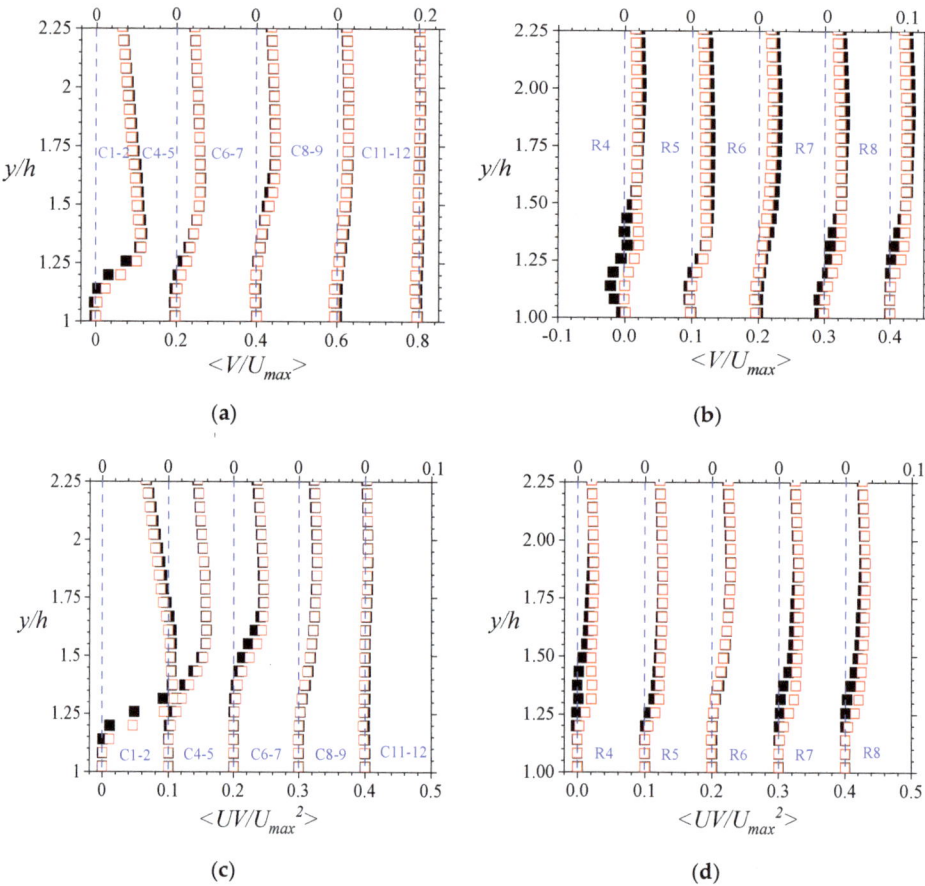

Figure 13. Turbulence quantities $<V/U_{max}>$, and $<UV/U_{max}^2>$ are respectively line-averaged mean wall-normal velocity, and momentum flux, normalized by the local maximum line-averaged streamwise mean velocity in open and porous regions. Here, the profiles for the flow in the open region are shown. Staggered plots (**a**,**c**) show the streamwise variations whereas (**b**,**d**) demonstrate the spanwise variations. The denominations C and R respectively stand for columns and rows, and are illustrated in Figure 1. Black-filled boxes are the results for Test *A*, and red-open boxes are for results for Test *B*.

To illustrate the rotational regions of flow above the porous medium, the line-averaged mean spanwise vorticity $<\Omega_z>$ ($<\partial V/\partial x - \partial U/\partial y>$) distributions are plotted in Figure 14. They underscore the complex vortical activity over the porous medium. These vortices are presumed to stem from Kelvin–Helmholtz (K–H) instabilities, and such are reported in other related studies [7,60]. In the current tests, the domination of these vortices were emphatic at the first seven columns of rods. For those domains, they were found to extend into increasing depth of the flow above the rods as they propagated through the flow. The location of the maximum vorticity also changed with the streamwise location. Compared with the flow through the porous medium, vortex motions over the rods were ~5 times more intense in magnitude, and more sustained. Moreover, the spatial variations along the span also showed that the presence of the side walls tended to heighten the vortex motions and more so at lower Reynolds number.

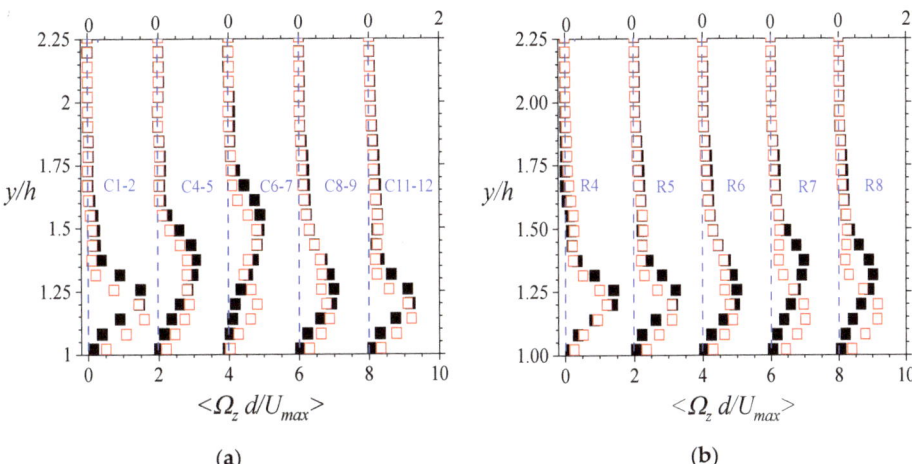

Figure 14. Line-averaged mean spanwise vorticity $<\Omega_z> = <\partial V/\partial x - \partial U/\partial y>$ plots normalized by the ratio of the local maximum line-averaged streamwise mean velocity and rod diameter. Plots in (**a**) are staggered distributions along the stream, and those in (**b**) are staggered across the span in the open region. The denominations C and R respectively stand for columns and rows, and are illustrated in Figure 1. Black-filled boxes are the results for Test *A*, and red-open boxes are the results for Test *B*.

3.2.3. Turbulence Intensities, Reynolds Stresses, and Turbulent Energy

The turbulent intensities of the flow over the porous surface were also examined. They are plotted as line-averaged distributions in Figure 15. The results indicate that compared with the local mean streamwise velocities, the maximum streamwise and wall-normal turbulent intensities were ~25% and ~12%, respectively. The Reynolds number effects were also apparent in the wall-normal turbulent intensities. The maximum turbulent intensities measured in the flow above the porous medium were much higher than those measured at the entry flow and within the porous medium. Indeed, the relative proportions of the streamwise and wall-normal components were found to be higher than that recorded in previous works [7,60]. It is also worth noting that the variation of the wall-normal locations of the maximum streamwise turbulent intensities was relatively small compared with other mean flow quantities (e.g., momentum flux). This indicates that the maximum turbulent intensities were not so much a factor of the mass and momentum dynamics of the turbulent flow as the static attributes of the flow section. Thus, it is reasonable to conclude that the turbulent intensities are mostly due to the porous boundary condition, and specifically the structural arrangements of the porous material.

The profiles of the line-averaged Reynolds normal and shear stresses are shown in Figure 16. The Reynolds normal stresses were found to be qualitatively similar to the turbulent intensities. Thus, here also, as the wall-normal components were expected to be curbed by the porous wall, these values were smaller than the streamwise components. Additionally, the wall-normal locations of the maximum Reynolds normal stresses did not very much, consolidating the likelihood of the porous boundary being the main factor of influence of their magnitudes. The Reynolds shear stresses, on the other hand, were quite distinctive. The profiles of flow over the first six columns of rods had negative maximum values, while those over the downstream rods had positive maximum values. This observation of negative Reynolds shear stress measurements of turbulent flows over porous media is rare in the literature. However, it should be conceded that the literature has barely any record of turbulent boundary layer flow developments over porous media. Nevertheless, it is notable that Lee and Sung [58] reported negative Reynolds shear stresses in their direct numerical simulations of the turbulent boundary layer over a rod-roughened wall. These negative values were also specifically located over the most upstream two-dimensional rod. Hence, this feature is not totally out of the bounds of possibility. This phenomenon requires further study. However, it is also reasonable to note that if the negative mean momentum flux $<-UV>$ is important in the production of the shear stress (as suggested in the streamwise momentum equation [57]), then large pockets of positive mean momentum flux $<UV>$ (Figure 13c,d) may be a factor in the sign change in $<-uv>$ [57].

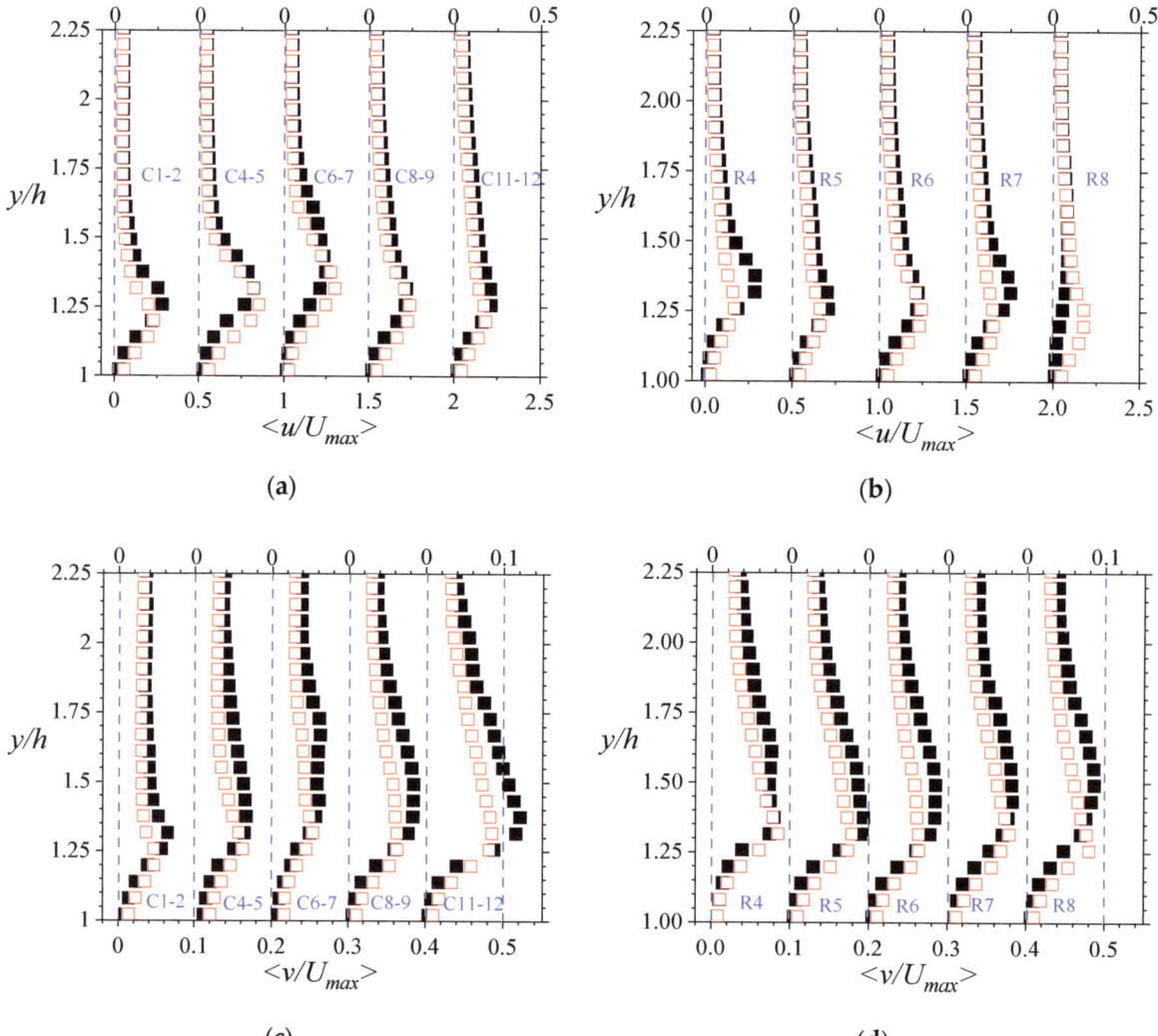

Figure 15. Parameters $<u/U_{max}>$ and $<v/U_{max}>$ are respectively line-averaged streamwise and wall-normal turbulence intensities normalized by the local maximum line-averaged streamwise mean velocity in both open and porous regions. Here, the profiles for the flow in the open region are shown. Staggered plots (**a**,**c**) show the streamwise variations whereas (**b**,**d**) demonstrate the spanwise variations. The denominations C and R respectively stand for columns and rows, and are illustrated in Figure 1. Black-filled boxes are the results for Test A, and red-open boxes are the results for Test B.

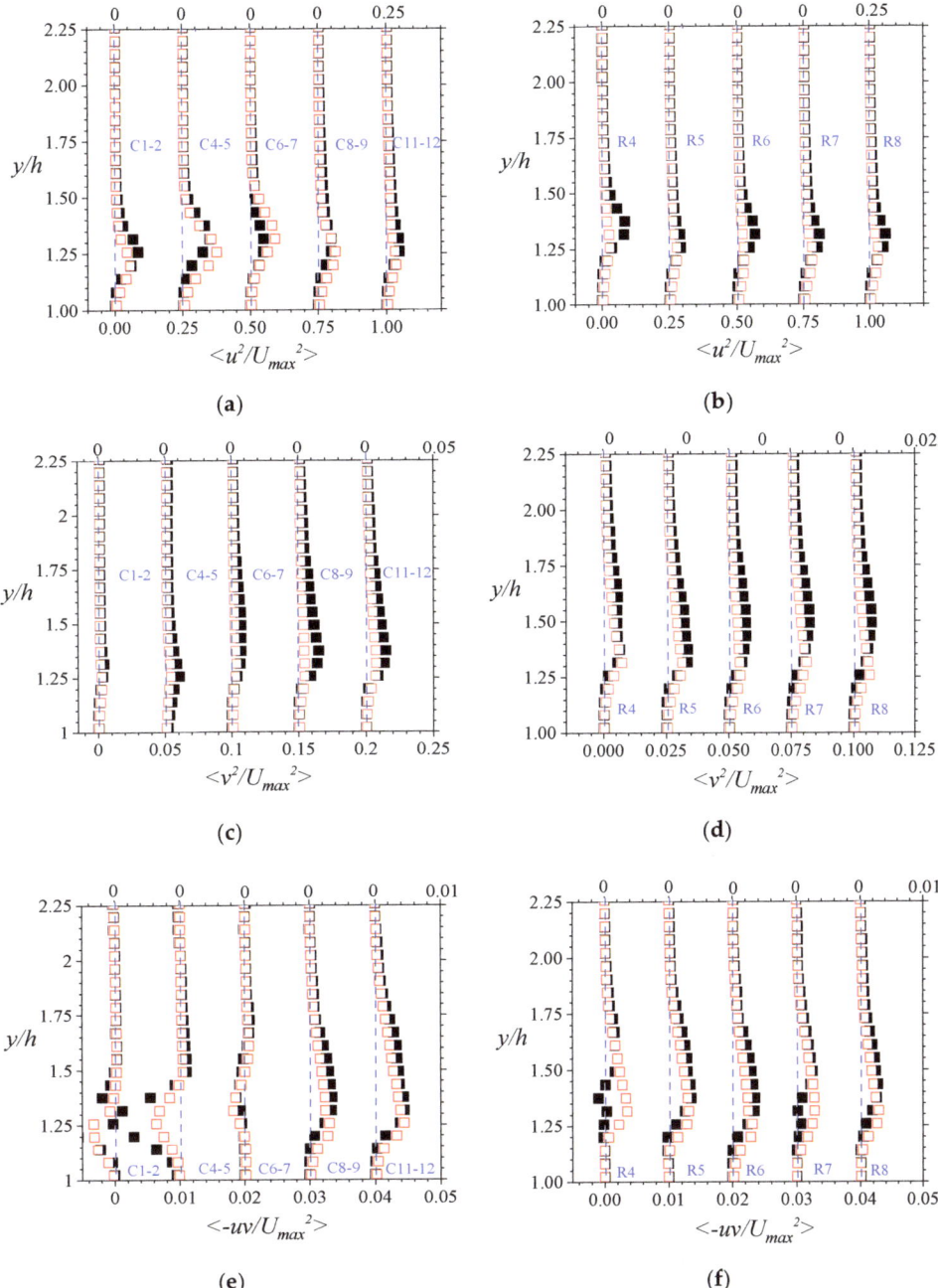

Figure 16. Turbulence quantities $\langle u^2/U_{max}^2 \rangle$, $\langle v^2/U_{max}^2 \rangle$, and $\langle -uv/U_{max}^2 \rangle$ are respectively line-averaged streamwise Reynolds normal stress, wall-normal Reynolds normal stress, and Reynolds shear stress in the streamwise-wall-normal plane, normalized by the local maximum line-averaged streamwise mean velocity in both open and porous regions. Here, the profiles for the flow in the open region are shown. Staggered plots (**a,c,e**) show the streamwise variations whereas (**b,d,f**) demonstrate the spanwise variations. The denominations C and R respectively stand for columns and rows, and are illustrated in Figure 1. Black-filled boxes are the results for Test A, and red-open boxes are the results for Test B.

To study large-scale anisotropy, ratios of the Reynold stresses were analyzed. The plots are presented in Figure 17. As shown, close to the porous boundary, the ratio of the normal stresses tended to decay and flatten to values other than unity. Further away from the walls, they increased to ~0.5. The ratios of the Reynolds shear stress and the normal stress, however, were complex. As the flow developed downstream, the ratio of the Reynolds shear stress and the streamwise component of the normal stress varied throughout the flow, with Reynolds effects apparent in the spanwise planes. As expected, the effects of negative Reynolds shear stresses were also present in the upstream rods. It has been reported that some researchers (e.g., Suga [61]) have been able to develop turbulence models for fully developed flows over porous media with varying degrees of success. The results showed herein, however, indicate that the use of models (e.g., k-ε and k-ω) that assume local isotropy do not apply to turbulent flows over compact porous media as tested in this work.

In another consideration, the turbulent kinetic energy k was assessed. As the current measurement system is planar, the Reynolds normal stress in the spanwise direction w^2 could not be measured directly. Thus, k was estimated as $0.75\,(u^2 + v^2)$. The normalized line-averaged profiles are shown in Figure 18. As the streamwise component of the Reynolds normal stress is the dominant contributor of k, the profiles are qualitatively similar to the streamwise Reynolds normal stress distributions in Figure 16. Clearly, there is significant turbulent kinetic energy at the domain where intense vortical activity is prevalent. The ratio of the Reynolds shear stress to the kinetic energy is an important parameter that is used to calibrate turbulence model coefficient C_k [62,63]. This constant (also called Townsend's structure parameter, $<-uv/(2k)>$ is used in the Kolmogorov–Prandtl expression associated with the eddy viscosity v_T (=$C_k\,k/L$, where L is a length scale). De Lemos [63] suggested a constant for turbulent flow over porous media. However, Figure 19 indicates that this coefficient may vary from ~-0.3 to 0.1 depending on the wall-normal location above the porous medium.

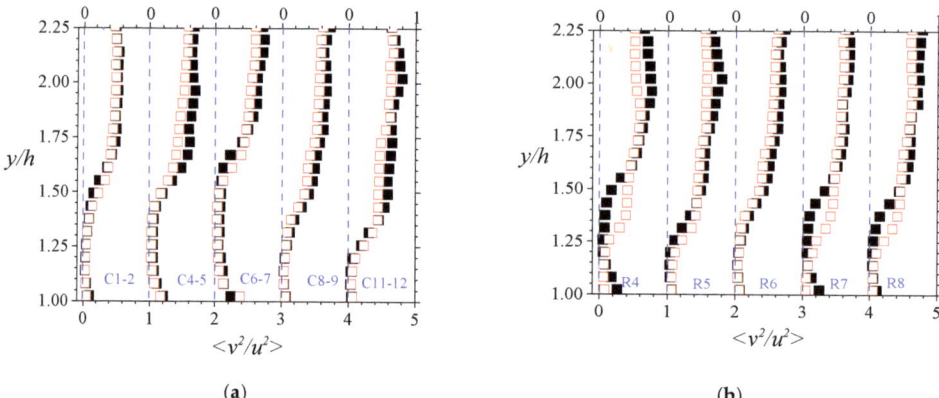

Figure 17. Cont.

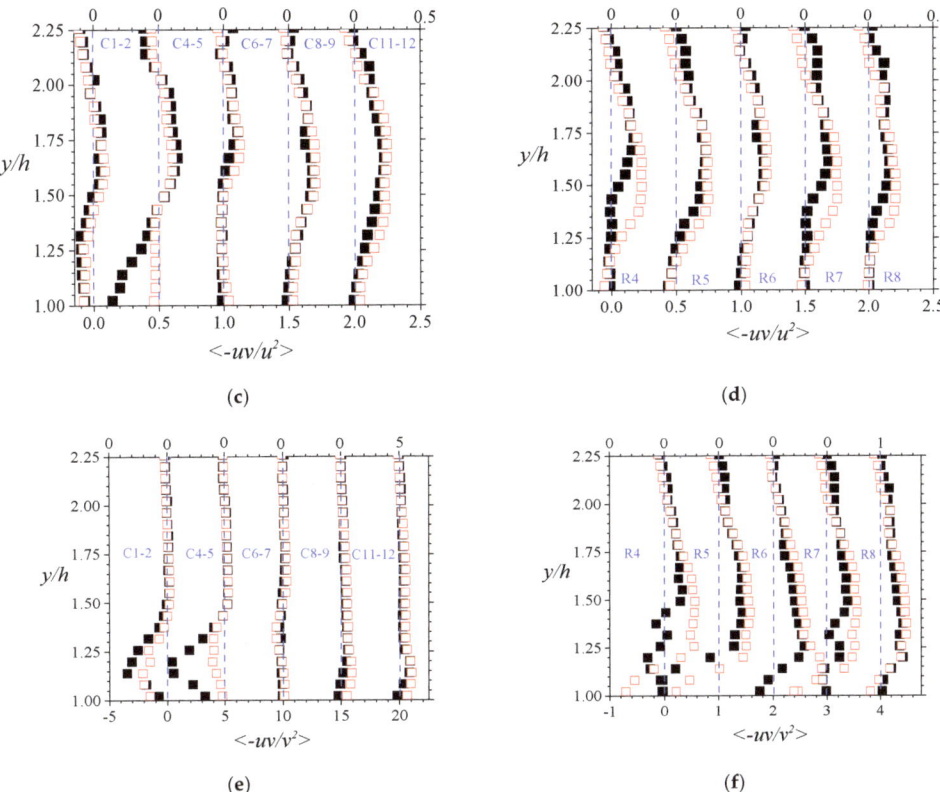

Figure 17. Reynolds stress ratios for the flow in the open region are shown. Staggered plots (**a,c,e**) show the streamwise variations whereas (**b,d,f**) demonstrate the spanwise variations. The denominations C and R respectively stand for columns and rows, and are illustrated in Figure 1. Black-filled boxes are the results for Test A, and red-open boxes are the results for Test B.

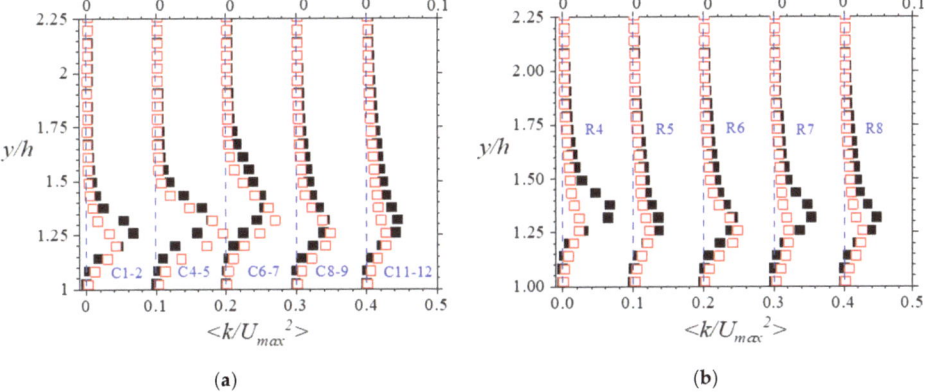

Figure 18. Turbulent kinetic energy profiles for the flow in the open region are shown. The plots are shown as normalized by the local maximum line-averaged streamwise mean velocity in both open and porous regions. Staggered plots (**a**) show the streamwise variations whereas staggered plots (**b**) demonstrate the spanwise variations. The denominations C and R respectively stand for columns and rows, and are illustrated in Figure 1. Black-filled boxes are the results for Test A, and red-open boxes are the results for Test B.

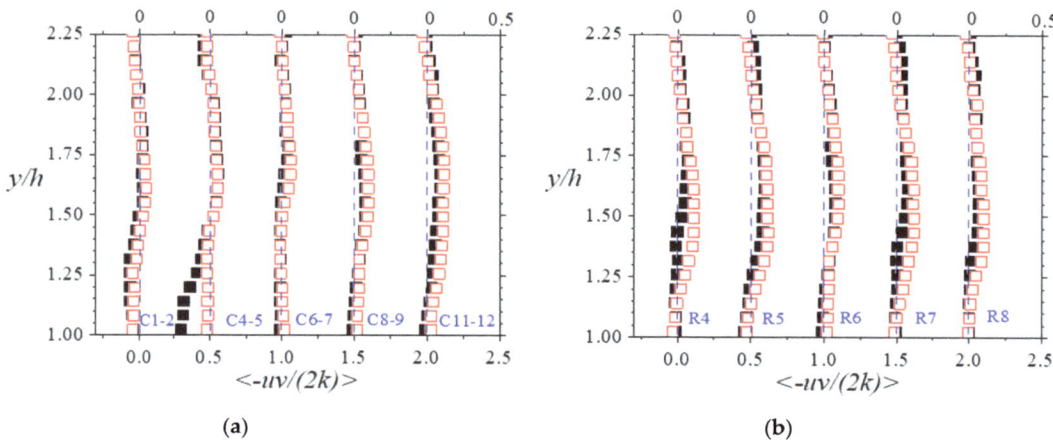

Figure 19. Evaluation of Townsend structure parameter for the flow in the open region is shown. Staggered plots (**a**) show the streamwise variations whereas staggered plots (**b**) demonstrate the spanwise variations. The denominations C and R respectively stand for columns and rows, and are illustrated in Figure 1. Black-filled boxes are the results for Test *A*, and red-open boxes are the results for Test *B*.

4. Conclusions

The focus of the present work was the investigation of a turbulent flow through and over a compact porous medium. This research was conducted to measure the development of the flow over the porous boundary, the penetration of the turbulent flow into the porous domain, the attendant three-dimensional effects, and Reynolds number effects. These objectives were achieved by conducting particle image velocimetry measurements in a test section with turbulent flow through and over a compact model porous medium of porosity 85%, and filling fraction of 21%. The porous medium model was made of a square array of cylindrical rods of twelve columns and nine rows. The tests were carried out at Reynolds numbers (based on the bulk flow through the entire channel, and the depth of the channel) = 14,338 and 24,510.

Overall, this work indicates that the flow through and over a compact porous medium exhibits phenomenon different and somewhat complex compared to that associated with fully developed domains. The flow above the compact porous medium showed a complex flow development pattern with the boundary layer thickening along the stream by about 90% at the trailing rods, increasing monotonically in momentum thickness, and growing non-monotonically in displacement thickness and shape parameter. The spanwise variations, on the other hand, were no more than 16% from the value at the mid-plane. The mean velocities showed that if existent, the logarithmic layer occupied a very limited region. The friction velocities were 9 to 18 times higher over the porous surface compared to the entry flow over the smooth wall. There were significant complex vortical activities over the porous medium, extending into an increasing depth of the flow above the rods as they propagate through the flow. The ratios of the Reynolds stresses showed a complex large-scale anisotropy in the flow over the porous medium. The implication then is that the use of turbulence models that assume local isotropy through and over porous media may not apply to turbulent flows over compact porous media. At the interface between the porous medium and the overlying turbulent flow, the slip velocities were no more than 6% of the maximum velocity. The shear rate obtained indicates that the average depth of such infiltration through the interfacial zone was only about 7% of the rod diameter. For the flow through the porous medium, the results showed that the streamwise velocities, in particular, had variations with negative gradients close to the lower wall and in the planes outside of the midspan. Pockets of variant sizes of the mean spanwise vortices appeared to be propagated from regions of shear through the porous medium, at the edge of the most upstream porous medium rods, and the interfacial region.

This work provides useful insight into the physics of flow pertinent to natural and industrial scenarios such as terrestrial or aquatic canopy flows over regions with small/confined arrangements of sparse vegetation, and flows associated with pin-fin arrays of rods for heat transfer enhancement. The data may serve as a useful resource for developing and validating numerical models.

Funding: This work was funded by the C Graydon and Mary E Rogers Fellowship.

Institutional Review Board Statement: Not applicable.

Informed Consent Statement: Not applicable.

Data Availability Statement: The data presented in this study are available on request from the corresponding author.

Acknowledgments: The author acknowledges with gratitude the financial support of Bucknell University through the C Graydon and Mary E Rogers Fellowship. The work of Owen Schiele in helping to complete the experiments, is also gratefully acknowledged.

Conflicts of Interest: The author declares no conflict of interest. The funders had no role in the design of the study; in the collection, analyses, or interpretation of data; in the writing of the manuscript, or in the decision to publish the results.

Nomenclature

English

A	Measured parameter: Test condition at Re_H = 14,338
B	Test condition at re_h = 24,510; logarithmic law constant
ΔB^+	Roughness function
b	Location in the wall-normal direction
C	Column
C_f	Skin-friction coefficient = 2 $(U_\tau/U_e)^2$
C_k	Townsend parameter = $<-uv/(2k)>$
d	Nominal diameter of rods (mm)
g	Acceleration due to gravity (m/s)
h	Average height of rods (mm)
H	Channel depth (mm); shape factor = δ^*/θ^*
k	Turbulent kinetic energy $\approx 0.75\ (u^2 + v^2)$ (m^2 s^{-2})
L	Length scale
l	Distance between centers of adjacent rods (mm)
n	Column number
P	Plane
R	Row, location in the spanwise direction
Re_H	Bulk Reynolds number = $u_b h/\nu$
Re_θ	Reynolds number based on momentum thickness and maximum velocity
U	Mean (time-averaged) streamwise velocity (m s^{-1})
u	Streamwise turbulence intensity (m s^{-1})
u^2	Streamwise Reynolds normal stress (m^2 s^{-2})
U_b	Bulk velocity in the channel (m s^{-1})
$U_{b,p}$	Average streamwise velocity within the porous region (m s^{-1})
U_e	Maximum velocity of entry flow (m s^{-1})
U_s	Streamwise mean slip velocity (m s^{-1})
U_{max}	Maximum time-averaged streamwise velocity in the flow section with a porous medium model (m s^{-1})
U_τ	Friction velocity (m s^{-1})
U^+	Time-averaged streamwise velocity in inner coordinates = U/U_τ

UV	Mean momentum flux (m^2 s^{-2})
$-uv$	Reynolds shear stress (m^2 s^{-2})
V	Mean (time-averaged) wall-normal velocity (m s^{-1})
v	Wall-normal turbulence intensity (m s^{-1})
v^2	Wall-normal Reynolds normal stress (m^2 s^{-2})
v_s	Settling velocity (m s^{-1})
x	Streamwise direction; streamwise distance (m)
y	Wall-normal direction; wall-normal distance (m)
y_o	Virtual origin
y^*	Wall-normal distance from the top plane of the rods (m)
y^+	Wall-normal direction in inner coordinates = $y\, U_\tau/\nu$
z	Spanwise direction; spanwise distance (m)
Greek	
δ	Boundary layer thickness (m)
δ^*	Displacement thickness (m)
ε	Porosity
η	Kolmogorov length scale $\approx 3\nu/U_\tau$ (m)
θ^*	Momentum thickness (m)
κ	Von Kármán constant
λ	Taylor microscale $\approx \sqrt{15(U_e\eta^2)}/\nu$ (m)
ν	Kinematic viscosity (m^2 s^{-1})
ρ_f	Fluid density (kg m^{-3})
ρ_p	Particle density (kg m^{-3})
τ_f	Viscous time scale = ν/U_τ^2 (s)
τ_R	Response time (s)
γ_0	Shear rate at the interface = $dU/dy|_{y=h}$ (s^{-1})
Ω_z	Mean spanwise vorticity = $\partial V/\partial x - \partial U/\partial y$ (s^{-1})
Other Symbols/Acronyms	
< >	Line-averaged parameter
K-H	Kelvin–Helmholtz
Nd:YAG	Neodymium: Yttrium Aluminum Garnet
PIV	Particle image velocimetry
RIM	Refractive index matching

References

1. Beavers, G.S.; Joseph, D.D. Boundary conditions at a naturally permeable wall. *J. Fluid Mech.* **1967**, *30*, 197–207. [CrossRef]
2. Neale, G.; Nader, W. Practical significance of Brinkman's extension of Darcy's law: Coupled parallel flows within a channel and a bounding porous medium. *Can. J. Chem. Eng.* **1974**, *52*, 475–478. [CrossRef]
3. Vafai, K.; Kim, S.J. Forced convection in a channel filled with a porous medium: An exact solution. *ASME J. Heat Transf.* **1989**, *111*, 1103–1106. [CrossRef]
4. Arthur, J.K.; Ruth, D.W.; Tachie, M.F. PIV measurements of flow through a model porous medium with varying boundary conditions. *J. Fluid Mech.* **2009**, *629*, 343–374. [CrossRef]
5. Florens, E.; Eiff, O.; Moulin, F. Defining the roughness sublayer and its turbulence statistics. *Exp. Fluids* **2013**, *54*, 1500. [CrossRef]
6. Kim, T.; Blois, G.; Best, J.L.; Christensen, K.T. Experimental study of turbulent flow over and within cubically packed walls of spheres: Effects of topography, permeability and wall thickness. *Int. J. Heat Fluid Flow* **2018**, *73*, 16–29. [CrossRef]
7. Suga, K.; Okazaki, Y.; Ho, U.; Kuwata, Y. Anisotropic wall permeability effects on turbulent channel flows. *J. Fluid Mech.* **2018**, *855*, 983–1016. [CrossRef]
8. Arthur, J.K. PIV study of flow through and over porous media at the onset of inertia. *Adv. Water Resour.* **2020**, *146*, 103793. [CrossRef]
9. Lyu, W.; Wang, X. Stokes–Darcy system, small-Darcy-number behaviour and related interfacial conditions. *J. Fluid Mech.* **2021**, *922*, 509. [CrossRef]
10. Angot, P.; Goyeau, B.; Ochoa-Tapia, J.A. A nonlinear asymptotic model for the inertial flow at a fluid-porous interface. *Adv. Water Resour.* **2021**, *149*, 103798. [CrossRef]
11. Sengupta, S.; De, S. Effect of the transition layer on the stability of a fluid-porous configuration: Impact on power-law rheology. *Phys. Rev. Fluids* **2021**, *6*, 063902. [CrossRef]
12. Hester, E.T.; Cardenas, M.B.; Haggerty, R.; Apte, S.V. The importance and challenge of hyporheic mixing. *Water Resour. Res.* **2017**, *53*, 3565–3575. [CrossRef]
13. Ghisalberti, M.; Nepf, H. Shallow flows over a permeable medium: The hydrodynamics of submerged aquatic canopies. *Transp. Porous Media* **2009**, *78*, 309–326. [CrossRef]

14. Belcher, S.E.; Finnigan, J.J.; Harman, I.N. Flows through forest canopies in complex terrain. *Ecol. Appl.* **2008**, *18*, 1436–1453. [CrossRef]
15. Segalini, A. Linearized simulation of flow over wind farms and complex terrains. *Philos. Trans. R. Soc. A Math. Phys. Eng. Sci.* **2017**, *375*, 20160099. [CrossRef] [PubMed]
16. Uzol, O.; Camci, C. Heat transfer, pressure loss and flow field measurements downstream of staggered two-row circular and elliptical pin fin arrays. *J. Heat Transf.* **2005**, *127*, 458–471. [CrossRef]
17. Prescott, P.J.; Incropera, F.P. The effect of turbulence on solidification of a binary metal alloy with electromagnetic stirring. *ASME* **1995**, *117*, 716–724. [CrossRef]
18. Manes, C.; Pokrajac, D.; McEwan, I.; Nikora, V. Turbulence structure of open channel flows over permeable and impermeable beds: A comparative study. *Phys. Fluids* **2009**, *21*, 125109. [CrossRef]
19. Raupach, M.; Thom, A.S. Turbulence in and above plant canopies. *Annu. Rev. Fluid Mech.* **1981**, *13*, 97–129. [CrossRef]
20. Finnigan, J. Turbulence in plant canopies. *Annu. Rev. Fluid Mech.* **2000**, *32*, 519–571. [CrossRef]
21. Ghisalberti, M.; Nepf, H.M. Mixing layers and coherent structures in vegetated aquatic flows. *J. Geophys. Res. Ocean.* **2002**, *107*, 3–11. [CrossRef]
22. Ruff, J.F.; Gelhar, L.W. Turbulent shear flow in porous boundary. *J. Eng. Mech. Div.* **1972**, *98*, 975–991. [CrossRef]
23. Zagni, A.F.; Smith, K.V. Channel flow over permeable beds of graded spheres. *J. Hydraul. Div.* **1976**, *102*, 207–222. [CrossRef]
24. Vollmer, S.; de los Santos Ramos, F.; Daebel, H.; Kühn, G. Micro scale exchange processes between surface and subsurface water. *J. Hydrol.* **2002**, *269*, 3–10. [CrossRef]
25. Breugem, W.P.; Boersma, B.J.; Uittenbogaard, R.E. The influence of wall permeability on turbulent channel flow. *J. Fluid Mech.* **2006**, *562*, 35–72. [CrossRef]
26. Pokrajac, D.; Manes, C. Velocity measurements of a free-surface turbulent flow penetrating a porous medium composed of uniform-size spheres. *Transp. Porous Media* **2009**, *78*, 367–383. [CrossRef]
27. Nikora, V.; McEwan, I.; McLean, S.; Coleman, S.; Pokrajac, D.; Walters, R. Double-averaging concept for rough-bed open-channel and overland flows: Theoretical background. *J. Hydraul. Eng.* **2007**, *133*, 873–883. [CrossRef]
28. Suga, K.; Nakagawa, Y.; Kaneda, M. Spanwise turbulence structure over permeable walls. *J. Fluid Mech.* **2017**, *822*, 186–201. [CrossRef]
29. Suga, K.; Matsumura, Y.; Ashitaka, Y.; Tominaga, S.; Kaneda, M. Effects of wall permeability on turbulence. *Int. J. Heat Fluid Flow* **2010**, *31*, 974–984. [CrossRef]
30. Pokrajac, D.; Manes, C.; McEwan, I. Peculiar mean velocity profiles within a porous bed of an open channel. *Phys. Fluids* **2007**, *19*, 098109. [CrossRef]
31. Sharma, A.; García-Mayoral, R. Turbulent flows over dense filament canopies. *J. Fluid Mech.* **2020**, *888*, A2. [CrossRef]
32. Szepessy, S.; Bearman, P.W. Aspect ratio and end plate effects on vortex shedding from a circular cylinder. *J. Fluid Mech.* **1992**, *234*, 191–217. [CrossRef]
33. Ostanek, J.K.; Thole, K.A. Effects of varying streamwise and spanwise spacing in pin-fin arrays. *Turbo Expo Power Land Sea Air* **2012**, *44700*, 45–57.
34. Anderson, C.D.; Lynch, S.P. Time-resolved stereo PIV measurements of the horseshoe vortex system at multiple locations in a low-aspect-ratio pin–fin array. *Exp. Fluids* **2016**, *57*, 5. [CrossRef]
35. Nepf, H.M. Flow and transport in regions with aquatic vegetation. *Annu. Rev. Fluid Mech.* **2012**, *44*, 123–142. [CrossRef]
36. Moroni, M.; Cushman, J.H. Statistical mechanics with three-dimensional particle tracking velocimetry experiments in the study of anomalous dispersion. II Experiments. *Phys. Fluids* **2001**, *13*, 81–91. [CrossRef]
37. Huang, A.Y.; Huang, M.Y.; Capart, H.; Chen, R. Optical measurements of pore geometry and fluid velocity in a bed of irregularly packed spheres. *Exp. Fluids* **2008**, *45*, 309–321. [CrossRef]
38. Raffel, M.; Willert, C.E.; Scarano, F.; Kähler, C.J.; Wereley, S.T.; Kompenhans, J. *Particle Image Velocimetry: A Practical Guide*; Springer: Berlin/Heidelberg, Germany, 2018.
39. Samimy, M.; Lele, S.K. Motion of particles with inertia in a compressible free shear layer. *Phys. Fluids A Fluid Dyn.* **1991**, *3*, 1915–1923. [CrossRef]
40. Prasad, A.K. Particle image velocimetry. *Curr. Sci.* **2000**, *79*, 51–60.
41. Wieneke, B. PIV uncertainty quantification from correlation statistics. *Meas. Sci. Technol.* **2015**, *26*, 074002. [CrossRef]
42. Kadivar, M.; Tormey, D.; McGranaghan, G. A review on turbulent flow over rough surfaces: Fundamentals and theories. *Int. J. Thermofluids* **2021**, *10*, 100077. [CrossRef]
43. Millikan, C.B. A critical discussion of turbulent flow in channels and circular tubes. In Proceedings of the 5th International Congress on Applied Mechanics, Wiley, New York, NY, USA, 12–16 September 1938; pp. 386–392.
44. Durst, F.; Fischer, M.; Jovanović, J.; Kikura, H. Methods to set up and investigate low Reynolds number, fully developed turbulent plane channel flows. *J. Fluids Eng.* **1998**, *120*, 496–503. [CrossRef]
45. Johansson, A.V.; Alfredsson, P.H. Effects of imperfect spatial resolution on measurements of wall-bounded turbulentbx shear flows. *J. Fluid Mech.* **1983**, *137*, 409–421. [CrossRef]
46. He, X.; Apte, S.V.; Finn, J.R.; Wood, B.D. Characteristics of turbulence in a face-centred cubic porous unit cell. *J. Fluid Mech.* **2019**, *873*, 608–645. [CrossRef]

47. Koch, D.L.; Ladd, A.J.C. Moderate Reynolds number flows through periodic and random arrays of aligned cylinders. *J. Fluid Mech.* **1997**, *349*, 31–66. [CrossRef]
48. Hill, R.J.; Koch, D.L. The transition from steady to weakly turbulent flow in a close-packed ordered array of spheres. *J. Fluid Mech.* **2002**, *465*, 59–97. [CrossRef]
49. Suekane, T.; Yokouchi, Y.; Hirai, S. Inertial flow structures in a simple-packed bed of spheres. *AIChE J.* **2003**, *49*, 10–17. [CrossRef]
50. Bernard, P.S. *Turbulent Fluid Flow*; John Wiley & Sons: Hoboken, NJ, USA, 2019.
51. Arthur, J.K. *Flow through and over Model Porous Media with or without Inertial Effects*; University of Manitoba: Winnipeg, MB, Canada, 2012.
52. Escudier, M.P.; Abdel-Hameed, A.; Johnson, M.W.; Sutcliffe, C.J. Laminarisation and re-transition of a turbulent boundary layer subjected to favourable pressure gradient. *Exp. Fluids* **1998**, *25*, 491–502. [CrossRef]
53. Narayanan, M.A.B.; Ramjee, V. On the criteria for reverse transition in a two-dimensional boundary layer flow. *J. Fluid Mech.* **1969**, *35*, 225–241. [CrossRef]
54. Purtell, L.P.; Klebanoff, P.S.; Buckley, F.T. Turbulent boundary layer at low Reynolds number. *Phys. Fluids* **1981**, *24*, 802–811. [CrossRef]
55. Tachie, M.F. Particle image velocimetry study of turbulent flow over transverse square ribs in an asymmetric diffuser. *Phys. Fluids* **2007**, *19*, 065106. [CrossRef]
56. Schultz, M.P.; Flack, K.A. Outer layer similarity in fully rough turbulent boundary layers. *Exp. Fluids* **2005**, *38*, 328–340. [CrossRef]
57. Ashrafian, A.; Andersson, H.I.; Manhart, M. DNS of turbulent flow in a rod-roughened channel. *Int. J. Heat Fluid Flow* **2004**, *25*, 373–383. [CrossRef]
58. Lee, S.; Sung, H.J. Direct numerical simulation of the turbulent boundary layer over a rod-roughened wall. *J. Fluid Mech.* **2007**, *584*, 125–146. [CrossRef]
59. Tachie, M.F.; Shah, M.K. Favorable pressure gradient turbulent flow over straight and inclined ribs on both channel walls. *Phys. Fluids* **2008**, *20*, 095103. [CrossRef]
60. Suga, K.; Mori, M.; Kaneda, M. Vortex structure of turbulence over permeable walls. *Int. J. Heat Fluid Flow* **2011**, *32*, 586–595. [CrossRef]
61. Suga, K. Understanding and modelling turbulence over and inside porous media. *Flow Turbul. Combust.* **2016**, *96*, 717–756. [CrossRef]
62. Pedras, M.H.; de Lemos, M.J. Macroscopic turbulence modeling for incompressible flow through undeformable porous media. *Int. J. Heat Mass Transf.* **2001**, *44*, 1081–1093. [CrossRef]
63. de Lemos, M.J. Turbulent kinetic energy distribution across the interface between a porous medium and a clear region. *Int. Commun. Heat Mass Transf.* **2005**, *32*, 107–115. [CrossRef]

Article
Small Alcohols as Surfactants and Hydrate Promotors

Bjørn Kvamme [1,2,3]

1. Hyzenenergy, 26701 Quail Creek, Laguna Hills, CA 92656, USA; bkvamme@strategic-carbonllc.com; Tel.: +47-934-51-95-6
2. Strategic Carbon LLC, 20 Ladd St., Suite 200, Portsmouth, NH 03801, USA
3. State Key Laboratory of Oil and Gas Reservoir Geology and Exploitation, Southwest Petroleum University, Xindu Road No.8, Chengdu 610500, China

Abstract: Many methods to produce hydrate reservoirs have been proposed in the last three decades. Thermal stimulation and injection of thermodynamic hydrate inhibitors are just two examples of methods which have seen reduced attention due to their high cost. However, different methods for producing hydrates are not evaluated thermodynamically prior to planning expensive experiments or pilot tests. This can be due to lack of a thermodynamic toolbox for the purpose. Another challenge is the lack of focus on the limitations of the hydrate phase transition itself. The interface between hydrate and liquid water is a kinetic bottle neck. Reducing pressure does not address this problem. An injection of CO_2 will lead to the formation of a new CO_2 hydrate. This hydrate formation is an efficient heat source for dissociating hydrate since heating breaks the hydrogen bonds, directly addressing the problem of nano scale kinetic limitation. Adding limited amounts of N_2 increases the permeability of the injection gas. The addition of surfactant increases gas/water interface dynamics and promotes heterogeneous hydrate formation. In this work we demonstrate a residual thermodynamic scheme that allows thermodynamic analysis of different routes for hydrate formation and dissociation. We demonstrate that 20 moles per N_2 added to the CO_2 is thermodynamically feasible for generating a new hydrate into the pores. When N_2 is added, the available hydrate formation enthalpy is reduced as compared to pure CO_2, but is still considered sufficient. Up to 3 mole percent ethanol in the free pore water is also thermodynamically feasible. The addition of alcohol will not greatly disturb the ability to form new hydrate from the injection gas. Homogeneous hydrate formation from dissolved CH_4 and/or CO_2 is limited in amount and not important. However, the hydrate stability limits related to concentration of hydrate former in surrounding water are important. Mineral surfaces can act as hydrate promotors through direct adsorption, or adsorption in water that is structured by mineral surface charges. These aspects will be quantified in a follow-up paper, along with kinetic modelling based on thermodynamic modelling in this work.

Keywords: hydrate; non-equilibrium; thermodynamics; carbon dioxide

1. Introduction

Natural gas hydrates are classes of composite structures in which water organizes to create cavities that enclathrate small non-polar molecules such as CH_4, C_2H_6, C_3H_8, and i-C_4H_{10}. Some small slightly polar components, such as H_2S and CO_2, also form hydrates. The molecules that enter cavities are called guest molecules. Hydrates in nature are created from two sources. Methane released through biogenic degradation of organic material in the upper crust is the most abundant source of known hydrates worldwide. These are almost pure methane hydrates and structure I hydrates. The smallest symmetrical unit of structure I hydrate is a cubic box containing 46 water molecules, 6 large cavities (24 water molecules), and 2 small cavities (20 water molecules) that can host molecules such as CH_4. Molecules that enter these cavities of hydrogen-bounded water molecules are called guest molecules. The size of the cubic unit cell varies with temperature. For typical temperatures above the water freezing point, a constant unit box length of 12.01 Å is generally accurate enough.

Many biogenic hydrate resources are connected to deeper sources of thermogenic hydrocarbons. These hydrocarbons are the result of long-term thermal degradation of organic matter and have a large variation in hydrocarbon composition. These resources typically contain sour gases such as H_2S, which is formed from the degradation of organic sulfur in the absence of oxygen. Utilization of the upper hydrates may lead to an inflow of thermogenic hydrate formers through the fracture systems below.

Structure II is different from structure I in two important ways. The largest cavity has 28 water molecules and can accommodate molecules such as C_3H_8 and i-C_4H_{10}. The small cavity in structure II is very similar to the small cavity of structure I. The second difference in structure II, as compared to structure I, is that the ratio of small cavities to large cavities in structure II is 2:1.

We will not discuss further details on the basic structures and properties of these two hydrate structures as this information is available in several books including Sloan and Koh [1] and Mokogon [2]. These books also contain more information on structure H, which can clathrate guest molecules up to heptane, but is exotic in terms of natural occurrence.

In this work we limit ourselves to structure I hydrates and pure CH_4 hydrates as the energy sources. All the methods and considerations in this work also apply to hydrates from thermogenic hydrocarbons and mixed sources of biogenic and thermogenic hydrates. This work is part of a larger research project that investigates combined safe long-term storage and energy production of CO_2. More specifically, we examine the addition of limited amounts of N_2 (roughly 20 mole%) to increase the permeability of injections of CO_2/N_2 mixtures into CH_4 hydrate-filled sediments. As a second additive to the CO_2/N_2 mixtures, a variety of different surfactants are evaluated experimentally and theoretically. Theoretical evaluations are mainly molecular dynamics (MD) simulations of model systems [3] used to investigate how surfactants affect the interface between liquid water and a separate phase containing hydrate forming components. This separate phase is denoted as gas later in the discussion, although it may be a liquid state hydrate in a former phase.

The primary purpose of the surfactant is to keep the liquid water/gas interface free of blocking hydrate films. Its secondary purpose is to increase the kinetics of mass transport into the liquid water of the interface. Finally, it is expected that the presence of a surfactant will increase the amount of hydrate formerly in the liquid water of the interface.

Small alcohols like methanol and ethanol have been used for many decades as thermodynamic inhibitors. Most experimental data for hydrate formation from water containing alcohols are in a concentration range in which the alcohols change the activity of water significantly. The low partial charge, relative to the size of the methyl group, results in a surfactant effect of methanol. Ethanol has greater surfactant properties due to its non-polar outer methyl group. These surfactant effects can be utilized in hydrate production using the CO_2/CH_4 hydrate exchange method for the simultaneous release of CH_4 from the hydrate and for the safe, long-term storage of CO_2 as a hydrate.

There is, however, a symbiosis between this project and some industrial hydrate problems caused by methanol. Methanol used to keep wells free of hydrate may result in various amounts of methanol remaining during multiphase transport and/or subsequent processing. This may result in situations in which remaining methanol promotes hydrate formation. A typical example was reported by STATOIL (now EQUINOR) several years ago [4].

The main objective of this work is to examine how limited amounts of ethanol affect water. Limited amounts here refer to amounts that are typically below the amounts needed for hydrate prevention. We facilitate the surfactant properties of ethanol without reducing hydrate formation ability substantially. The focus here is how the addition of small amounts of alcohols affects thermodynamic properties responsible for phase transitions (Gibbs free energy) and associated changes in enthalpies. From a technical point of view, we seek alcohols that can easily be mechanically mixed in with the injection gas (CO_2 containing N_2). This excludes the high viscosity glycols and the main focus here is on methanol and ethanol.

A secondary objective is to shed light on some important factors related to design of a combined scheme for safe long-term storage of carbon dioxide, and simultaneous release of natural gas for energy. This also includes aspects related to the non-equilibrium nature of hydrates in sediments and different routes to hydrate formation. There are many experimental papers on carbon dioxide hydrate/methane hydrate swapping. It is beyond the scope of this work to discuss any of them. This also excludes our own experiments over a period of almost two decades.

Hydrate phase transition kinetics are implicit functions of thermodynamic control (Gibbs free energy changes), associated mass transport, and associated heat transport. The kinetic models will be derived and discussed in a separate follow-up paper. However, it is important in the context of this work to describe how various thermodynamic properties can be calculated.

A third objective is to demonstrate how the thermodynamic properties related to hydrate phase transitions can be calculated in a consistent way. It is critical in multiphase systems to utilize a uniform reference system. This will facilitate direct comparison of phase stability in terms of Gibbs free energies of the different co-existing phases.

To our knowledge, we are the only group that utilize residual thermodynamics for all phases, including hydrates. For that reason, there are few references to publications from other groups. There are certainly many good publications from other groups. However, within the limited theoretical focus in this work they may not fit into the main objectives discussed above.

Many of the calculated data provided in this work, like for instance Gibbs free energies, are not directly available experimentally. However, model calculations can be verified indirectly towards experimental data, for instance hydrate stability limits in pressure and temperature. Enthalpies of hydrate formation are hard to measure accurately. Additionally, I have not found any experimental data for hydrate formation from dissolved hydrate formers in water [5]. To my knowledge, the method utilized in this work for calculating enthalpies [5–8] is the only general method that can be applied to any hydrate phase transition. Methods that utilize gradient along phase co-existence curves, like Clausius and Clausius–Clapeyron, are at best applicable to pure hydrate formers along pressure temperature stability limit projection. Empirical concepts that utilize a definition of hydrate fugacity are not thermodynamically consistent. Fugacity is defined on a component basis and related to a specific, component based, reference state. In residual thermodynamic formulation fugacity at reference state is pure components pressure. For symmetric excess thermodynamics the reference state is pure liquid component fugacity. Additionally, for asymmetric excess thermodynamics the reference state is infinite dilution fugacity for the component in a specific solvent.

The concept we present, and use, in this work is thermodynamically consistent. The reason for the consistency is that the equation for enthalpy is derived from Gibbs free energy using fundamental thermodynamic relationships. Practically this also implies that the corresponding entropy change for hydrate formation is consistent. It can therefore be expected that the final hydrate has the correct hydrate structure.

No similar studies on water containing ethanol have been found in the accessible literature.

The paper is organized as follows. The traditional use of alcohols for thermodynamic hydrate inhibitor is briefly discussed in Section 2. Since this is not an important focus in this work, there are no references to any publications from my research group or other international research groups. In Section 3, I give a brief overview of hydrate production philosophy. The main focus of this section is on and kinetic challenges related to the various hydrate production methods. The use of CO_2 for exchange with in situ CH_4 in hydrate is a priority in that section. Scientific methods utilized in this work are briefly described in Section 4. The non-equilibrium thermodynamic nature of hydrates in sediments is discussed in Section 5. In particular it is argued that hydrates in sediments cannot reach thermodynamic equilibrium because there are too many independent thermodynamic variables in the co-existing phases compared to constraints on these variables. These constraints are mass conservations and thermodynamic equilibrium equations. Hydrates in sediments are normally in a stationary state. The rest of the paper is devoted to thermodynamic analysis of the consequences of adding small amounts of ethanol to CO_2 and mixtures of CO_2 with N_2. The purpose of the analysis is to investigate if the ability to form a new hydrate from CO_2 or CO_2/N_2 is significantly reduced by adding up to 3 mole percent ethanol to the pore water. The effects of adding ethanol on heterogeneous gas/liquid formation are discussed in Section 6. The main focus in this section is on hydrate stability. A secondary focus in this section is the ability of injection gas (CO_2 or CO_2/N_2 mixture) to form a new hydrate which releases enough heat to dissociate in situ CH_4 hydrate. A similar analysis for homogenous hydrate formation from water and dissolved hydrate former is given in Section 7. The paper is concluded with a discussion in Section 8 and conclusions in Section 9.

2. The Use of Alcohols for Hydrate Prevention in Industrial Settings

Problems related to the formation of hydrate from hydrocarbon mixtures and water have been important motivations for industrial hydrate research during the latest seven decades. The interest for natural gas hydrates as energy source has increased substantially during the latest four decades. Research activities on hydrates in sediments, and hydrate production technologies, have increased substantially.

Historically, small alcohols like methanol and ethanol have been used to prevent hydrate formation. Addition of alcohols to liquid water results in a reduction (more negative) of liquid water chemical potential. A consequence of this reduced liquid water chemical potential is that a higher pressure is needed in order to form hydrate. This will be discussed in more detail later.

Methanol totally dominates the market among these two alcohols. Exceptions are countries, such as Brazil, which produce ethanol at reasonable prices from a waste product in sugar production.

Thermodynamic hydrate inhibition as discussed above is based on the effect of dissolved alcohols on "bulk" water [9,10]. Most efficient concentrations of alcohol are concentrations when alcohol is the solvent for water, rather than the opposite. The transition from water as solvent, over to methanol as solvent, is clearly visible as a transition in the liquid mixture dielectric constant as function of alcohol concentration [10].

There are many different ways that a hydrate can form in a multiphase system. Hydrate formation on the interface between liquid water, and a separate phase for guest molecules. Guest molecules are the hydrate forming molecules that enter the cavities created by hydrogen bonded water. Methane and natural gas are examples of phases containing guest molecules. Hydrate can only form if Gibbs free energy of the hydrate is lower than Gibbs free energy for the same amounts of water and guest molecules in the original phases. Technically a hydrate equilibrium calculation involves solving for the same water chemical potential in liquid water and hydrate water when also chemical potentials for guest molecules are the same in hydrate and in gas phase. Addition of significant amounts of alcohols to the water reduces the chemical potential of liquid water and higher pressure is needed in order to create hydrate.

However, hydrate can also form from dissolved hydrate formers in liquid water, from water dissolved in gas and towards mineral surfaces like for instance rust. As will be discussed here, all these hydrates are different in composition and stability (Gibbs free energy).

Hydrate inhibition is not a primary objective in this work. However, many of the thermodynamic calculations presented here can also provide valuable extensions in hydrate risk analysis. In particular, the multi-phase hydrate stability analysis should be incorporated. An analysis based on temperature and pressure alone is not sufficient. There are many different ways that hydrate can dissociate again even in temperature and pressure is inside hydrate formation region.

Since there are several routes to hydrate formation it is important to be able to analyze the relative stabilities of the different hydrates than are formed in presence of alcohols if the local inhibitor concentrations are insufficient for total thermodynamic inhibition of hydrate.

3. Small Alcohols as Surfactants in Hydrate Production

Countries such as China and Japan are very actively working towards full scale hydrate production schemes.

A kinetically critical element in hydrate dissociation is the transport of guest molecules across a thin (roughly 1.2 nm) interface between a hydrate and its surroundings [8,9,11–13]. During the hydrate dissociation process, there will be liquid water facing the dissociating hydrate. The interface between liquid water and the hydrate will reproduce itself continuously. This is determined by the molecular physics of water. Partial charges on the surface of a hydrate are fairly fixed except for small translational and rotational movements relative to a minimum energy situation. Liquid water molecules, on the other hand, are highly dynamic. However, liquid water molecules have to relate to the hydrate water molecules in order to optimize entropy.

Methanol has surfactant properties because of the large methyl group. The low partial charge on the methyl group results in a small charge per atomistic group surface. The concentration of methanol, on the surface of water that faces the gas, will therefore be higher than the average concentration of methanol in "bulk" water. Carbon dioxide has a significant quadrupole moment. Solubility of carbon dioxide in water is therefore higher than solubility of small hydrocarbons in water.

Efficient hydrate dissociation depends on a favorable Gibbs free energy change. Temperature and pressure are two of all the independent thermodynamic variables. Concentrations of all molecules in all the co-existing are the other independent thermodynamic variables. If temperature and pressure are outside hydrate formation limits then gas and liquid water is more stable than the hydrate and Gibbs free energy change is favorable for hydrate dissociation. This is, however, only one of the thermodynamic conditions that has to be fulfilled. The first law of thermodynamics also requires that enough the enthalpy of hydrate dissociation can be supplied.

Reducing pressure to a condition outside hydrate stability is considered as a low-cost method for producing natural gas from a hydrate. The challenge is that the natural supply of heat may not be sufficient. Pressure reduction results in a temperature gradient that can supply heat. Local geological depth involves a geothermal gradient. Both of these possible sources of heat are low temperature sources and may not be efficient for breaking hydrogen bonds. The duration of early stages pilot tests [14,15] were extremely short. The results from these pilot studies are far too short to provide significant learning about hydrate production. The first offshore test [16] ended after 6 days. Officially, the reasons were massive sand production and reservoir freezing. The second test [17] was supposed to last for 6 months but the sediments froze after 24 days [17–19].

The enthalpy needed to dissociate CH_4 hydrate is roughly 58 kJ/mole CH_4 at 273.15 K [5–9]. Heat is transported fast through water phases [20]. Injection of steam or hot water is expensive. Heat will also be lost to minerals. However, it still addresses the breaking of hydrogen bonds. Injection of alcohols or salts is also very efficient in breaking hydrogen bonds.

Natural gas hydrates in sediments are unable to touch mineral surfaces due to extremely low chemical potential of the structured adsorbed water [21–26]. Another way to look at this fact is to consider distribution of atomistic charges in mineral surfaces. Distribution of atomistic charges on mineral surfaces are noy compatible with average partial charges in liquid water. The result is rather extreme water densities in the first adsorbed water layers on mineral surfaces. Many theoretical studies are available in the literature. A more limited number of experimental studies are also available. Water densities in the order of 3 times liquid water density is quite common. The paper by Geissbühler et al. [27] is a representative experimental report. Geo-mechanical models based on hydrate "cementing" pores are therefore physically wrong. Practically there will be minimum distances between mineral surfaces and hydrate. This minimum distance is controlled by the mineral/water structures discussed above, and also hydrate/liquid water interfaces. Molecular diffusion and induced hydrodynamic flow increase the distance between mineral surfaces and hydrate.

Fracture systems lead to hydrodynamic flow in offshore hydrates. Fractures that connect the seafloor to hydrate filled sediments lead to hydrate dissociation. The reason for this dissociation is the seawater concentration of guest molecules, which are almost at infinite dilution in the seawater. Limits for hydrate stability, in terms of guest concentration in surrounding liquid water, are discussed in the section describing homogeneous hydrate formation. Fracture systems below the hydrate can lead to inflow of new hydrocarbons, and can lead to subsequent hydrate formation when the hydrocarbons enter hydrate forming conditions. Hydrates in sediments are not thermodynamically stable (see Section 5). The system of a hydrate, liquid water, and gas will establish a situation that can be close to stationary. The dynamics of this stationary situation varies from one hydrate reservoir to another. The stationary situation is not constant. A net hydrate dissociation, which is higher than the creation of new hydrate formation, will over time lead to reduced hydrate saturation. Even in the permafrost in Alaska the average hydrate saturation is around 75% and it is rare to find hydrate deposits that exceed 85% of hydrate pore filling volume.

Injection of carbon dioxide is efficient because there is available free pore water that can create carbon dioxide hydrate. Enthalpy of hydrate formation for CO_2 hydrate is roughly -68 kJ/mole CO_2 at 273.15 K[5–9]. Addition of up to 30 mole% N_2 to the CO_2 increases this value to around -66 kJ/mole guest. The net difference between what is needed to dissociate the in situ CH_4 hydrate is still sufficient.

The first question is whether addition of limited amounts of alcohol will significantly disturb the hydrate formation ability (Gibbs free energy). The second question is whether the released enthalpy from formation of new hydrate from CO_2 or CO_2/N_2 mixture is sufficient to dissociate the in situ CH_4 hydrate.

4. Methodology

The primary scientific tool in this work is classical thermodynamics. There are elements of statistical mechanics in the historical development of the theory for chemical potential of water in hydrate, and interactions between water and guest in the hydrate lattice. As will be discussed in Section 6, our model for chemical potential of water in hydrate differs from older models. Using molecular dynamics simulations, we were able to incorporate effects of guest movements on water lattice destabilization. For example, carbon dioxide movements interfere with water movements in the hydrate lattice. The interference affects specific water movement frequencies and leads to water lattice destabilization. The result is a difference of roughly 1 kJ/mole between a rigid hydrate water lattice and a dynamic lattice. Ethane in a large cavity also disturbs water movements but less so than carbon dioxide. For methane as a guest, the difference between a dynamic water lattice and a rigid lattice is almost zero.

5. Non-Equilibrium Nature of Hydrates in Sediments

The early hydrate equilibrium experiments started around 1940. With CH_4 gas, liquid water, and a hydrate phase the researchers knew from Gibbs phase rule that only one independent thermodynamic variable could be fixed. Despite this fact, these hydrate equilibrium curves are also used as equilibrium curves when both temperature and pressure are fixed. As will be discussed later, hydrate stability also depends on a minimum concentration of guest molecules in the surrounding liquid water. Hydrates can form from liquid water solutions of guest molecules. The concentration range for this hydrate formation is between liquid water solubility concentration and minimum concentration for hydrate stability. Hydrates formed along this route have different composition and density. They are therefore thermodynamic definition.

As discussed above, solid mineral surfaces are very active relative to hydrate in two ways. In simplified terms, mineral surfaces are hydrate inhibitors because hydrates cannot exist close to mineral surfaces due to low chemical potential of adsorbed water. However, mineral surfaces can adsorb hydrate formers directly (polar guest molecules) [28–33] or indirectly in structured water [26,28–36] slightly outside mineral surface.

Hydrates formed from different phases are different because composition, density, and Gibbs free energy are different. The reason for this is that there is a mathematical imbalance between the number of independent thermodynamic variables and the associated constraint (mass conservation and equilibrium equations). If the thermal (same temperatures in all phases) equilibrium and the mechanical equilibrium (same pressures in all phases) is fulfilled, we can return to the simple system of CH_4, liquid water, and a hydrate. There are six remaining mole-fractions in the three phases. The constraints are the conservation of mole-fractions in each of the three phases. In addition, there are 4 independent equations of equal chemical potential for CH_4 and water in the three phases. With 7 equations in 6 variables there is no unique mathematical solution. The first and second laws of thermodynamics reduce to the following form when temperatures and pressures are the same for all phases:

$$\min \underline{G} = \min \left[\sum_{j=1}^{p} G^j N^j \right] = \min \left[\sum_{j=1}^{p} N^j \sum_{j=1}^{n} \mu_i^j(T, P, \overline{x}^j) x_i^j \right] \quad (1)$$

in which the line below G denotes extensive property (unit Joule), j is a phase index, and N is the number of moles in each phase, as indicated by counter j. Total number of co-existing phases is p. i is a component index and x is mole-fraction. The line above x in the chemical potential dependency denotes the vector of all mole-fractions in the specific phase. This also includes mineral surfaces since they serve as catalysts for hydrate nucleation while at the same time not being able to "host" the hydrate on the mineral surface. It also includes all different hydrate phases and in principle also adsorbed phases on hydrate surfaces. The abbreviation min denotes minimum.

Number of degrees of freedom is the number of thermodynamic independent variables that must be specified for a system to be able to reach equilibrium. I prefer to use discrete counting rather than the compact Gibbs phase rule formula. One reason for this is that mixtures of guest molecules in a gas phase increases the complexity. The result can be that the system is brought even more out of the possibility to read the thermodynamic equilibrium. Many of these aspects will not be captured by Gibbs phase rule. As an example, consider an equimolar mixture of carbon dioxide and methane. Carbon dioxide is closer to condensation on water than methane. Carbon dioxide has a stronger attraction to the liquid water surface. In more rigorous physics, one example of a two-dimensional adsorption theory is described by Kvamme [37]. If the system is closed, then the result of selective adsorption is that many different hydrates can form. Carbon dioxide will have a higher mole-fraction in the adsorbed layer on liquid water. This also implies a higher mole-fraction than methane of the liquid side of the interface. The first hydrates that form will be dominated by carbon dioxide. Gradually the carbon dioxide content of the gas decreases and the final hydrates that form will be pure methane hydrate. Similar physical differences apply to other ways that hydrate can form, such as hydrate formation towards solid surfaces.

In summary, there is no way to establish thermodynamic equilibrium for hydrate in sediments. A theoretical philosophy about phases that can be totally consumed and then disappear from the balance will generally not be realistic for real hydrate reservoirs. As mentioned above, offshore hydrates are normally in a stationary balance of hydrate dissociation and formation of new hydrates from upcoming gas. On top of this, there are active phases that are normally not counted for, including mineral adsorption and hydrate nucleation towards mineral surfaces

In view of the above, the thermodynamic equilibrium is not generally possible for hydrates in sediments. A similar line of arguments can be established for hydrates in industrial processing units and in pipeline transport of gas containing dissolved water. Multiphase transport of gas and water is another example.

Due to general non-equilibrium situation, I will use the term stability limits rather than equilibrium curves. For heterogeneous hydrate formation on gas/liquid water interface the set natural set of independent thermodynamic variables are temperature and pressure. For homogeneous hydrate formation from liquid water and dissolved gas components, the independent variables are mole-fractions of the guest molecules in water.

6. Heterogeneous Hydrate Formation in Systems Containing Alcohols

20 mole% N_2 seems feasible based on earlier studies in open literature. In order to limit the number of figures, I investigate hydrate free energy for pure CH_4, pure CO_2, and an injection mixture with 20 mole% N_2 in CO_2. Most natural gas hydrates are formed from biogenic sources and almost pure CH_4 hydrates. There are, however, no limits in the analysis here. The same calculations can be applied to thermogenic gas and mixed structure I and structure II hydrates.

From an environmental point of view, and also in terms of surfactant properties, ethanol is more interesting than methanol. I will therefore mainly focus on ethanol. In Section 6.1, I focus on the free energy changes needed to form hydrates and in Section 6.2 I examine changes in released enthalpy during hydrate formation.

The main goal is to establish knowledge on the stability of CH_4 and CO_2 hydrates in real thermodynamic variables (Gibbs free energy) and the ability of hydrate formation from injection gas (CO_2 and CO_2/N_2 mixtures) to deliver enough heat to dissociate in situ CH_4 hydrate.

The thermodynamic models are described in Section 6.1 after some examples for model verification purposes. The model for water activity coefficients used in Equation (2) is given in Table 1. Note that this is a simplified model. For NaCl it only applies to calculations of water activity. For situations in which salinity in the water is important for interactions with mineral reactions then a totally different model is needed.

$$\gamma_{H_2O}(T, x_{H_2O}) = a_0 + a_1 x_{H_2O} + a_2 x_{H_2O}^2 + a_3 x_{H_2O}^3, \quad a_k = c_{0,k} + \frac{c_{1,k}}{T_R} + \frac{c_{2,k}}{T_R^2}, \quad T_R = \frac{T}{273.15}$$

Table 1. Model and parameters for water activity coefficients for Equation (1).

a_k	Methanol			Ethanol		
	c_0	c_1	c_2	c_0	c_1	c_2
a_0	0.74821	0.52077	−0.59936	0.73743	0.51369	−0.58176
a_1	0.54174	−0.47388	0.54721	0.53390	−0.45996	0.53981
a_2	−0.53859	0.56767	−0.52552	−0.52277	0.55995	−0.51009
a_3	0.35068	−0.53117	0.37324	0.34566	−0.51558	0.36817
a_k	Ethylene Glycol (MEG)			Triethylene Glycol (TEG)		
	c_0	c_1	c_2	c_0	c_1	c_2
a_0	0.59006	0.44750	−0.49897	0.10473	0.19307	−0.12389
a_1	0.47286	−0.39499	0.47783	0.15496	0.17274	−0.15457
a_2	−0.44892	0.49855	−0.43804	0.08770	0.08731	0.11456
a_3	0.37487	−0.44859	0.36800	0.37947	0.80576	−0.82038
a_k	Glycerol			NaCl		
	c_0	c_1	c_2	c_0	c_1	c_2
a_0	0.10453	0.08003	0.35094	0.14486	0.15079	0.26256
a_1	−0.00151	0.32200	0.21219	0.09071	−0.09709	−0.63019
a_2	0.04988	0.04965	0.00515	0.13827	−0.28171	0.57169
a_3	0.44152	0.56533	−1.25419	0.13053	1.11504	−0.55605

6.1. Hydrate Phase Transition and Free Energy Changes

Ethanol and methanol only have one hydroxyl group. It can be of interest to compare hydrates formed from liquid water containing glycols and CH_4 with hydrates formed from water containing ethanol or methanol. MEG is mono-ethylene glycol and TEG is tri-ethylene glycol. In Figure 1a, I plot hydrate stability pressure limits as function of temperature and concentration of MEG or TEG in the water. Gibbs free energies for the formed hydrates are plotted in Figure 1b.

Similar plots for water containing ethanol and methanol are plotted in Figure 2a,b.

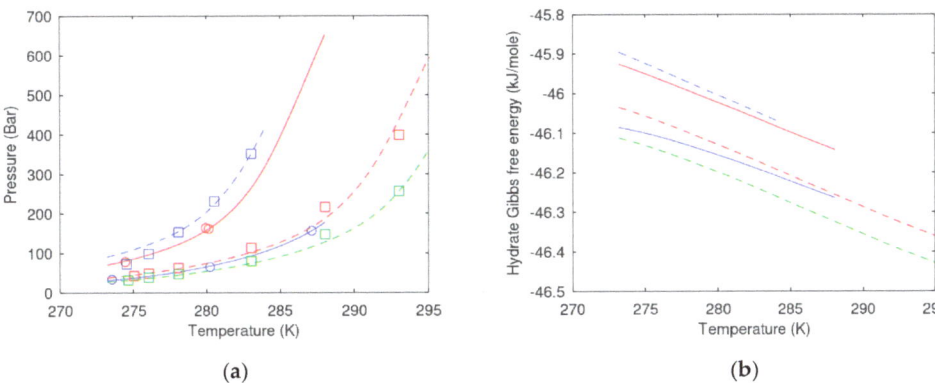

Figure 1. (**a**) Temperature pressure stability limits for CH_4 hydrate formed from CH_4 and water containing MEG (circles are experimental data [38] and solid curves are calculated) or TEG (squares are experimental data [39] and dashed curves are calculated). Blue curves and circles are for 10 wt% MEG in water. Red circles and curve are for 30 wt% MEG. Green squares and dashed curve are for 10 wt% TEG. Red squares and dashed curve are for 20 wt% TEG and blue squares and dashed curve are for 40 wt% TEG. (**b**) Gibbs free energy for hydrates formed along the stability curves in Figure 1a. Same line notations as in Figure 1a.

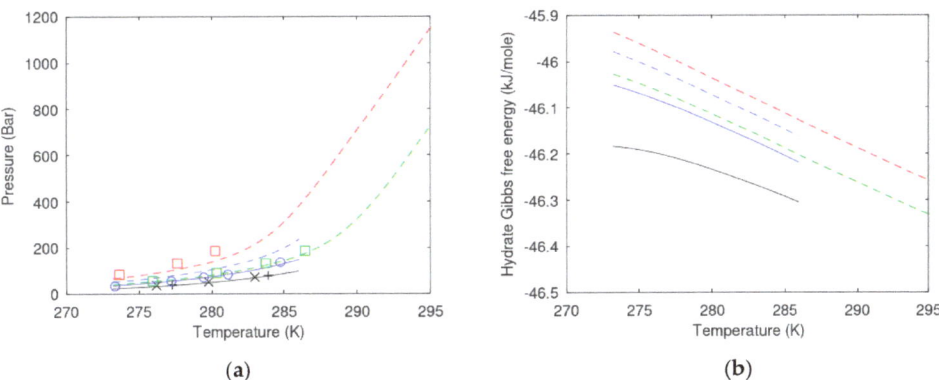

Figure 2. (**a**) Temperature pressure stability limits for CH_4 hydrate formed from CH_4 and water containing ethanol (circles are experimental data [40] and solid curves are calculated) and methanol (squares are experimental data [41] and dashed curves are calculated). Blue circles and solid blue curves are for 15 wt% ethanol. Green squares and dashed curve are for 10 wt% methanol. Dashed blue is for 15 wt% methanol. Red squares and dashed are for 20 wt% methanol. Solid black curve is calculated data for pure water. Crosses are experimental data from Sabil et al. [42] and plusses are experimental data from Tumba et al. [43]. (**b**) Gibbs free energy for hydrates formed from CH_4 and water containing various concentrations of ethanol and methanol. Same notations as in Figure 2a.

Stability of hydrates formed from water containing glycols, and hydrates formed from water containing ethanol and methanol are very similar.

Fugacity formulations dominate hydrate stability limit calculations in open literature on hydrates. Academic and commercial hydrate software is also based on the so-called reference scheme from around 1970. It might be useful to provide the residual thermodynamic model behind the curves in Figures 1 and 2. Liquid water chemical potential is formulated by symmetric excess thermodynamics as:

$$\mu_i^{(p,mix)}(T^{(p)}, P^{(p)}, \vec{x}^{(p)}) = \mu_i^{(p,pure)}(T^{(p)}, P^{(p)}) + RT \ln x_i + RT \ln \gamma_i^{(p)}(T^{(p)}, P^{(p)}, \vec{x}^{(p)}) \quad (2)$$

in which p is an indicator of liquid phase and p is water in this case. Subscript i denotes component index and i is water in this case. Superscript mix denote mixture and superscript pure denotes pure component. x is liquid mole-fraction and the arrow on x denote vector of all mole-fractions in the liquid mixture. $\gamma_i^{(p)}$ is the activity coefficient for component i. Asymptotic limit is unity when mole-fraction of component i approaches unity. Activity correction is in excess thermodynamics. The whole quantity is, however, in residual thermodynamics since the pure water term (first term on right hand side) is derived from Molecular Dynamics (MD) simulations for ice. Consistent chemical potential for pure liquid water is derived from enthalpy of ice dissociation at 0 °C and extended in the liquid region using experimental heat capacities. The second term on the right-hand side is the ideal mixing term due to entropy of mixing. The final term corrects for the deviations from ideal mixing. A simple model for water activity coefficients in mixtures with some alcohols is given in Table 1. These are based on 1 bar pressure. Correction from 1 bar to actual pressure is included through a Poynting correction for pure liquid water in Equation (2). A Poynting correction for water in empty hydrate is also included in Equation (3) below.

Chemical potential for water in hydrate is given by:

$$\mu_{H_2O}^H = \mu_{H_2O}^{O,H} - \sum_{k=1,2} RTv_k \ln\left(1 + \sum_i h_{ki}\right) \quad (3)$$

$\mu_{H_2O}^{O,H}$ is the chemical potential for water in an empty clathrate. Number of cavities is v, with subscripts k for large and small cavities, respectively. For structure I, which is the main focus in this overview on thermodynamics, $v_{large} = 3/24$ and $v_{small} = 1/24$. Within the scope of this work, I will assume that only one guest molecule can enter a cavity. The harmonic oscillator approach model can then be expressed as:

$$h_{ki} = e^{\beta[\mu_{ki} - \Delta g_{ki}]} \quad (4)$$

Chemical potential for molecule type i in cavity type k. I will assume that small and large cavities are at equilibrium so that:

$$\mu_{large_i} = \mu_{small_i} \quad (5)$$

For a system at equilibrium the chemical potential for a guest molecule in a cavity is equal to the chemical potential for the same molecule in the equilibrium phase. Δg_{ki} is the free energy change for inclusion for guest molecule i in a cavity of type k.

The most classical example is a hydrate former phase (gas, liquid, or supercritical) in contact with liquid (or ice) water which forms a hydrate at the interface. For these three phases there are 12 independent thermodynamic variables, 3 mass conservation equations, and 8 conditions of equilibrium. The difference between the number of independent variables minus mass conservations and equilibrium conditions leaves only one independent thermodynamic variable that can be fixed for equilibrium to be achieved. For this particular case of equilibrium Equation (4) can be written as:

$$h_{ki} = e^{\beta[\mu_i^{gas}(T,P,\vec{x}) - \Delta g_{ki}]} \quad (6)$$

For the heterogeneous hydrate formation, the phase index p in Equation (7) is gas. $\phi_i^{(p)}$ is the fugacity coefficient for CH_4. In this work I utilize the Soave–Redlich–Kwong (SRK) equation of state [14]. The first term on the right-hand side is the ideal gas chemical potential as function of temperature and density. Ideal gas chemical potential is trivial to calculate using the momentum space of the canonical partition function. The methane model is a sphere and the methane ideal gas chemical potential is easy to calculate. The analytical expression can be found in any textbook on physical chemistry. Guest molecules with rotational degrees of freedom also require the moments of inertia for the rotational degrees of freedom ideal gas chemical potential. In this work, all models are rigid and there are no ideal gas contributions from intramolecular degrees of freedom. The second term is the ideal gas mixing terms that comes from the entropy of mixing ideal gas components at constant pressure.

$$\mu_i^{(p,mix)}(T^{(p)}, P^{(p)}, \vec{x}^{(p)}) = \mu_i^{(p,pure,ideal\ gas)}(T^{(p)}, P^{(p)}) + RT \ln x_i \\ + RT \ln \phi_i^{(p)}(T^{(p)}, P^{(p)}, \vec{x}^{(p)}) \tag{7}$$

Since methane is pure in the first systems illustrated in Figures 1 and 2 mole-fraction is unity. Equation (7) enters into Equation (6). The free energy change of inclusion for methane in small and large cavity is given elsewhere [9,13,26,44–46]. With Equation (5) in Equation (2) and solving Equation (2) equal to Equation (3) for the same chemical potential of water in liquid water as in the hydrate. Chemical potential for water in empty clathrate is also given by Kvamme and Tanaka [46]. I have solved the equations iteratively for pressures, with temperature as the chosen fixed variable. The chemical potential for the gas changes significantly with pressure. I therefore also plot the composition of the formed hydrate as well as chemical potential of CH_4.

The filling fractions are trivially available from the semi grand canonical ensemble used for derivation of (3). This part is the same for the van der Waal and Platteeuw [47] model and our model [46]:

$$\theta_{ki} = \frac{h_{ki}}{1 + \sum_j h_{kj}} \tag{8}$$

θ_{ki} is the filling fraction of component i in cavity type k. Also:

$$x_{i,large}^H = \frac{\theta_{large,i} \nu_{large}}{1 + \theta_{large,i} \nu_{large} + \theta_{small,i} \nu_{small}} \tag{9}$$

$$x_{i,small}^H = \frac{\theta_{small,i} \nu_{small}}{1 + \theta_{large,i} \nu_{large} + \theta_{small,i} \nu_{small}} \tag{10}$$

where ν is the fraction of cavity per water for the actual cavity type, as indicated by subscripts. The corresponding mole-fraction water is then given by:

$$x_{H_2O}^H = 1 - \sum_i x_{i,large}^H - \sum_i x_{i,small}^H \tag{11}$$

and the associated hydrate free energy is then:

$$G^{(H)} = x_{H_2O}^H \mu_{H_2O}^H + \sum_i x_i^H \mu_i^H \tag{12}$$

Additionally, note that the results in Figures 1a and 2a are verification of the model system since these figures are solutions to:

$$\Delta G^{(H)} = x_{H_2O}^H \left[\mu_{H_2O}^H - \mu_{H_2O}^{liquid} \right] + \sum_i x_i^H \left[\mu_i^H - \mu_i^{gas} \right] = 0 \tag{13}$$

The molecular weight of ethanol and methanol is very different. The weight% alcohol used in industry is not convenient for comparison of thermodynamic effects since these are on a molar basis.

The changes in chemical potential for CH_4 as guest varies substantially along the stability limits of temperature and pressure, and also as function of increasing ethanol content. The composition of the hydrates also varies substantially with amount of ethanol added to the water.

Addition of ethanol as a surfactant to CO_2/N_2 system should rarely exceed 2 mole% but I also include 3 mole% as a possible upper limit and 1.5 mole% as minimum addition of ethanol. CO_2 hydrate is generally a more stable hydrate than CH_4 hydrate. This is illustrated in Figure 3b below. The density of CO_2 is also higher than CH_4 for the same range of conditions. There is a CO_2 phase transition to a higher density that results in a (relative) steep change in hydrate formation pressure as function of temperature. The stability limits of the hydrate are not substantially shifted for ethanol concentrations up to 3 mole% in water. In Figure 4a, we plot mole-fractions CO_2 in the formed hydrates as function of ethanol mole-fractions up to 3 mole%.

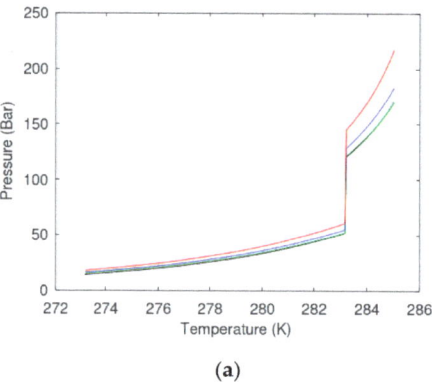
(a)

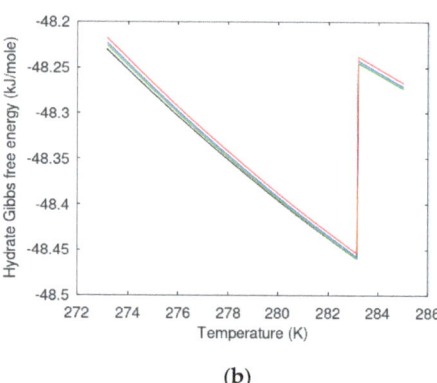
(b)

Figure 3. (a) Temperature pressure stability limits for CO_2 hydrate formed from CO_2 and water containing ethanol. Black curve is for 0 mole-fraction ethanol, green is for 0.015 mole-fraction ethanol, blue is for 0.02 mole-fraction ethanol, and red is for 0.03 mole-fraction ethanol. (b) Gibbs free energy for hydrates formed from CO_2 and water containing various concentrations of ethanol. The color codes are the same as in Figure 3a.

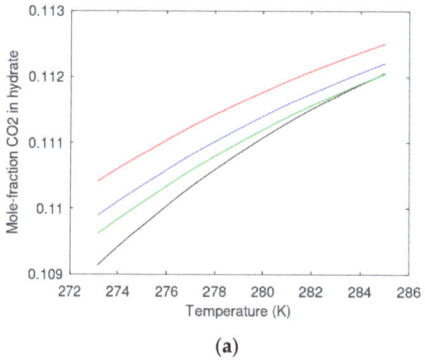
(a)

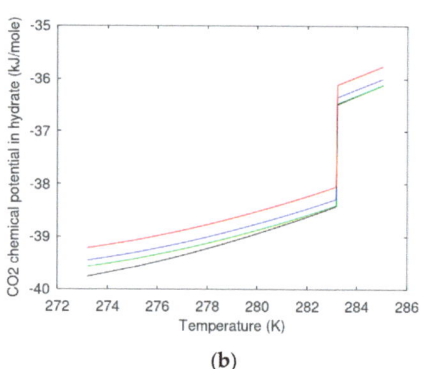
(b)

Figure 4. (a) Mole-fraction CO_2 in CO_2 hydrate formed from CO_2 and water containing ethanol. Black curve is for 0 mole-fraction ethanol, green curve is for 0.015 mole-fraction ethanol, blue curve is for 0.02 mole-fraction ethanol, and red curve is for 0.03 mole-fraction ethanol. (b) Chemical potential for CO_2 in CO_2 hydrate formed from CO_2 and water containing various concentrations of ethanol. Color codes are the same as in Figure 4a.

Based on Figure 3, it is considered feasible to add up to 3 mole% ethanol to the water phase. Thermodynamic stability of the formed hydrate is not affected substantially for ethanol concentrations below this limit. The changes in hydrate composition and chemical potential of CO_2 in hydrate are also small, as is illustrated in Figure 5. Changes in temperature and pressure hydrate formation conditions are also limited. Technically the best way is to add the ethanol to the injection gas. This will likely keep the surfactant effect for longer times before ethanol distributes and dissolve into groundwater.

The same calculations presented in Figures 3 and 4 are also conducted for 20 mole% N_2 in CO_2/N_2 mixture. Results are presented in Figures 5 and 6.

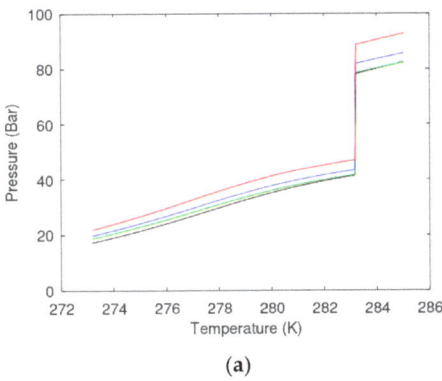
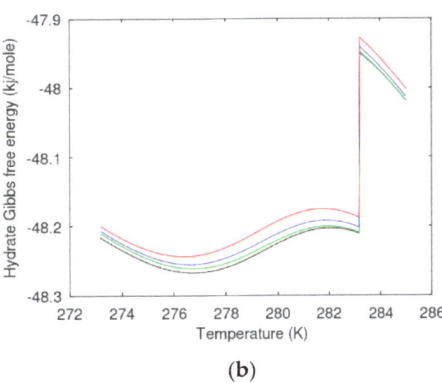

(a)   (b)

Figure 5. (a) Temperature pressure stability limits for mixed CO_2/N_2 hydrate formed from a CO_2/N_2 mixture with 20 mole percent N_2 and water containing ethanol. Black curve is for 0 mole-fraction ethanol, green curve is for 0.015 mole-fraction ethanol, blue curve is for 0.02 mole-fraction ethanol, and red curve is for 0.03 mole-fraction ethanol. (b) Gibbs free energy for hydrates formed from a CO_2/N_2 mixture with 20 mole percent N_2 and water containing various concentrations of ethanol. The color notation is the same as in Figure 5a.

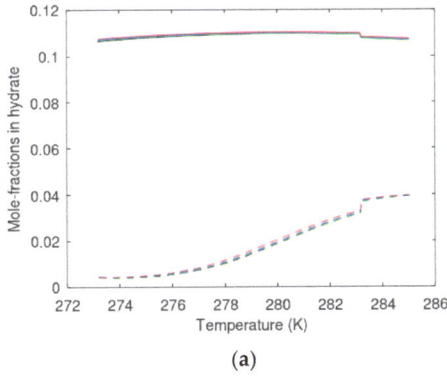
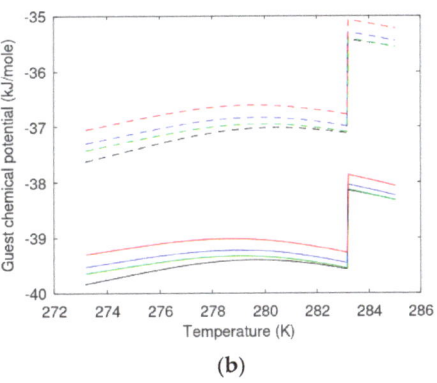

(a)   (b)

Figure 6. (a) Hydrate composition in mixed CO_2/N_2 hydrate formed from a CO_2/N_2 mixture containing 20 mole percent N_2 and water containing ethanol. Dashed curves are N_2 and solid curves are CO_2. Black curves are for 0 mole-fraction ethanol, green curves are for 0.015 mole-fraction ethanol, blue curves are for 0.02 mole-fraction ethanol, and red curves are for 0.03 mole-fraction ethanol. (b) Chemical potentials for CO_2 (solid) and N_2 (dashed curves) in hydrates formed from a CO_2/N_2 mixture with 20 mole percent N_2 and water containing various concentrations of ethanol. Color codes are the same as in Figure 6a.

There is no experimental evidence that CO_2 enters the small cavity of structure I at *liquid water conditions*. Hydrate formation mechanisms in the ice range is different and can likely lead to trapping of CO_2 in small cavity. There are some references to work conducted in the group of Werner Kuhs [48–50] that provides evidence of small cavity filling of CO_2 for temperatures far below zero and atmospheric pressures. We have conducted theoretical studies on different levels. These range from quantum mechanics to molecular dynamics simulations in the classical regime. All conclusions are the same. There is no free energy contribution to hydrate stability from CO_2 filling in small cavities at liquid water temperatures. At some conditions of full filling of large cavities with CO_2, a limited portion of CO_2 in small cavities is not enough to result in destabilization of the hydrate as a whole. However, that does not mean that CO_2 "willingly" enters a small cavity. According to Equation (5), the canonical partition function for CO_2 in small cavity is zero at liquid water conditions.

Another challenge in the experiments that are conducted at atmospheric pressure and 1 bar [48–51] is that the conversion between CH_4 hydrate and CO_2 hydrate is totally different since there is no liquid side interface. Specifically, a solid-state mechanism has been proposed [51]. A solid-state mechanism is not in accordance with kinetic rates we have observed during 2 decades of CH_4/CO_2 exchange experiments in liquid water regime. As mentioned in the introduction, this is a theoretical paper and we are therefore not reporting from our experiment or any other experiments on this specific issue.

For CO_2/N_2 mixtures at liquid water state conditions the hydrate formation from injection gas will facilitate from N_2 filling of the small cavities. This can be seen from a different stability limit window in Figure 5a. However, the free energies of the formed hydrates are less stable than hydrates created from pure CO_2, as can be seen by comparing Figures 3b and 5b. This is due to the change in CO_2 chemical potential though dilution with N_2. The reduction in stability is, however, limited and stability of the mixed hydrates formed from up to 3 mole% ethanol in water is significantly more stable than the CH_4 hydrate.

Composition of the hydrate is plotted in Figure 6a. Chemical potentials of the guests in the hydrate are plotted in Figure 6b. The values are reasonable in view of the influence of CO_2 on chemical potential for N_2 through the density of the mixture and the fugacity coefficient.

Thermodynamically it is feasible to use ethanol in CO_2/N_2 with 20 mole% N_2, and an amount of ethanol that would result in maximum 3 mole% ethanol in the pore water. Higher mole-fractions of ethanol in water have not been investigated since 3 mole% ethanol in water is equivalent to 7.33 weight%.

6.2. Enthalpies of Hydrate Formation

A consistent way to calculate ΔH^{Total} is given by the general thermodynamic relationship:

$$\frac{\partial \left[\frac{\Delta G^{Phasetransition}}{RT} \right]_{P,\vec{N}}}{\partial T} = - \left[\frac{\Delta H^{Phasetransition}}{RT^2} \right] \quad (14)$$

The route to enthalpy changes, from the residual thermodynamic scheme described in this work, is fairly straightforward utilizing Equation (3) for hydrate water. Liquid water chemical potential from Equation (2) as basis for corresponding enthalpy of liquid water is trivial and available in thermodynamic textbooks.

I now use H_1 to denote heterogeneous hydrate formation from liquid water (or ice) and methane gas. The same proportionality as in Equation (14) applies to relationship between chemical potential and partial molar enthalpy:

$$\frac{\partial \left[\frac{\Delta \mu_{H_2O}^{(H_1),Phasetransition}}{RT} \right]_{P,\vec{N}}}{\partial T} = - \left[\frac{\Delta \overline{H}_{H_2O}^{(H_1),Phasetransition}}{RT^2} \right] \quad (15)$$

in which $\Delta \overline{H}_{H_2O}^{(H_1), Phasetransition}$ is the partial molar enthalpy change for water during the phase transition.

Differentiation of Equation (3) gives the following results for partial molar enthalpy of water in hydrate [6,12–15]:

$$\overline{H}_{H_2O}^{H} = -RT^2 \frac{\partial \left[\frac{\mu_{H_2O}^{0,H}}{RT}\right]_{P,\vec{N}}}{\partial T} + \left[\sum_{k=1,2} v_k \frac{\sum_i h_{ki}\left[(\mu_{ki} - \Delta g_{ki}) - T(\frac{\partial \mu_{ki}}{\partial T} - \frac{\partial \Delta g_{ki}}{\partial T})\right]}{\left(1 + \sum_i h_{ki}\right)}\right] \quad (16)$$

and from Equation (2) for liquid water:

$$\overline{H}_{H_2O}^{water} = -RT^2 \frac{\partial \left[\frac{\mu_{H_2O}^{pure,H_2O}}{RT}\right]_{P,\vec{N}}}{\partial T} - -RT^2 x_{H_2O}^{water} \left.\frac{\partial \ln \gamma_{H_2O}^{water}}{\partial T}\right|_{P,\vec{N}} \quad (17)$$

and the enthalpy of hydrate formation is then:

$$\Delta \overline{H}_{H_2O}^{(H_1), Phasetransition} = x_{H_2O}^{(H_1)} \left[\overline{H}_{H_2O}^{H_1} - \overline{H}_{H_2O}^{water}\right] + \sum_{i=1}^{n} x_i^{(H_1)} \left[\overline{H}_i^{H_1} - \overline{H}_i^{gas}\right] \quad (18)$$

The last term is the partial molar enthalpy of the guest molecules in the gas (or liquid) hydrate former phase. This is trivially given by the Equation of state using the same type of relationship as Equation (15) with chemical potential for methane from Equation (7). The analytical expression for the SRK [52] version of this term is available from engineering handbooks and does not need space here. The partial molar enthalpy of each guest molecule in the hydrate, $\overline{H}_i^{H_1}$ needs some more attention.

The ideal gas enthalpy depends only on temperature and will be the same for the guest molecule inside the cavities and in the gas mixture. As such we only need the residual contributions for these two terms. Entropy related properties like chemical potentials and free energies are sensitive to possible interactions between water movements and guest movements. Average interaction energy is less sensitive to these effects. I have therefore used a Monte Carlo approach [53,54] with fixed water lattice to evaluate average interaction energies between guest and water for the various cavity types in structure I. I also sample the efficient volume for the guest in the cavity in order to calculate a compressibility factor for the guest molecule inside the two types of cavities.

$$z_{ki} = \frac{PV_{ki}}{RT} \quad (19)$$

where V_{ki} is the sampled molar volume of guest molecule i in cavity type k.

$$H_{ki}^{R} = U_{ki}^{R} + (z_{ki} - 1)RT \quad (20)$$

$$x_i^{H_1} \overline{H}_i^{H_1} - x_i^{H_1} \overline{H}_i^{gas} = x_i^{H_1} \overline{H}_i^{H_1,R} - x_i^{H_1} \overline{H}_i^{gas,R}$$
$$= x_{large,i}^{H_1} \overline{H}_{large,i}^{H_1,R} + x_{lsmall,i}^{H_1} \overline{H}_{small,i}^{H_1,R} - x_i^{H_1} RT^2 \left.\frac{\partial \ln \phi_i^{gas}(T,P,\vec{x}^{gas})}{\partial T}\right| \quad (21)$$

In Figure 7 below, I plot enthalpies of hydrate formation for pure CO_2 and water containing various concentrations of ethanol up to mole-fraction 0.03. These are plotted as per mole hydrate in (a) as well as per guest in (b) in order to illustrate that the temperature dependency of the coordination number (moles water per mole guest in the formed hydrate) changes significantly when adding N_2. As an example: for the pure CO_2 hydrate formed form water containing 3 mole% ethanol the coordination number changes from around 7.0 for the lowest temperature range and down to 6.5 after the CO_2 phase transition temperature. The coordination number is actually the numbers in Figure 8b divided by corresponding numbers in Figure 7a. Similar comparison for CO_2 with 20 mole% N_2 is illustrated is Figure 7c,d. As can be seen from Figure 6a, there is a significant change in the nitrogen filling of the hydrate with temperature. The coordination number changes substantially with temperature as compared to the pure CO_2 case. However, as can be seen by comparing Figure 7b,d, the available enthalpy that can be used to dissociate in situ CH_4 hydrate is significantly decreased by the addition of N_2.

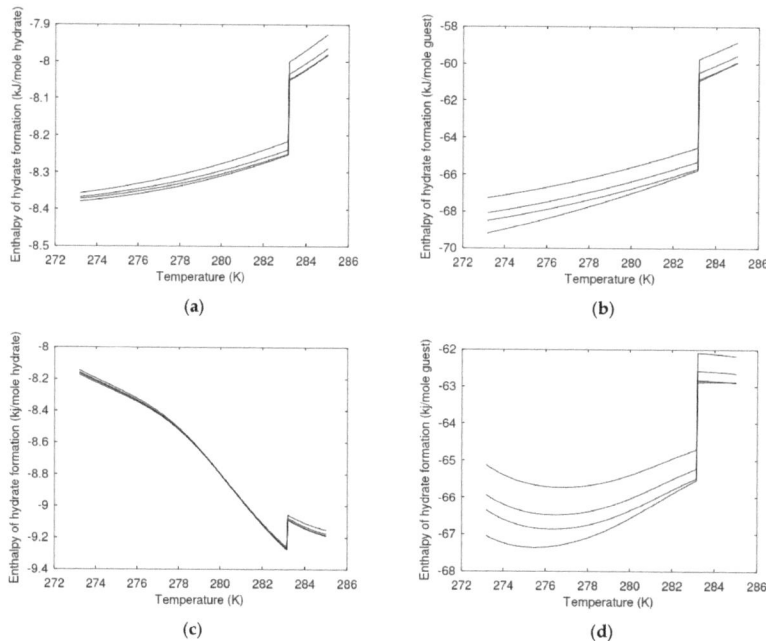

Figure 7. (a) Enthalpy of hydrate formation per mole hydrate as function of temperature and mole-fraction ethanol in water for pure CO_2 hydrate. The lowest curve is for no ethanol in water, then 0.015 mole-fraction ethanol in water, then 0.02 mole-fraction ethanol in water, and upper curve 0.03 mole-fraction ethanol in water. (b) Enthalpy of hydrate formation per mole guest in hydrate as function of temperature and mole-fraction ethanol in water for pure CO_2 hydrate. The lowest curve is for no ethanol in water, then 0.015 mole-fraction ethanol in water, then 0.02 mole-fraction ethanol in water, and upper curve 0.03 mole-fraction ethanol in water. (c) Enthalpy of hydrate formation per mole hydrate as function of temperature and mole-fraction ethanol in water for hydrate formed from 20 mole percent N_2 in CO_2. The lowest curve is for no ethanol in water, then 0.015 mole-fraction ethanol in water, then 0.02 mole-fraction ethanol in water, and upper curve 0.03 mole-fraction ethanol in water. (d) Enthalpy of hydrate formation per mole guest in hydrate as function of temperature and mole-fraction ethanol in water for hydrate formed from 20 mole percent N_2 in CO_2. The lowest curve is for no ethanol in water, then 0.015 mole-fraction ethanol in water, then 0.02 mole-fraction ethanol in water, and upper curve 0.03 mole-fraction ethanol in water.

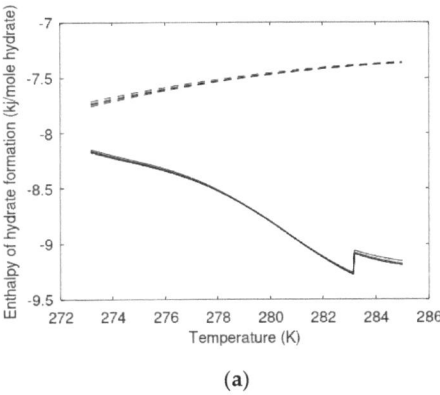 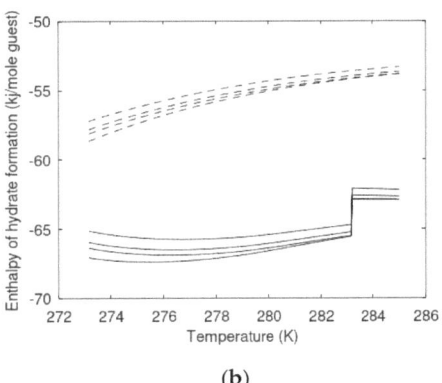

Figure 8. (a) Enthalpy of hydrate formation per mole hydrate and (b) enthalpy of hydrate formation per mole guest. Both figures are plotted as function of ethanol concentration in water. Dashed curves are for CH_4 hydrates formed from water with various mole-fractions ethanol in water. Solid curves are for hydrates formed from CO_2 with 20 mole% N_2. Lowest curves are for no ethanol, then 0.015 mole-fraction ethanol in water, then 0.02 mole-fraction ethanol in water, and upper curves for 0.03 mole-fraction ethanol in water.

How much ethanol that will dissolve into "bulk" pore water during the average residence time of the injection gas in the pore is unknown. As such, it makes sense to also compare enthalpies per guest for methane as function of mole% ethanol in water as well. This will enable us to illustrate the range between maximum and minimum available formation enthalpy that can be used for dissociation of the in situ CH_4 hydrate.

In Figure 8, I plot the same curves as in Figure 7 but now also with CH_4 hydrate formation enthalpy included. In the worst case scenario, the available net enthalpy of hydrate formation from the CO_2/N_2 mixture is roughly 6 kJ/mole more than what is needed for dissociation of the in situ CH_4 hydrate. These are all values at pressure temperature phase stability boundaries.

Heat transport is at least 2–3 times faster than mass transport through water systems [20]. The hydrate formation enthalpy for hydrates formed from CO_2/N_2 mixtures containing up to 20 mole% N_2 is therefore sufficient to dissociate in situ CH_4 hydrate for up to 3 mole% ethanol in the pore water. As illustrated in the previous section, the free energy of the hydrates formed from CO_2/N_2 mixture is thermodynamically more stable than CH_4 hydrates. This is illustrated by comparing Figures 2b and 5b. The difference is in the order of 2 kJ/mole hydrate. Formation of a new hydrate from the injection gas (CO_2/N_2/surfactant) will result in higher salinity of the remaining liquid pore water. Eventually, the resulting changes in liquid water chemical potential will lead to dissociation of hydrate in the pore. Additionally, since the stability of CH_4 hydrate is lower than stability of CO_2 hydrate then the CH_4 hydrate will dissociate first.

7. Homogeneous Hydrate Formation in Systems Containing Alcohols

CH_4 from dissociating hydrates is expected to distribute as gas bubbles escaping upwards in the reservoir due to buoyancy. Some methane will dissolve rapidly in water and can generate hydrate cores for further growth elsewhere. However, it is also important to investigate hydrate stability limits towards concentration of CH_4 in surrounding water. This is also a route to hydrate dissociation even for other production schemes. Pumping out water during pressure reduction in a hydrate reservoir that has no free gas leads to change in water surrounding the hydrate. If the surrounding water contains less CH_4 than hydrate stability limit concentration then the hydrate will dissociate.

Of course, we need to know the enthalpy of homogeneous hydrate formation since the released enthalpy of hydrate formation will generate local temperature gradients.

Homogeneous CH_4 hydrate formation is discussed in Section 7.1 and homogeneous CO_2 hydrate formation is discussed in Section 7.2.

7.1. Methane Hydrate Formation from Water Solution

Solubility of CO_2 in water is significantly larger than solubility of CH_4 in water. Both of these hydrate formers can form hydrate homogeneously in water. They can also form hydrate heterogeneously since dissolved hydrate formers in water can adsorb on a hydrate film separating liquid water from the guest phase. Density of CH_4 hydrate is lower that water density and will reside on top of liquid water. A hydrate film which has been formed heterogeneously, as discussed in Section 6, might already exist.

Generally, enthalpies of hydrate formation from guest molecules dissolved in water are smaller in absolute value than enthalpies of hydrate formation from gas and liquid water. The reason is that contributions to enthalpy changes for guest molecules are smaller. Partial molar volumes of these guest molecules in water are significantly smaller than the molar volumes for the same molecules in gas phase. The guest interactions with surrounding water are also closer to that of interactions between guest and water in a closed hydrate cavity [5,53,54].

A residual thermodynamic formulation for methane dissolved in water was developed by Kvamme [5] and briefly the resulting equation is:

$$\mu_{CH_4}^{aqueous}(T,P,\vec{x}) = \mu_{CH_4}^{\infty,Residual}(T,P,\vec{x}) + \mu_{CH_4}^{ideal\ gas}(T,P,\vec{x}) + RT \ln\left[x_{CH_4}\gamma_{CH_4}^{\infty}(T,P,\vec{x})\right] \quad (22)$$

$$\mu_{CH_4}^{\infty,Residual} = 3.665 + \frac{40.667}{T_R} - \frac{48.860}{T_R^2} \quad (23)$$

where T_R is temperature divided by critical temperature for methane. The maximum usable temperature for Equation (23) is 325 K, which is of course far above most hydrate forming condition. The ideal gas term on right hand side is calculated using infinite dilution partial molar volume for CH_4 as function of temperature. Pressure dependency is negligible since water is almost incompressible. The ideal gas term is therefore calculated using the analytical solution for translational movement of CH_4 in momentum space since CH_4 is approximated as a sphere.

The activity coefficient for methane in water is based on the asymmetric excess convention. In this convention, the activity coefficient for CH_4 in water approaches unity when mole-fraction CH_4 goes to zero. A fitted equation, and corresponding parameters, is given in Table 2.

$$\ln \gamma_{CH_4}^{\infty}(T,P,\vec{x}) = \sum_{i=1,2}^{39}\left[a_0(i) + \frac{a_1(i+1)}{T_R}\right](x_{CH_4})^{[0.05+\frac{i-1}{40}]} \quad (24)$$

$T_R = \frac{T}{T_{C,CH_4}}$. The critical temperature for CH_4 is 190.6 K.

The lower summation 1,2 indicates starting from 1 and counting in steps of 2.

Parameters for Equation (24).

The density of CH_4 at ideal gas reference state is almost constant. With constant density the ideal gas chemical potential can be approximated by the following fit:

$$\mu_{CH_4}^{ideal\ gas} = -73.901 + \frac{129.925}{T_R} + \frac{-70.024}{T_R^2} \quad (25)$$

The homogeneous hydrate formation region is limited by solubility of methane in water, and minimum mole-fraction of CH_4 in water needed to keep hydrate stable. Solubility in liquid water is calculated by solving gas chemical potential, Equation (7), being equal to dissolved CH_4 chemical potential for CH_4 according to Equation (22).

Table 2. Activity model for CH_4 dissolved in water.

i	a_0	a_1
1	1.360608	3.796962
3	0.033630	−0.703216
5	0.656974	−12.441339
7	1.763890	−21.119318
9	5.337858	−33.298760
11	−0.024750	12.387276
13	48.353808	17.261174
15	−11.580192	16.384626
17	−0.087295	13.171333
19	−0.558793	13.556732
21	−23.753020	16.573197
23	−10.128675	13.591099
25	−41.212178	5.060082
27	−31.279868	31.289978
29	−23.855418	31.720767
31	−35.125907	37.064849
33	−33.675110	41.544360
35	−27.027285	57.609882
37	−19.026786	54.961702
39	−37.872252	57.204781

Minimum mole-fraction for keeping the hydrate stable is calculated by inserting Equation (22) into Equation (4) and then assuming equilibrium (or rather, the stability limit) between hydrate CH_4 and CH_4 in solution. For defined temperature and pressure, we can then solve for equal water chemical potentials in liquid, Equation (2), and hydrate, Equation (3). In Figure 9, I plot CH_4 solubility and CH_4 hydrate stability limit as function of mole-fraction CH_4 in surrounding water. These are plotted for various mole-fractions ethanol in water. Solubility plots are shown for the same range of conditions of homogeneous hydrate formation.

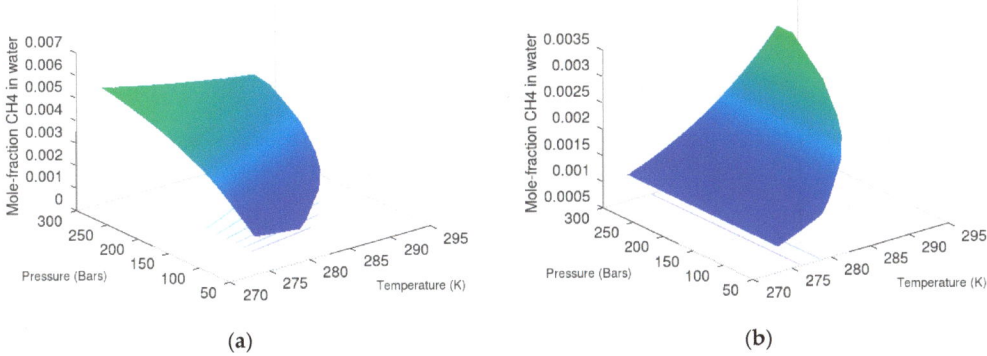

(a) (b)

Figure 9. *Cont.*

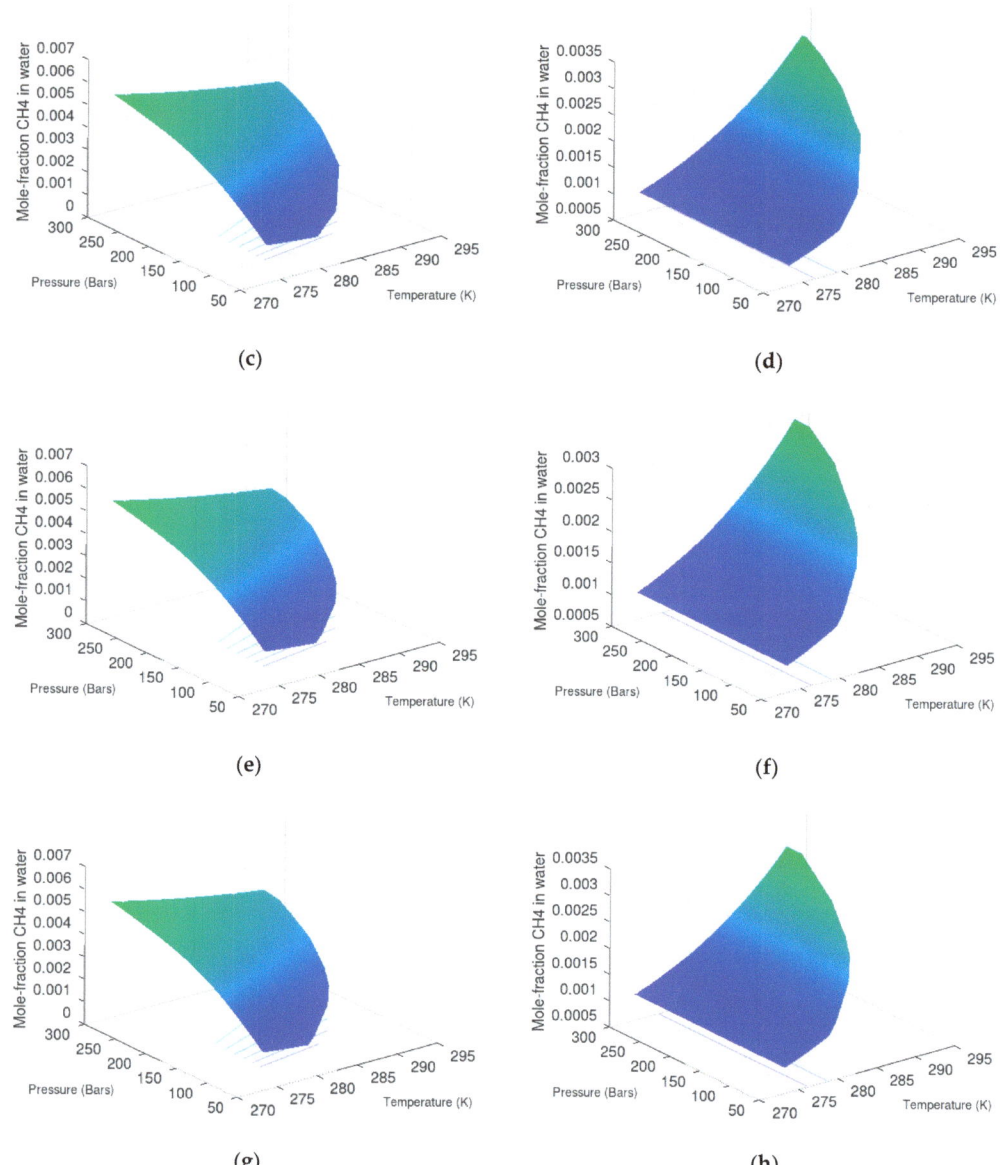

Figure 9. Liquid water solubility and hydrate stability limit mole-fractions of hydrate for various mole-fractions of ethanol. (**a**) Solubility for 0 ethanol, (**b**) hydrate stability limit for 0 ethanol, (**c**) solubility for 0.015 mole-fraction ethanol in water, (**d**) hydrate stability limit for 0.015 mole-fraction ethanol in water, (**e**) solubility for 0.02 mole-fraction ethanol in water, (**f**) hydrate stability limit for 0.02 mole-fraction ethanol in water, (**g**) solubility for 0.03 mole-fraction ethanol in water, and (**h**) hydrate stability limit for 0.03 mole-fraction ethanol in water.

These 3D plots are not easy to read even with grids on. However, in the context of this work they are mainly intended as illustrations of the concentration window for possible hydrate grow from dissolved CH_4 in water. As mentioned earlier, the hydrates formed from solution of CH_4 are different from the heterogeneous hydrate discussed in Section 6. This can be seen by comparing Figure 10 below and gas CH_4 chemical potentials from Figure 3b. This will result in different cavity partition functions, e.g., Equation (4). Compositions of hydrate will therefore be different and then also Gibbs free energies are different.

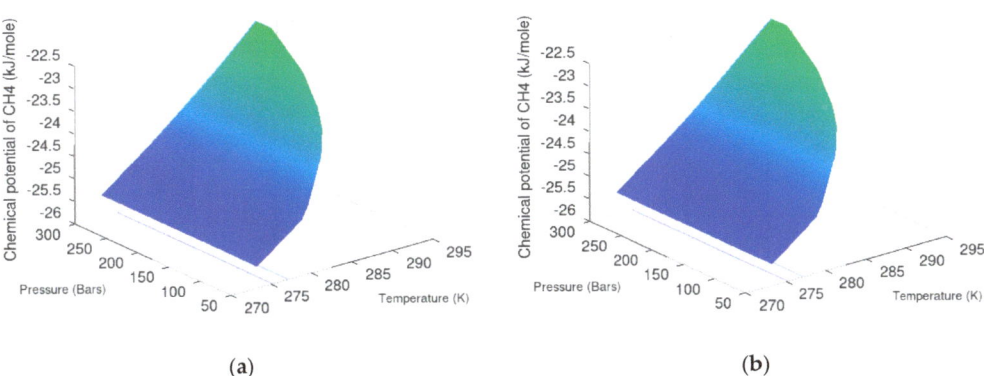

Figure 10. Chemical potential of CH_4 entering hydrate as function of temperature pressure and mole-fractions CH_4 (see Figure 10) for: (**a**) 0 ethanol and (**b**) 3 mole percent ethanol in water.

The differences between heterogeneous and homogenous hydrate can be seen by comparing Figure 10 and gas CH_4 chemical potentials from Figure 3b. Since the gas chemical potential and the liquid water chemical potential of CH_4 are different in a non-equilibrium system then cavity partition functions, such as in Equation (3), are also different. As a consequence, Gibbs free energies for heterogeneous and homogeneous hydrates are different. By thermodynamic definition hydrates with different composition and different Gibbs free energy are a different phase. In Figure 11, I plot free energy for the homogeneously formed hydrates for 0 ethanol and 3 mole percent ethanol in water. Hydrate can form in the whole concentration range between solubility and hydrate stability limits. Each mole-fraction guest in water will result in a unique hydrate composition. Additionally, note from the plots in Figures 9–11 that pressure dependency in the hydrate plots is very small since these are phase transitions that only involve condensed phases.

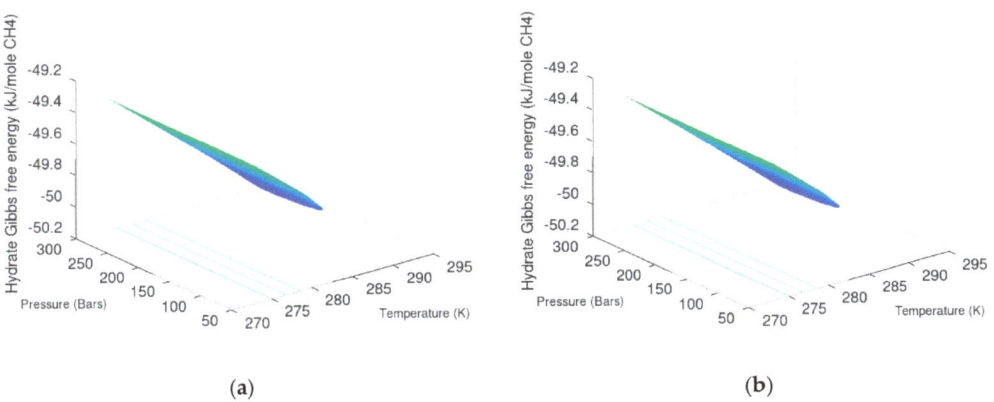

Figure 11. Gibbs free energy for homogeneously formed CH_4 hydrate as function of temperature pressure and mole-fractions CH_4 (see Figure 10) for: (**a**) 0 ethanol and (**b**) 3 mole percent ethanol in water.

A very important practical implication of the hydrate stability limits towards outside water methane concentration is the role it plays in destabilizing offshore hydrate. Most offshore hydrate reservoirs worldwide are in a stationary balance between formation of new hydrate from upcoming gas, and dissociation towards incoming seawater. Fracture systems in offshore hydrate frequently connect hydrate filled sections to the seafloor. The CH_4 concentration in this seawater is close to infinite dilution. The chemical potential for CH_4 at infinite dilution in water is lower than chemical potential for CH_4 in hydrate. As a result, the hydrate will dissociate due to a more beneficial situation for CH_4 in the outside water. The result of this mechanism for hydrate dissociation is observed worldwide as fluxes of methane from natural gas hydrate deposits. The flux rates can range from invisible bubbles to massive gas fluxes that bubble through the water column. Another example is all the hydrate mounds that sit on top of fractures that connect from seafloor and downwards towards conventional hydrocarbon reservoirs. Hydrate forms from upcoming gas and seawater, due to the temperature and pressure. The formed hydrate dissociates from the top due to chemical potential gradients of CH_4 between hydrate and seawater. The hydrate formation/dissociation couplings are kinetically limited by the hydrate dissociation and these mounds can seem stationary. Dissociation fluxes can be monitored by measuring local CH_4 concentrations in surrounding water. Since the CH_4 is an attractive source for several biological ecosystems the biological activity on these mounds is normally high. A complete dynamic modelling of CH_4 fluxes from these mounds will therefore include dynamics related the thermodynamics of hydrate dissociation towards seawater, as well as dynamics of biological consumption of CH_4.

Homogeneous hydrate formation will happen from injection gas when $CO_2/N_2/$ surfactants are used for hydrate production. This can be significant but is likely not very important. This route to hydrate formation is likely more important for CH_4. Release of CH_4 from a mole-fraction of roughly 0.14 in a hydrate will result in CH_4 bubbles. However, it is also likely that the water surrounding the dissociating hydrate will be super-saturated with CH_4. Possible CH_4 reformation nucleation can therefore potentially happen heterogeneously as well as homogeneously. Examples for enthalpies of hydrate formation for two conditions of temperature and pressure are plotted in Figure 12 below.

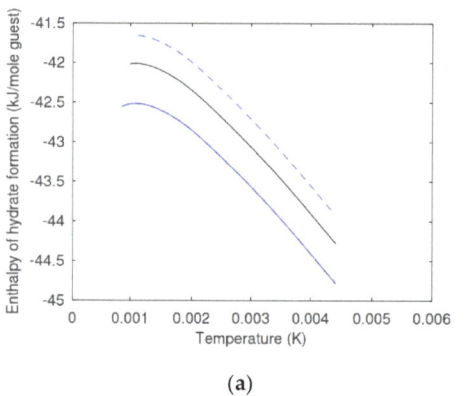 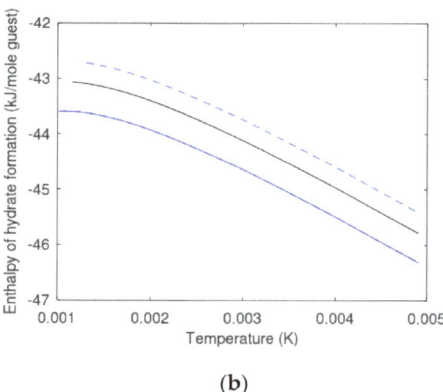

(a)   (b)

Figure 12. Enthalpies for hydrate formation (vertical axis) from dissolved methane at two conditions. (**a**) Temperature 273.16 K and pressure 200 bar. Black solid curve is for pure water and CH_4. CH_4 solubility mole-fraction in water is 4.42×10^{-3} and hydrate stability limit is 8.40×10^{-4}. Solid blue curve is for 3 mole percent ethanol. CH_4 solubility mole-fraction in water is 4.42×10^{-3} and hydrate stability limit is 9.90×10^{-4}. Dashed curve is for 3 mole percent methanol. CH_4 solubility mole-fraction in water is 4.42×10^{-3} and hydrate stability limit is 1.12×10^{-3}. (**b**) Temperature 276 K and pressure 300 bar. Black solid curve is for pure water and CH_4. CH_4 solubility mole-fraction in water is 4.42×10^{-3} and hydrate stability limit is 1.02×10^{-3}. Solid blue curve is for 3 mole percent ethanol. CH_4 solubility mole-fraction in water is 4.92×10^{-3} and hydrate stability limit is 1.16×10^{-3}. Dashed curve is for 3 mole percent methanol. CH_4 solubility mole-fraction in water is 4.92×10^{-3} and hydrate stability limit is 1.31×10^{-3}.

7.2. Carbon Dioxide Hydrate Formation from Water Solution

As discussed in the previous section, hydrate mounds on the seafloor can be connected to hydrate deposits. These mounds can, however, also be connected to conventional hydrocarbons sources that leak towards seafloor through fractures. When these hydrocarbons enter the seafloor at hydrate forming conditions, they create a dynamically unstable situation. These hydrate mounds are dynamically controlled by biological activity and also by chemical potential for CH_4 in surrounding seawater. As a result of the biological activity also CO_2 will be generated. CO_2 hydrate formation from released CO_2 will often be slow enough to dissolve in water. Seawater enriched with CO_2 is heavier than regular seawater and will sink. In either case, CO_2 hydrate can form locally from the dissolved CO_2 in water. Although the main focus herein is on hydrate production using CO_2/N_2 mixtures modelling here, and the follow-up paper on kinetic modelling will also be useful for other groups involved in studies of hydrocarbon fluxes and the resulting changes in the ocean carbon cycle.

The solubility of N_2 in water is extremely low and of no practical importance within the scope of this work. With 20 mole% N_2 in the CO_2/N_2 mixture the chemical potential for CO_2 "bulk" gas can be calculated using SRK [14]. On the other hand, the properties of CO_2 and N_2 are extremely different in terms of selective adsorption on liquid water surface [32]. It might therefore be a fair approximation to calculate solubility based on almost pure CO_2 in the adsorbed water layer.

The density of CO_2 as dissolved in water will correspond to the partial molar volume of CO_2 at infinite dilution. The density of CO_2 at infinite dilution is simply the inverse of the partial molar volume. Chemical potential for ideal gas is a trivial analytical function of the center of mass and molecular weight and the moments of inertia for rotation. It is the Boltzmann factor integrals of the momentum space in the canonical partition function. The integrals end up as analytical simple expressions that can be found in any textbook on physical chemistry or statistical mechanics. The infinite dilution ideal gas chemical potential is not very sensitive to pressure. The following fitted approximation to only temperature dependency is considered as adequate:

$$\mu_{CO_2}^{\infty, ideal\ gas} = -130.006 + \frac{163.818}{T_{0,R}} - \frac{64.898}{T_{0,R}^2} \qquad (26)$$

where $T_{0,R}$ is 273.15 K divided by the actual temperature. Equation (26) does not apply to temperatures above 303 K due to the limited range of temperatures for which infinite partial molar volumes are used and for temperatures above 273.15 K.

Many methods for production of hydrate reservoirs have been proposed during the latest three decades. Thermal stimulation, or injection of thermodynamic hydrate inhibitors are just two examples of methods that have more or less lost attention due to the high cost. One problem is, however, that different methods for producing hydrates are not evaluated thermodynamically prior to planning expensive experiments or even more expensive pilot tests. This can be due to lack of a thermodynamic toolbox for the purpose. Another challenge is the lack of focus on the limitations of the hydrate phase transition itself. The interface between hydrate and liquid water is a kinetic bottle neck. Reducing pressure does not address this problem. Injection of CO_2 will lead to formation of a new CO_2 hydrate. This hydrate formation is an efficient heat source for dissociating hydrate since heating will break the water hydrogen bonds and directly address the important nano scale kinetic limitation. Adding limited amounts of N_2 increases permeability for the injection gas. The addition of surfactant increases gas/water interface dynamics and promote heterogeneous hydrate formation. In this work, we demonstrate a residual thermodynamic scheme that opens up for thermodynamic analysis of different routes to hydrate formation and dissociation. It is demonstrated that 20 mole% N_2 added to the CO_2 is thermodynamically feasible for generating a new hydrate in the pores. The available hydrate formation enthalpy from CO_2/N_2 hydrate formation is also feasible for dissociating in situ CH_4 hydrate. Up to 3 mole% ethanol in the free pore water is also thermodynamically feasible. The addition of alcohol will not significantly disturb the ability to form new hydrate from the injection gas. Homogeneous hydrate formation from dissolved CH_4 and/or CO_2 is limited in amount and not important. However, the hydrate stability limits related to concentration of hydrate former in surrounding water is important. Mineral surfaces can act as hydrate promotors through direct adsorption, or adsorption in water which is structured by the mineral surface charges. These aspects will be quantified in a follow-up paper, along with kinetic modelling based on thermodynamic modelling in this work.

For CO_2 in water, I utilize the following function:

$$\ln \phi_{CO_2}^{water}(T, P, \vec{x}) = \sum_{i=1,2}^{39} \left[a_0(i) + \frac{a_1(i+1)}{T_R} \right] (x_{CO_2})^{[0.05 + \frac{i-1}{40}]} \quad (27)$$

where T_R is reduced temperature and defined as actual T in Kelvin divided by critical temperature for CO_2 (304.35 K). The lower summation 1, 2 indicates starting from 1 and counting in steps of 2. Parameters are given in Table 3 below. The arrow on top of x denotes the vector of all mole-fractions in the actual phase.

Table 3. Parameters for Equation (26).

i	a_0	a_1
1	−139.137483	−138.899061
3	−76.549658	−72.397006
5	−20.868725	−14.715982
7	18.030987	24.548835
9	44.210433	52.904238
11	63.353037	71.596515
13	74.713278	82.605791
15	80.411175	88.536302
17	82.710575	90.262518
19	82.017332	89.094887
21	79.373137	85.956670
23	75.429910	81.519167
25	70.680932	76.270320
27	65.490785	70.551406
29	60.126698	64.683147
31	54.782421	58.865478
33	49.592998	53.235844
35	44.500001	47.728622
37	39.869990	42.730831
39	35.597488	38.125674

The chemical potential for CO_2 that applies to Equation (23) for an equilibrium case is then given as:

$$\mu_{CO_2}^{aqueous}(T, P, \vec{y}) = \mu_{CO_2}^{\infty, ideal\ gas}(T, P, \vec{y}) + RT \ln \left[x_{CO_2} \phi_{CO_2}(T, P, \vec{x}) \right] \quad (28)$$

In Figure 13a, I plot solubility of CO_2 in pure water. Hydrate stability limits for CO_2 as function of mole-fractions CO_2 in surrounding water is plotted in Figure 13b. Similar results for water containing 3 mole% ethanol are plotted in Figure 14.

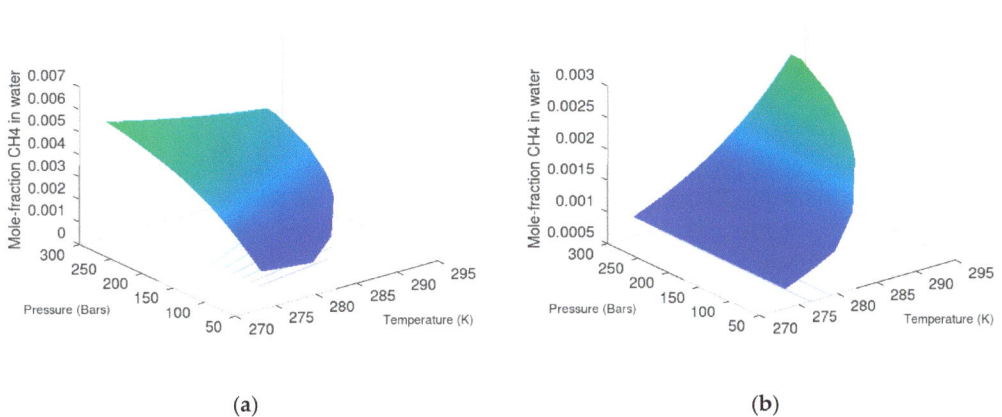

Figure 13. Pure water solubility for CO_2 and hydrate stability limits for CO_2. (**a**) Solubility of CO_2 in water for the range of hydrate forming conditions. (**b**) Hydrate stability limit concentrations for CO_2.

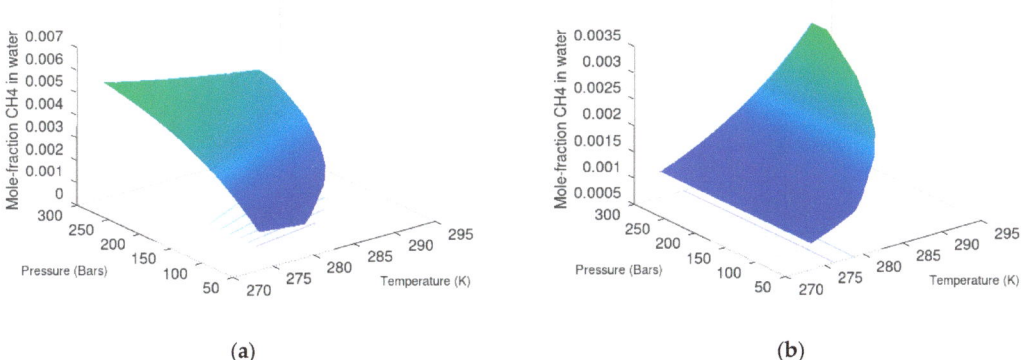

Figure 14. Solubility of CO_2 in water containing 3 mole percent ethanol and hydrate stability limits for CO_2. (**a**) Solubility of CO_2 in water for the range of hydrate forming conditions. (**b**) Hydrate stability limit concentrations for CO_2.

In order to get a picture of the mole-fraction range for homogeneous hydrate growth from CO_2 dissolved in water, I plot solubility in water minus hydrate stability limit in water as a function of mole-fraction CO_2 in water from solubility mole-fraction and down to hydrate stability limit mole-fraction. These are plotted in Figure 15 as function of solubility and temperature. Solubility axis reflects a pressure range from 90 bar to 300 bar.

The stability of hydrate formed homogeneously from dissolved CO_2 in water is lower than heterogeneously formed CO_2 hydrate, as illustrated in Figure 16.

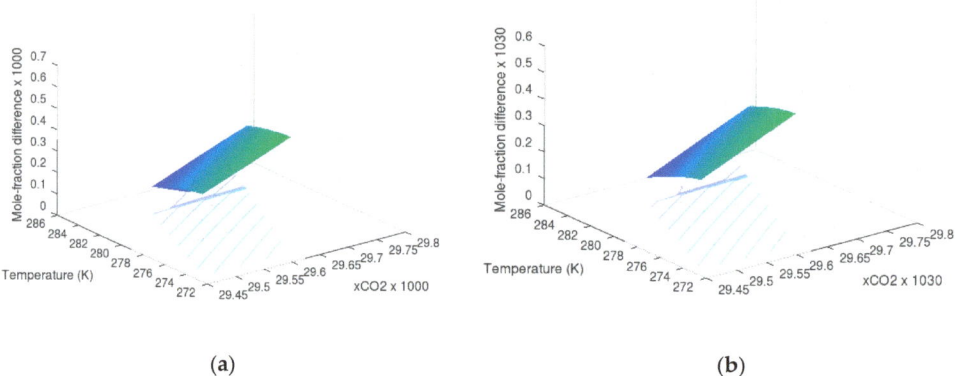

Figure 15. Hydrate growth potential in terms of mole-fraction, from solubility down to hydrate stability limit (z-axis), as function of solubility mole-fraction (x-axis) and temperature (y-axis) and pressure. For each temperature on the temperature axis, solubilities and growth potentials are calculated for pressures from 90 bar to 300 bar This is reflected in the increased solubility along the x-axis. (**a**) Pure water and (**b**) 3 mole% ethanol in water.

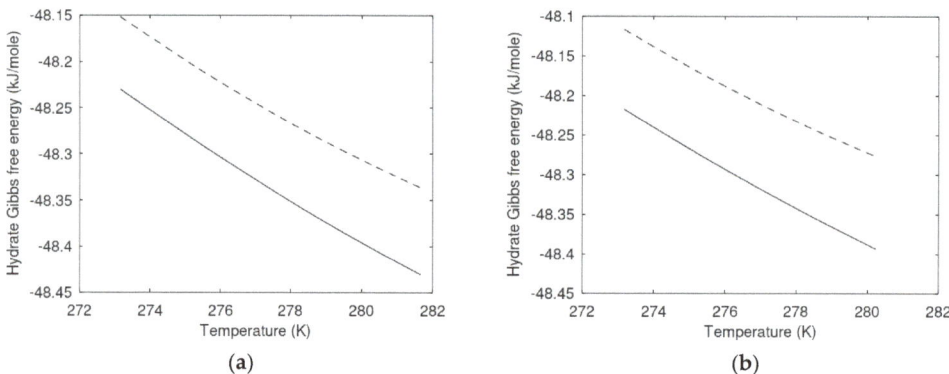

Figure 16. Hydrate Gibbs free energy for hydrate formed heterogeneously from CO_2 and liquid water (solid) and homogeneously formed hydrate from dissolved CO_2 in water (dashed). (**a**) Pure water. (**b**) Water containing 3 mole percent ethanol.

The change in the enthalpy model for CO_2 is the same that was done for CH_4 (but not shown here). It simply involves replacing the partial molar enthalpy from a gas phase as in the heterogeneous case to partial molar enthalpy for guest molecules dissolved in water. Specifically for CO_2:

$$H^R_{CO_2} = -RT^2 \left[\frac{\partial \ln \phi^{water}_{CO_2}}{\partial T} \right]_{P, y_{j \neq CO2}} \tag{29}$$

and

$$H^{water}_{CO_2} = -RT^2 \left[\frac{\partial \left(\frac{\mu^{\infty, ideal\ gas}_{CO_2}}{RT} \right)}{\partial T} \right]_{P, y_{j \neq CO2}} + \left[\frac{\partial \ln \phi^{water}_{CO_2}}{\partial T} \right]_{P, y_{j \neq CO2}} \tag{30}$$

Examples for two conditions of temperatures and pressures are plotted in Figure 17.

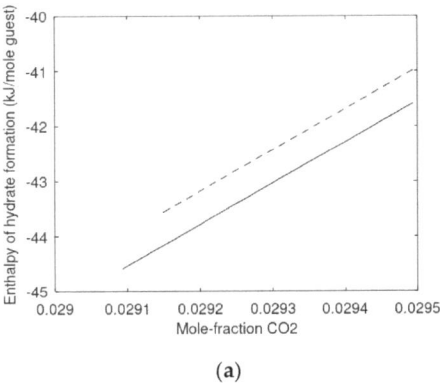 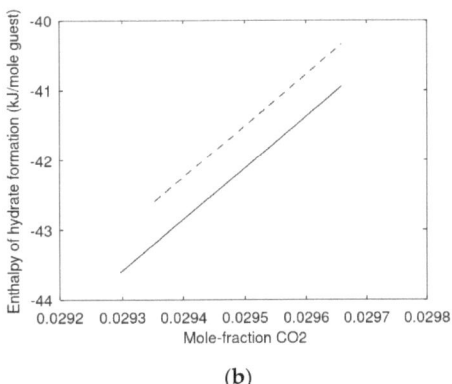

Figure 17. Enthalpy of hydrate formation from dissolved CO_2 in water. (**a**) Temperature 274 K and pressure 100 bar. Solid curve is for pure water and dashed curve is for water containing 3 mole percent ethanol. (**b**) Temperature 280 K and pressure 180 bar. Solid curve is for pure water and dashed curve is for water containing 3 mole percent ethanol.

The enthalpies of hydrate formation are small due to smaller changes for guest molecules as compared to changes for a gas molecule that gets trapped into a higher density. Practically this homogeneous hydrate formation may not be critically important for the hydrate production planning. The hydrate stability limits towards CO_2 in surrounding water may, however, be more important in terms of long terms stability of the stored CO_2. Nevertheless- storing CO_2 as solid hydrate in a reservoir with proven sealing is a safe option. The in situ CH_4 hydrate may have been trapped for millions of years due to sealing formations (clay and shale).

8. Discussion

Interest for using CO_2 in a concept for combined safe long-term storage of CO_2, and simultaneous release of CH_4 is increasing worldwide. Associated thermodynamic properties and knowledge of conversion mechanisms are critical. I have many years of experience on the water/gas interface dynamics and how it affects conditions for hydrate nucleation on liquid water side of the interface. All the papers in the PhD theses [49–52] and other references here [5,7,53] support the physical picture that heterogeneous hydrate nucleation is facilitated by water structures on the liquid water side of the interface. Recent studies on surfactant stimulated water/carbon dioxide interfaces underscore the importance of surfactants for hydrate nucleation [3,10,55,56]. There is a fundamental difference between heterogeneous hydrate nucleation in liquid water region, and at conditions far into the ice region. CH_4 hydrates made from powdered ice and CH_4, and then exposed to CO_2 at ice conditions do not have a liquid phase. There will for sure be some sort of interface between ice and hydrate, but conversion from CH_4 hydrate to CO_2 hydrate does not have much flexibility. It is understandable that experimental cells in atomistic imaging equipment cannot handle high pressures. It is therefore no surprise that hydrate exchange experiments at atmospheric conditions and very low temperatures [51] reveal a solid-state exchange mechanism. There is no liquid water and a totally different ice/hydrate interface as compared to the dynamic and broad (relative) hydrate/liquid water interface, and also a dynamic interface between liquid water and gas. I have been quite purposeful in not referring to experimental data, including our own data. Kvamme et al. [11] and Tegze et al. [12] are referred to in this work for interface details and modelling aspects. However, these references also contain information from experiments. A solid-state mechanism is a totally impossible explanation of our experiments over 15 years. These experiments were all conducted at liquid water conditions. An important element of Baig's study [57] was to shed light on conversion mechanisms. The results revealed a fast conversion mechanism based on creation of new CO_2 hydrate as long as free water was available. After the stage when all free water was consumed over to CO_2 hydrate then a very slow solid state conversion mechanism was applied to the rest of the simulation period.

I have deliberately not examined any higher content of N_2 than 20 mole% in a CO_2/N_2 mixture. Based on earlier studies [37] it is known that hydrate formation for CO_2/N_2 mixtures containing more than 25 mole% N_2. However, it may not be thermodynamically feasible. I have demonstrated that the addition of both N_2 (20 mole% in gas mixture) and ethanol up to 3 mole% (in local liquid water) reduces the available released heat from formation of a new hydrate from the injection gas. The reduction is, however, limited and will not affect the ability to dissociate in situ CH_4 hydrate substantially.

9. Conclusions

Ethanol is an interesting surfactant for addition to CO_2 and CO_2/N_2 mixtures. In this work I have examined the effect of ethanol on thermodynamic properties, and the ability for CO_2/N_2 mixtures to form a new hydrate with free pore water when injected in CH_4 hydrate filled sediments. It is found that ethanol injection as part of CO_2/N_2 mixture is thermodynamically feasible for pure CO_2 as well as CO_2 with up to 20 mole% N_2. The stability of the formed hydrate from the mixture with N_2 (20 mole%) is significantly higher than stability of the CH_4 hydrate. For injection of pure CO_2, the released enthalpy of hydrate formation is between 8 and 10 kJ/mole guest higher than what is needed to dissociate CH_4 hydrate. This potential is decreased by roughly 2 kJ/mole guest for the most unfavorable situation of 20 mole% N_2 and 3 mole% ethanol in water. Homogeneous hydrate formation from CH_4 and CO_2 dissolved in water is thermodynamically feasible. The amount of hydrate that can be formed through this route is very limited. Released enthalpy is less than from heterogeneous hydrate formation from gas and liquid water. The hydrate stability limit towards surrounding water concentration of guest molecules, however, needs to be taken into consideration.

Funding: This research received no external funding.

Conflicts of Interest: The author declares no conflict of interest.

References

1. Sloan, E.D.; Koh, C.A. *Clathrate Hydrates of Natural Gases*, 3rd ed.; CRC Press: Boca Raton, FL, USA, 2007.
2. Mokogan, Y.F. *Hydrates of Hydrocarbons*, 1st ed.; PennWell Books: Tulsa, OK, USA, 1997; ISBN 978-0878147182.
3. Kvamme, B.; Selvåg, J.; Aromada, S.A.; Saeidi, N.; Kuznetsova, T. Methanol as hydrate inhibitor and hydrate activator. *Phys. Chem. Chem. Phys.* **2018**, *20*, 21968–21987. [CrossRef]
4. Austvik, T.; Hustvedt, E.; Gjertsen, L.H. Formation and Removal of Hydrate Plugs—Field Trial at Tommeliten. In Proceedings of the 76 Annual Meeting of the Gas Processors Association (GPA), San Antonio, TX, USA, 10–12 March 1997; p. 249.
5. Kvamme, B. Enthalpies of Hydrate Formation from Hydrate Formers Dissolved in Water. *Energies* **2019**, *12*, 1039. [CrossRef]
6. Kvamme, B.; Aromada, S.A.; Gjerstad, P.B. Consistent Enthalpies of the Hydrate Formation and Dissociation Using Residual Thermodynamics. *J. Chem. Eng. Data* **2019**, *64*, 3493–3504. [CrossRef]
7. Aromada, S.A.; Kvamme, B.; Wei, N.; Saeidi, N. Enthalpies of Hydrate Formation and Dissociation from Residual Thermodynamics. *Energies* **2019**, *12*, 4726. [CrossRef]
8. Kvamme, B. Consistent Thermodynamic Calculations for Hydrate Properties and Hydrate Phase Transitions. *J. Chem. Eng. Data* **2020**, *65*, 2872–2893. [CrossRef]
9. Kvamme, B.; Zhao, J.; Wei, N.; Sun, W.; Zarifi, M.; Saeidi, N.; Zhou, S.; Kuznetsova, T.; Li, Q. Why Should We Use Residual Thermodynamics for Calculation of Hydrate Phase Transitions? *Energies* **2020**, *13*, 4135. [CrossRef]
10. Kvamme, B. Thermodynamic properties and dielectric constants in water/methanol mixtures by integral equation theory and molecular dynamics simulations. *Phys. Chem. Chem. Phys.* **2002**, *4*, 942–948. [CrossRef]
11. Kvamme, B.; Graue, A.; Buanes, T.; Kuznetsova, T.; Ersland, G. Storage of CO_2 in natural gas hydrate reservoirs and the effect of hydrate as an extra sealing in cold aquifers. *Int. J. Greenh. Gas Control* **2007**, *1*, 236–246. [CrossRef]
12. Tegze, G.; Pusztai, T.; Tóth, G.; Gránásy, L.; Svandal, A.; Buanes, T.; Kuznetsova, T.; Kvamme, B. Multiscale approach to CO_2 hydrate formation in aqueous solution: Phase field theory and molecular dynamics. Nucleation and growth. *J. Chem. Phys.* **2006**, *124*, 234710. [CrossRef]
13. Kvamme, B.; Coffin, R.B.; Zhao, J.; Wei, N.; Zhou, S.; Li, Q.; Saeidi, N.; Chien, Y.-C.; Dunn-Rankin, D.; Sun, W.; et al. Stages in the Dynamics of Hydrate Formation and Consequences for Design of Experiments for Hydrate Formation in Sediments. *Energies* **2019**, *12*, 3399. [CrossRef]

14. Bybee, K. Natural Gas Technology/Monetization: Overview of the Mallik Gas-Hydrate Production Research Well. *J. Pet. Technol.* **2004**, *56*, 53–54. [CrossRef]
15. Moridis, G.J.; Collett, T.S.; Dallimore, S.R.; Satoh, T.; Hancock, S.; Weatherill, B. Numerical studies of gas production from several CH_4 hydrate zones at the Mallik site, Mackenzie Delta, Canada. *J. Pet. Sci. Eng.* **2004**, *43*, 219–238. [CrossRef]
16. Konno, Y.; Fujii, T.; Sato, A.; Akamine, K.; Naiki, M.; Masuda, Y.; Yamamoto, K.; Nagao, J. Key Findings of the World's First Offshore Methane Hydrate Production Test off the Coast of Japan: Toward Future Commercial Production. *Energy Fuels* **2017**, *31*, 2607–2616. [CrossRef]
17. Yamamoto, K.; Wang, X.-X.; Tamaki, M.; Suzuki, K. The second offshore production of methane hydrate in the Nankai Trough and gas production behavior from a heterogeneous methane hydrate reservoir. *RSC Adv.* **2019**, *9*, 25987–26013. [CrossRef]
18. Oyama, A.; Masutani, S.M. A Review of the Methane Hydrate Program in Japan. *Energies* **2017**, *10*, 1447. [CrossRef]
19. Tenma, N. Recent Status of Methane Hydrate R&D Program in Japan. Presented at the 11th IMHRD, Corpus Christie, TX, USA, 7 December 2017.
20. Svandal, A.; Kvamme, B.; Granasy, L.; Pusztai, T. The influence of diffusion on hydrate growth. In Proceedings of the 1st International Conference on Diffusion in Solids and Liquids DSL-2005, Aveiro, Portugal, 6–8 July 2005.
21. Kvamme, B.; Kivelæ, P.-H.; Kuznetsova, T. Adsorption of water and carbon dioxide on hematite and consequences for possible hydrate formation. *Phys. Chem. Chem. Phys.* **2012**, *14*, 4410–4424. [CrossRef]
22. Olsen, R.; Leirvik, K.; Kvamme, B.; Kuznetsova, T. A molecular dynamics study of triethylene glycol on a hydrated calcite surface. *Langmuir* **2015**, *31*, 8606–8617. [CrossRef] [PubMed]
23. Olsen, R.; Leirvik, K.N.; Kvamme, B. Effects of glycol on adsorption dynamics of idealized water droplets on LTA-3A zeolite surfaces. *AIChE J.* **2019**, *65*, e16567. [CrossRef]
24. Olsen, R.; Leirvik, K.N.; Kvamme, B.; Kuznetsova, T. Effects of Sodium Chloride on Acidic Nanoscale Pores Between Steel and Cement. *J. Phys. Chem. C* **2016**, *120*, 29264–29271. [CrossRef]
25. Olsen, R.; Leirvik, K.N.; Kvamme, B. Adsorption characteristics of glycols on calcite and hematite. *AIChE J.* **2019**, *65*, e16728. [CrossRef]
26. Kvamme, B.; Zhao, J.; Wei, N.; Sun, W.; Saeidi, N.; Pei, J.; Kuznetsova, T. Hydrate Production Philosophy and Thermodynamic Calculations. *Energies* **2020**, *13*, 672. [CrossRef]
27. Geissbühler, P.; Fenter, P.; DiMasi, E.; Srajer, G.; Sorensen, L.; Sturchio, N. Three-dimensional structure of the calcite–water interface by surface X-ray scattering. *Surf. Sci.* **2004**, *573*, 191–203. [CrossRef]
28. Phan Van, C. Transport and Adsorption of CO_2 and H_2O on Calcite and Clathrate Hydrate. Ph.D. Thesis, University of Bergen, Bergen, Norway, 2012.
29. Van Cuong, P.; Kvamme, B.; Kuznetsova, T.; Jensen, B. Adsorption of water and CO_2 on calcite and clathrate hydrate: The effect of short-range forces and temperature. *Int. J. Energy Env.* **2012**, *6*, 301.
30. Van Cuong, P.; Kvamme, B.; Kuznetsova, T.; Jensen, B. Molecular dynamics study of calcite, hydrate and the temperature effect on CO_2 transport and adsorption stability in geological formations. *Mol. Phys.* **2012**, *110*, 1097–1106. [CrossRef]
31. Van Cuong, P.; Kvamme, B.; Kuznetsova, T.; Jensen, B. Adsorption energy and stability of H_2O and CO_2 on calcite effect by short-range force field parameters and temperature. In *Recent Researches in Applied Mathematics and Economics*; WSEAS Press: Athens, Greece, 2012; pp. 66–72, ISBN 978-1-61804-076-3.
32. Kuznetsova, T.; Jensen, B.; Kvamme, B.; Sjøblom, S. Water-wetting surfaces as hydrate promoters during transport of carbon dioxide with impurities. *Phys. Chem. Chem. Phys.* **2015**, *17*, 12683–12697. [CrossRef] [PubMed]
33. Jensen, B. Investigations into the Impact of Solid Surfaces in Aqueous Systems. Ph.D. Thesis, University of Bergen, Bergen, Norway, 2016.
34. Mohammad, N. Heterogeneous Hydrate Nucleation on Calcite {1014} and Kaolinite {001} Surfaces: A Molecular Dynamics Simulation Study. Master's Thesis, University of Bergen, Bergen, Norway, 2016.
35. Nesse Knarvik, A.B. Examination of Water and Methane Structuring at a Hematite Surface in the Presence of MEG. Master's Thesis, Department of Physics and Technology, University of Bergen, Bergen, Norway, 2017.
36. Austrheim, M.H. Evaluation of Methane and Water Structure at a Hematite Surface—A Hydrate Prevention Perspective. Master's Thesis, Department of Physics and Technology, University of Bergen, Bergen, Norway, 2017.
37. Kvamme, B. Thermodynamic Limitations of the CO_2/N_2 Mixture Injected into CH_4 Hydrate in the Ignik Sikumi Field Trial. *J. Chem. Eng. Data* **2016**, *61*, 1280–1295. [CrossRef]
38. Robinson, D.B.; Ng, H.-J. Hydrate Formation and Inhibition in Gas or Gas Condensate Streams. *J. Can. Pet. Tech.* **1986**, *25*, 26–30. [CrossRef]
39. Ross, M.J.; Toczylkin, L.S. Hydrate Dissociation Pressures for Methane or Ethane in the Presence of Aqueous Solutions of Triethylene Glycol. *J. Chem. Eng. Data* **1992**, *37*, 488–491. [CrossRef]
40. Kobayashi, R.; Withrow, H.J.; Williams, G.B.; Katz, D.L. Gas hydrate formation with brine and ethanol solutions. *Proc. NGAA* **1951**, *27*, 57–66.
41. Ng, H.-J.; Robinson, D.B. Hydrate formation in systems containing methane, ethane, propane, carbon dioxide or hydrogen sulfide in the presence of methanol. *Fluid Phase Equilibria* **1985**, *21*, 145. [CrossRef]
42. Sabil, K.M.; Nashed, O.; Lal, B.; Ismail, L.; Japper-Jaafar, A. Experimental investigation on the dissociation conditions of methane hydrate in the presence of imidazolium-based ionic liquids. *Thermodyn. J. Chem.* **2015**, *84*, 7–13.

43. Tumba, K.; Tumba, P.; Reddy, P.; Naidoo, D.; Ramjugernath, A.; Eslamimanesh, A.; Mohammadi, H.; Richon, D. Phase equilibria of methane and carbon dioxide clathrate hydrates in the presence of aqueous solutions of tributylmethyl-phosphonium methylsulfate ionic liquid. *J. Chem. Eng. Data* **2011**, *56*, 3620–3629. [CrossRef]
44. Kvamme, B.; Zhao, J.; Wei, N.; Saeidi, N. Hydrate—A Mysterious Phase or Just Misunderstood? *Energies* **2020**, *13*, 880. [CrossRef]
45. Kvamme, B.; Zhao, J.; Wei, N.; Sun, W.; Zarifi, M.; Saeidi, N.; Zhou, S.; Kuznetsova, T.; Li, Q. Thermodynamics of hydrate systems using a uniform reference state. *Asia-Pac. J. Chem. Eng.* **2021**, e2706. [CrossRef]
46. Kvamme, B.; Tanaka, H. Thermodynamic Stability of Hydrates for Ethane, Ethylene, and Carbon Dioxide. *J. Phys. Chem.* **1995**, *99*, 7114–7119. [CrossRef]
47. Van der Waals, J.H.; Platteeuw, J.C. Clathrate solutions. In *Advances in Chemical Physics*, 1st ed.; Prigogine, I., Ed.; John Wiley & Sons, Inc.: Hoboken, NJ, USA, 1958; pp. 1–57.
48. Falenty, A.; Salamatin, A.N.; Kuhs, W.F. Kinetics of CO_2-Hydrate Formation from Ice Powders: Data Summary and Modeling Extended to Low Temperatures. *J. Phys. Chem. C* **2013**, *117*, 8443–8457. [CrossRef]
49. Salamatin, A.N.; Falenty, A.; Hansen, T.C.; Kuhs, W.F. Guest Migration Revealed in CO_2 Clathrate Hydrates. *Energy Fuels* **2015**, *29*, 5681–5691. [CrossRef]
50. Falenty, A.; Genov, G.; Hansen, T.C.; Kuhs, W.F.; Salamatin, A.N. Kinetics of CO_2 Hydrate Formation from Water Frost at Low Temperatures: Experimental Results and Theoretical Model. *J. Phys. Chem. C* **2011**, *115*, 4022–4032. [CrossRef]
51. Lee, H.; Seo, Y.; Seo, Y.-T.; Moudrakovski, I.L.; Ripmeester, J.A. Recovering Methane from Solid Methane Hydrate with Carbon Dioxide. *Angew. Chem.* **2003**, *115*, 5202–5205. [CrossRef]
52. Soave, G. Equilibrium constants from a modified Redlich-Kwong equation of state. *Chem. Eng. Sci.* **1972**, *27*, 1197–1203. [CrossRef]
53. Kvamme, B.; Lund, A.; Hertzberg, T. The influence of gas-gas interactions on the Langmuir constants for some natural gas hydrates. *Fluid Phase Equilibria* **1993**, *90*, 15–44. [CrossRef]
54. Kvamme, B.; Førrisdahl, O.K. Polar guest-molecules in natural gas hydrates. *Fluid Phase Equilibria* **1993**, *83*, 427–435. [CrossRef]
55. Selvåg, J.; Kuznetsova, T.; Kvamme, B. Molecular dynamics study of surfactant-modified water–carbon dioxide systems. *Mol. Simul.* **2017**, *44*, 128–136. [CrossRef]
56. Selvåg, J. Molecular simulations for surfactants at aqueous-nonpolar liquid interfaces. Ph.D. Thesis, University of Bergen, Bergen, Norway, May 2021.
57. Baig, K. Nano to Micro Scale Modeling of Hydrate Phase Transition Kinetics. Ph.D. Thesis, University of Bergen, Bergen, Norway, 2017.

Article

Features of the Hydrocarbon Distribution in the Bottom Sediments of the Norwegian and Barents Seas

Inna A. Nemirovskaya * and Anastasia V. Khramtsova

Shirshov Institute of Oceanology of Russian Academy of Sciences, 117997 Moscow, Russia; asya-medvedeva95_16@mail.ru
* Correspondence: nemir44@mail.ru

Abstract: The results of the study of hydrocarbons (HCs): aliphatic (AHCs) and polycyclic aromatic hydrocarbons (PAHs) in bottom sediments (2019 and 2020, cruises 75 and 80 of the R/V Akademik Mstislav Keldysh) in the Norwegian-Barents Sea basin: Mohns Ridge, shelf Svalbard archipelago, Sturfiord, Medvezhinsky trench, central part of the Barents Sea, Novaya Zemlya shelf, Franz Victoria trough are presented. It has been established that the organo-geochemical background of the Holocene sediments was formed due to the flow of sedimentary material in the coastal regions of the Barents Sea on shipping routes. The anthropogenic input of HCs into bottom sediments leads to an increase in their content in the composition of C_{org} (in the sandy sediments of the Kaninsky Bank at an AHC concentration up to 64 µg/g, when its proportion in the composition of C_{org} reaches 11.7%). The endogenous influence on the of the Svalbard archipelago shelf in Sturfiord and in the Medvezhinsky Trench determines the specificity of local anomalies in the content and composition of HCs. This is reflected in the absence of a correlation between HCs and the grain size composition of sediments and C_{org} content, as well as a change in hydrocarbon molecular markers. At the same time, the sedimentary section is enriched in light alkanes and naphthalene's that may be due to emission during point discharge of gas fluid from sedimentary rocks of the lower stratigraphic horizons and/or sipping migration.

Keywords: hydrocarbons (aliphatic and polycyclic aromatic hydrocarbons); organic matter; bottom sediments; alkanes; fluid flows; Norwegian Sea; Barents Sea

1. Introduction

The Norwegian-Barents Sea basin is one of the most promising areas for the development of shelf resources [1]. Anomalies of the OM components in distribution and of hydrocarbons (HCs) in their composition can be direct indicators of their origin [2–4].

Besides, the study of the composition, distribution and genesis of (HCs) in bottom sediments is necessary for subsequent geoecological control during exploration and mining [5–8].

OM and HCs of waters and bottom sediments usually show a complex composition [5,9]. These are autochthones and allochthones components with different origin. The former are syngenetic to the environment and consist of products of bio- and geochemical processes of OM transformation taking place in the water column during sedimentogenesis and at the beginning stages of burial in sediments [2,10]. The second epigenetic category of HCs is even more diverse. It includes products migrating from sedimentary strata, where their formation takes place during catagenesis and in the harsh conditions of metamorphism [11]. Furthermore, the HCs could contain anthropogenic components that enter the aquatic environment (especially in shallow waters) and bottom sediments with oil and oil products polluting water areas [5,7,12].

In different years, the research on the cruises of the R/V Akademik Mstislav Keldysh in the Norwegian and Barents Seas covered water and sediments in hydrothermal fields located within the Jan Mayenne axial volcanic uplift of the Mohns Ridge; in places of outcrops of cold methane seeps on the continental margins of the Spitsbergen archipelago; the Medvezhinsky Trench, the Novaya Zemlya shelf, and craters in the central part of the Barents Sea were examined [13–16] (Figure 1).

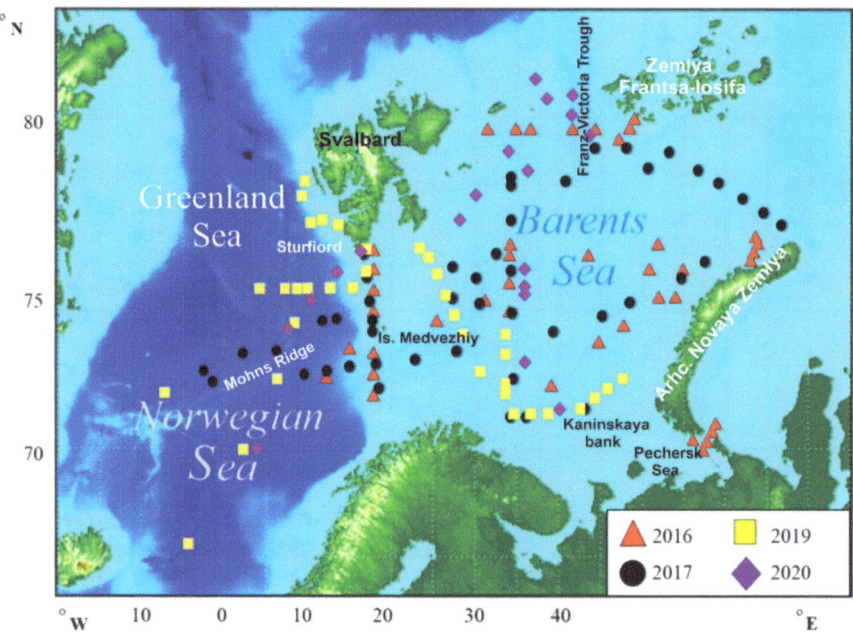

Figure 1. Scheme of stations in different years of research.

The purpose of our research is to obtain new data (2019 and 2020) on the spatial distribution and composition of aliphatic (AHCs) and polycyclic aromatic hydrocarbons (PAHs) in the bottom sediments of the Norwegian and Barents Seas, to establish contribution of HCs vertical migration to their total hydrocarbon pool in bottom sediments.

2. Methods of the Studies

The upper layer of bottom sediments was sampled with a Russian-made Okean-25 bottom grab, and undisturbed cores were sampled with a multicore, Mini Muc. K/MT410, KUM, Germany. To determine the moisture content of sediments, the samples was dried at 100 °C in a drying oven to constant weight. To determine Eh, a portable pH 3110 ionometer (WTW, Germany, Frankfurt) with selective electrodes was used—WTW Electrode Sen Tix ORP.

All solvents were of high purity grade. Methylene chloride was used to extract lipids from bottom sediments. The individual AHCs fractions were separated with hexane by means of column chromatography on silica gel. The concentrations of lipids and AHCs (before and after the chromatography, respectively) were determined by IR spectroscopy using an IR Affinity 1 instrument (Shimadzu, Japan, Kyoto). A mixture of isooctane, hexadecane, and benzene (37.5, 37.5, and 25 vol%, respectively) was used as a standard [5,8,17,18]. The sensitivity of the procedure amounted to 3 µg/mL of the extract [5].

The content and composition of PAHs were determined by the method of high performance liquid chromatography (HPLC) on a LC-20 Prominence liquid chromatograph (Shimadzu, Japan, Kyoto). An Envirosep PP column was used at 40 °C in a thermostat under gradient conditions (up to 50 to 90% in a volume of acetonitrile in water). A 1 cm^3/min flow rate of the eluent was used, and a RF 20A fluorescent detector with programmed wavelengths of absorption and excitation. The calculations were performed by means of LC Solution software. The equipment was standardized with individual

PAHs and their mixtures manufactured by Supelco Co (Sigma-Aldrich, Germany, Darmstadt). As a result, the key polyarenes recommended for studying the pollution of marine objects [12,19] were identified: naphthalene (Naph), 1-methylnaphthalene (1-MeNaph), 2-methylnaphthalene (2-MeNaph), acenaphthene (ACNF), fluorene (FL), phenanthrene (PHEN), anthracene (ANTR), fluoranthene (FLT), pyrene (PYR), benzo(a)anthracene (BaA), chrysene (CHR), benzo(e)pyrene (BeP), benzo(a)pyrene (BaP), benzo(b)fluoranthene (BbF), benzo(k)fluoranthene (BkF) dibenzo(a,h)anthracene (DbhA), indeno(1,2,3-c,d)pyrene (INP), and benzo(g,h,i)perylene (BPl).

The organic carbon in the samples of the SPM was determined by dry combustion with the TOC-L, (Shimadzu, Japan, Kyoto). The sensitivity amounted to 6 µg of carbon in a sample at a precision of 3–6%. The AHCs concentrations were converted into C_{org} a factor of 0.86 [5,20].

3. Results

3.1. 2019 (The 75th Cruise of the R/V Akademik Mstislav Keldysh)

The sediments of the Jan-Mayen fault on the surface were represented by hydrothermal agglomerate silt with lumps of dark gray and bluish color. Red spots of ferruginization, rock fragments, shells, and pebbles were visible on the surface of the soft sediment. The AHC concentrations varied in the range of 5–51 µg/g (Supplementary Materials, Table S1), with the maximum value in the finely dispersed sediment of st. 6131 (Figure 2).

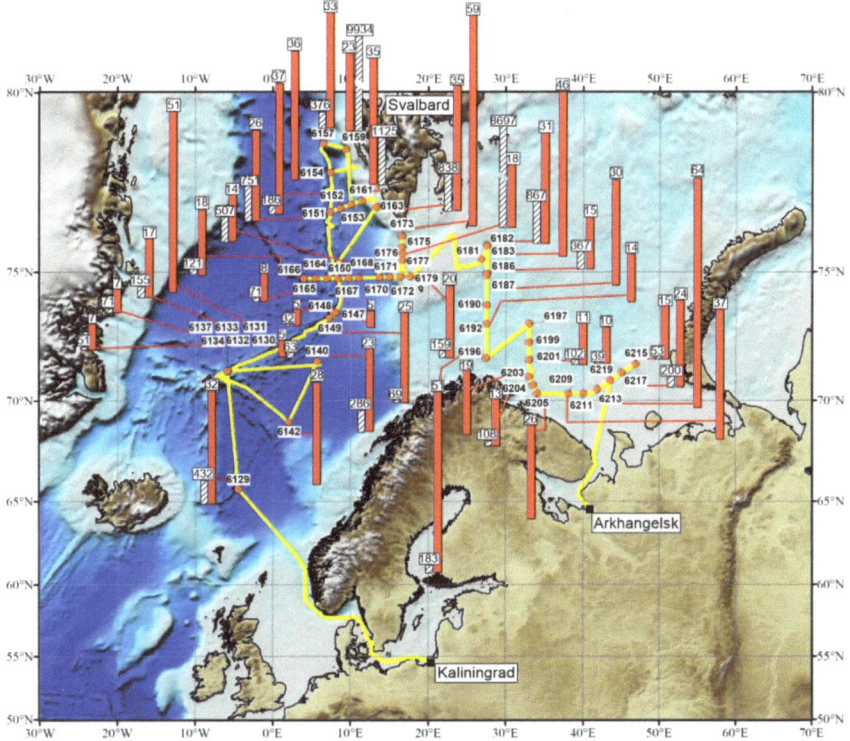

Figure 2. Distribution in the surface layer of bottom sediments of AHC concentrations (µg/g, values are shown above red columns) and PAHs (ng/g, values are above light columns). Squares 6129-6219 show the station numbers (the 75th cruise of the R/V Akademik Mstislav Keldysh).

The bottom sediments of the Barents Sea could be divided into three layers, which reflect three main stages of postglacial sedimentogenesis [21]. At the early stage, Holocene

BS formed, showing the predominance of pelite fraction formed due to fine products of glacier melting supplied to the basin. The second stage corresponds to the time of Atlantic climatic optimum with its maximal transgression level, active shore abrasion, and increasing role of sand-aleurite fraction. The third stage—the deposition of modern Late Holocene BS—falls on the subboreal–subatlantic period of basin regression, corresponding to an increase in the amount of pelite particles because of the stabilization of the rate of shore abrasion, supplying coarsegrained material.

In the Barents Sea, a high content of AHCs was found on the Eastern shelf of the Spitsbergen archipelago (51 µg/g, station 6196), and the maximum concentration was found in the southern part (64 µg/g, station 6213) on the North of the Kaninsky Bank (Figure 2). Light homologues slightly prevailed in the composition of n-alkanes of the surface layer of bottom sediments in most areas with the ratio L/H = $\sum(C_{12}-C_{24})/\sum(C_{25}-C_{37})$, averaged 1.22 (Table S2), which may indicate the intensity of autochthonous processes. The lowest values of this ratio were found on the Scandinavian shelf (stations 6203, 0.22) and in the northern part of the Pechora Sea (station 6217, 0.28), the maximum (station 6179, 2.06) was revealed in the eastern part of the latitudinal section in the Kveitola Trough.).

The area of the Mohns Ridge (station 6131), the Lofoten Basin (station 6142) and the area of the Knipovich Ridge (station 6154) shows microbial even alkanes domination in the low molecular weight area and a series of odd C_{25}–C_{31} homologues prevails in the high molecular weight area (Figure 3). Low CPI values (1.40–2.06, average 1.64) and close concentrations of iso-compounds (pristane/phytane ratio—1.01 on average) may indicate a slight transformation of alkanes.

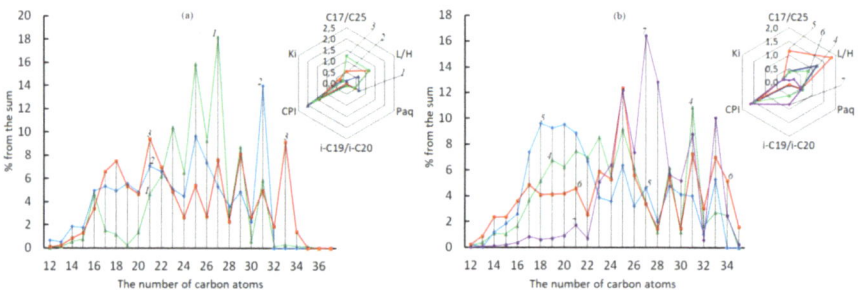

Figure 3. Composition of n-alkanes in the surface layer of bottom sediments at individual stations (75 cruise R/V Akademik Mstislav Keldysh, 2019)–(**a**), *1*–station 6131, *2*–station 6142, *3*–station 6154, (**b**), *4*–station 6190, *5*–station 6192, *6*–station 6213, *7*–station 6217. The inset shows the distribution of the main markers:. L/H = $\sum(C_{12}-C_{24})/\sum(C_{25}-C_{37})$; Paq = $(C_{23}+C_{25})/(C_{23}+C_{25}+C_{29}+C_{31})$; CPI = $\sum(odd)/\sum(even)$; K_i = $(i-C_{19}+i-C_{20})/(C_{17}+C_{18})$.

3.2. 2020 (The 80th Cruise of the R/V Akademik Mstislav Keldysh)

The range of AHCs concentrations in bottom sediments was significantly larger than in 2019: 3–186 µg/g (Figure 4). Nevertheless, in the area of the Mohns Ridge, a rather low content of both AHC (on average 14 µg/g) and C_{org} was found. (average 0.44%). The sediments here are represented by sandy-silty-pelitic silt from dark boggy to almost black color with a small admixture of gravel and pebble material of volcanic origin (pyroclastic material). On the surface of the sediment, Fe-Mn crusts ranging in size from 1 to 10 cm thickness are noted.

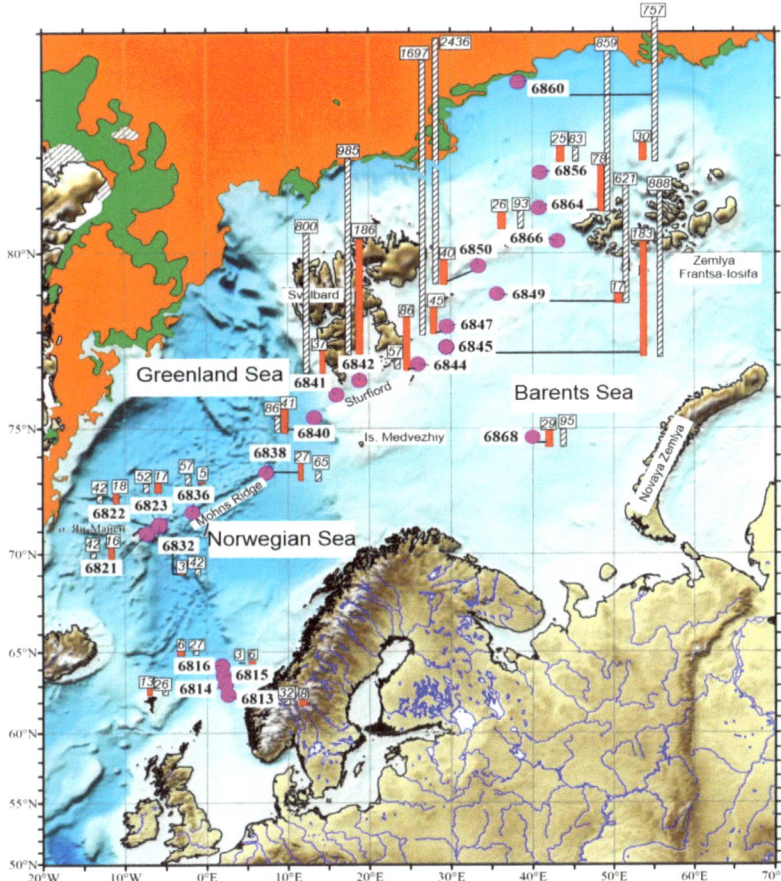

Figure 4. Distribution of AHC concentrations (at the top of the red columns, µg/g) and PAHs (at the top of the light columns, ng/g) in the surface layer of bottom sediments in the area of the 80th cruise of the R/V Akademik Mstislav Keldysh (2020); purple color shows the location of stations (station numbers in white squares).

The highest AHC concentrations in 2020 were found in Sturfiord (on average 90 µg/g, Table S1), with the maximum content at station 6842. Concentrations of AHCs were significantly lower on the eastern shelf of Svalbard (average 52 µg/g, Table S1).

Favorable ice conditions provided a unique opportunity to conduct research within the Franz Victoria Trench from depths of 3700 m (station 6860) to the shelf with a depth of 593 m (station 6864) and 403 m (station 6866), i.e., within the water area usually covered with ice. Here, the content of AHC (average 25 µg/g) in the surface layer of sediments was the lowest (Table S1).

The composition of alkanes in the surface layer of sediments was quite varied (Figure 5). On the Svalbard shelf, most samples were dominated by light homologues (stations 6844, 6845, L/H = 1.24–1.40) with low CPI values (1.23–1.69). On the other hand, in the Franz Victoria Trough (station 6860), the amount of high-molecular-weight alkanes increased (L/H = 0.53), but the CPI values (1.97) only slightly increased.

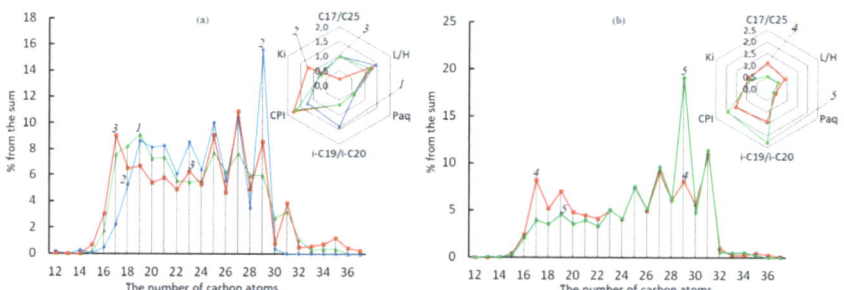

Figure 5. Composition of n-alkanes in the surface layer of bottom sediments at individual stations (80 cruise R/V Akademik Mstislav Keldysh, 2020) (**a**), *1*—station 6844, *2*—station 6845, *3*—station 6847, (**b**), *4*—station 6849, *5*—station. 6860. Inset: the distribution of markers in their composition.

The distribution of PAHs in surface sediments is mosaic, and their total content in 2019 varied in the range of 32–9934 ng/g, and in 2020—3–2430 ng/g (Figure 4). Such a wide concentration range is apparently due not only to different lithological-facial conditions of sedimentation, but also to the relative variability of the main geochemical parameters in the sedimentary strata.

The concentrations of PAHs found in the Mohns Ridge area averaged 50 ng/g were rather low. In Sturfiord, the PAH content at station 6842 (985 ng/g) with the maximum AHC content was also higher than at station 6841 (800 ng/g). However, in contrast to AHCs, the highest PAH concentrations are confined to the shelf of Svalbard. Therefore, there is no correlation between these hydrocarbon classes in the distribution of their concentrations in the surface layer: $r = 0.30$, $n = 21$.

In the northern part of the Barents Sea, in the sediments of the Franz Victoria Trench, their concentrations in the surface layer varied from 83 to 859 ng/g, with a maximum at station 6864.

In the bottom sediments thikness, the concentrations of both C_{org} and HCs decreased with the depth of burial only at some stations only (Figure 6a). However, in most areas, there was an increase in their content in individual horizons. An example of such a distribution is the Sturfiord sedimentary stratum at station 6841 (Figure 6a). The sediment on the surface consisted of olive-brown silt-pelitic, which became darker with the depth of burial. When collecting the sediment, the smell of hydrogen sulfide was felt. In the sedimentary sequence (>2 cm), a large number of hydrotroilite smears and authigenic carbonate crusts were observed. The maximum concentration (218 µg/g) was established with a change in the redox potential in the mountains 6–7 cm (Eh = −80), where the content of C_{org} sharply decreased, while AHCs, on the contrary, increased, that is, the formation of AHCs occurred due to the decomposition of C_{org}. At the same time, there were no connections in the distribution of C_{org} and AHC ($r = -0.16$), but a dependence was observed in the distribution of AHCs and PAHs ($r = 0.56$).

The composition of alkanes at station 6841 sharply differed from station 6840 in terms of the content and distribution of homologues (Figure 7). Low molecular weight homologues dominated in their composition, and the L/H ratio increased with the burial depth. The latter testifies to the intensity of autochthonous processes in the sedimentary strata even at a horizon of 22–26 cm. The CPI value at station 6840 (on average 1.8, maximum 2.6) was higher than at station 6841 (on average 1.4, maximum 1.8) may indicate a lesser transformation of high molecular weight homologues directly in the sedimentary sequence. The CPI values usually increase with the depth of burial, since a series of odd more stable homologues increases in the composition of alkanes during the transformation of AHCs [9,22,23]. For comparison, in the Holocene shelf sediments of the Kara Sea, the CPI C_{22-33} values varied in the range of 2.5–8.1, with an average of 5.2 [24].

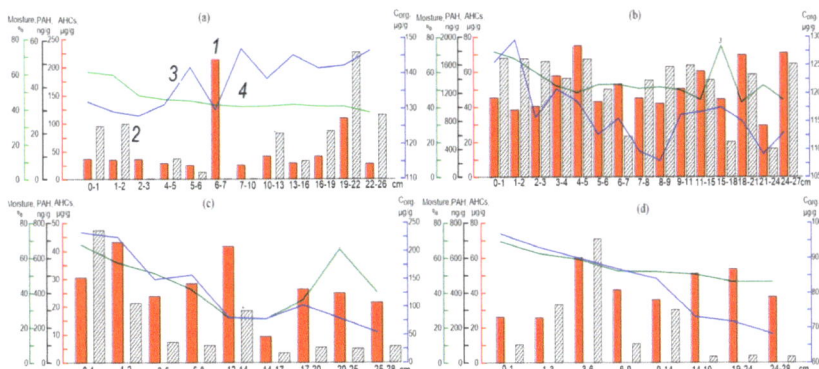

Figure 6. Changes in the concentrations of AHCs (1), PAHs (2), C_{org} (3) and moisture (4) precipitation with the depth at stations 6841 (**a**), 6847 (**b**), 6860 (**c**) and 6866 (**d**). The location of the stations is shown in Figure 4.

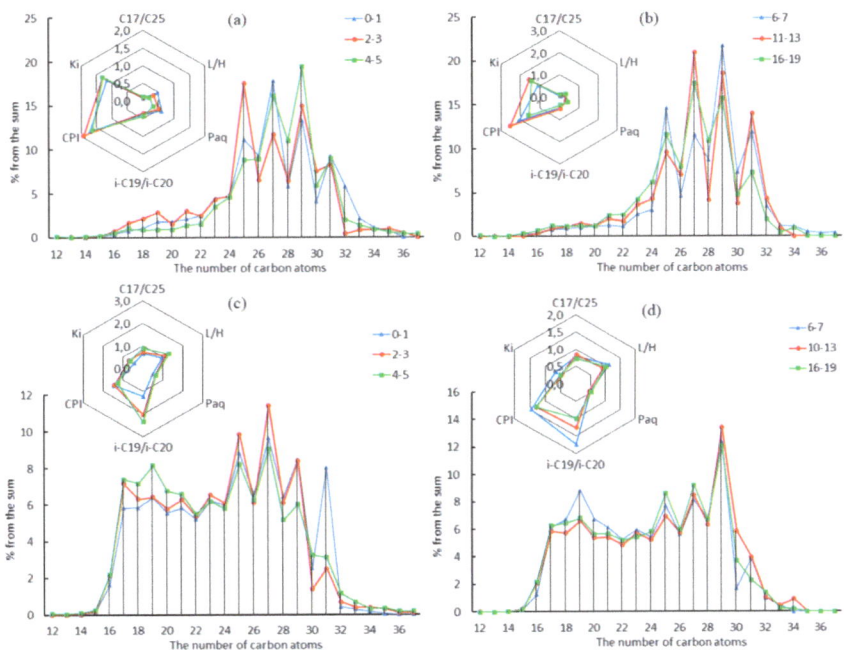

Figure 7. Composition of alkanes with burial depth at stations 6840 (**a,b**) and 6841 (**c,d**) and the distribution of the main markers in their composition. The location of the stations is shown in Figure 4.

At the stations in Sturfiord, the composition of PAHs was dominated by 2, 3-ring arenas: naphthalene, 2-methylnaphthalene (27–43% of the total), and phenanthrene (from 22 to 40%) (Figure 8). Naphthalenes are the least stable compounds and should degrade during sedimentation [25], therefore, their rather high content may be due to the formation of sediments directly in the strata. At the same time, at station 6841, the PAH content in the lower core horizon was higher than in the surface one (2633 and 2164 ng/g, respectively).

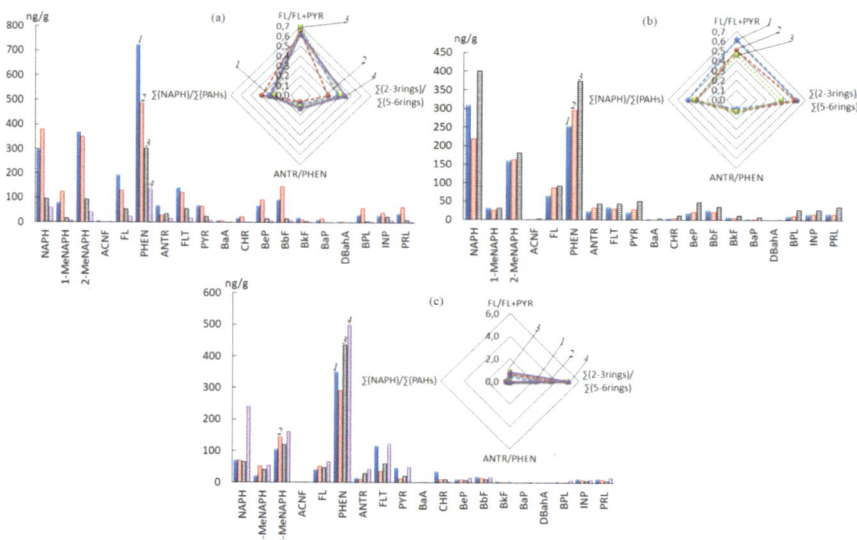

Figure 8. Changes in PAH composition with burial depth (**a**) at station 6841: *1*—0-1, *2*—3-4, *3*—5-6, *4*—7-8 cm; (**b**) station 6847: *1*—1–2, *2*—5–6, *3*—24–27 cm; (**c**) station 6864: *1*—0–1 cm, *2*—3–4 cm, *3*—13–16 cm, *4*—24–26 cm. The location of the stations is shown in Figure 4.

Sediments of the Eastern shelf arch. Spitsbergen, represented by strongly bioturbated silty-pelitic silts of a dark yellowish-brown color, oxidized to 8 cm. They also did not show a gradual decrease in the concentrations of C_{org}, AHCs and PAHs. The increased concentrations of naphthalene's and phenanthrene are confined not to the surface, but to the 19–22 cm horizon (Figure 6b). In particular, in the sedimentary stratum, there was no dependence in the distribution of AHCs and C_{org} (r = 0.07), and, in contrast to C_{org}, the AHCs distribution did not depend on the granulometric composition of sediments–r(AHCs-MOI.). = −0.57, while r (C_{org}-MOI) = 0.58.

It is known that the porosity and moisture content of the sediment characterizes its granulometric composition [26]. Sediments with a high moisture content (up to 90% and more) are formed, as a rule, by a finely dispersed suspension of biogenic origin (for example, fragments of dying planktonic organisms). On the contrary, low moisture values (less than 40%) are characteristic of coarse bottom sediments formed by lithogenic material entering water bodies as a result of erosion of the coastal zone and with slope water runoff.

In the Franz Victoria Trench in the deepest part at station 6860 (depth 3703 m) in the sediment core (Figure 6c), there was a sharp decrease in the content of AHCs (by 4.5 times) and PAHs (by 5.4 times) at the 14–17 cm while the concentration of C_{org} increased almost 2 times (from 1.142 up to 2.107%). At the same time, the composition of the sediment changed, and lenses of dense sandy material appeared in the 13 cm layer, and bioturbation took place from 19 cm. On the shelf in this area (stations 6864 and 6866 with a depth of 594–403 m), an uneven decrease in AHC concentrations was observed in the sediment mass against the background of a decrease in the C_{org} content. At a horizon of 3–6 cm, the AHC content increased more than twofold (up to 60 µg/g) and PAHs (up to 638 ng/g). A similar distribution of organic compounds was observed in the sediment layer at station 6864, where, in the transition from the oxidized to the reduced layer, the amount of naphthalene's increased from 29 to 36% in the PAH composition (Figure 8c).

4. Discussion

The data obtained in 2019–2020 show that the anthropogenic input of HCs into bottom sediments were limited by the coastal areas, where their content in the composition of C_{org} increases. In particular, in 2019, with an AHC content of 64 µg/g and PAH 600 ng/g

(Figure 2) in the sandy sediments of the Kaninsky Bank (Moisture 17.4%), their concentration reached an anomalously high value in the C_{org} composition up to –11.7%. Usually, in bottom sediments, the AHC content in the C_{org} composition was less than 0.5%, and the PAH content was < 0.002% (Figure 9). It is believed that the background concentration of AHCs in coarsely dispersed sediments were 10 µg/g, and in finely dispersed sediments—50 µg/g [5,27]. Only in water areas with anthropogenic oil inflows, mud volcanism, and endogenous migration did the AHC concentration increase and exceed 1% in the C_{org} composition. [5,25,27] Therefore, the increase in the AHC value relative to the background concentration, both in terms of dry sediment and in the composition of C_{org}, is most likely due to anthropogenic sources.

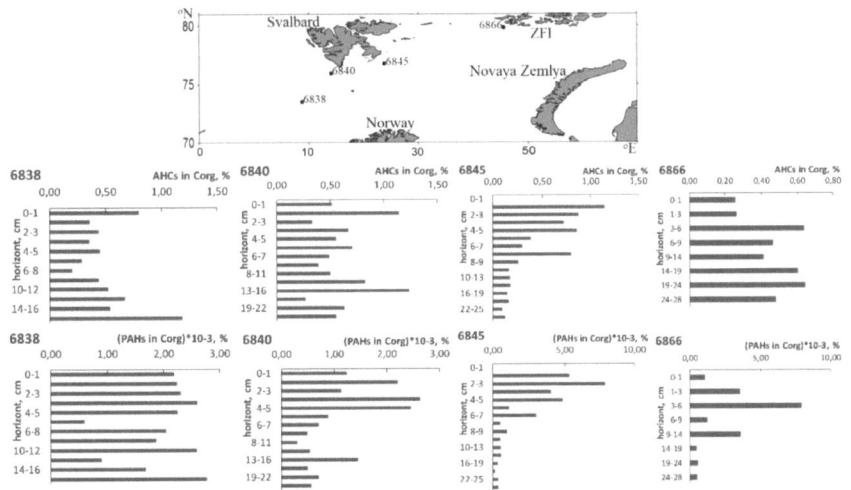

Figure 9. Change in the content of AHCs and PAHs in the composition of C_{org} (%) with the depth of burial at individual stations. The location of the stations is shown in Figure 4.

According to our data of 2020, in the coastal sandy sediments (moisture content 11–14%) in the Murmansk region with AHC content of 54–73 µg/g, their proportion in C_{org} varied in a range 3.4–3.6%. However, the composition of alkanes did not correspond to the smooth petroleum distribution of homologues [5,7], since; in the low-molecular-weight region odd homologues of n-C_{15}–C_{19} dominated, while in the high-molecular region prevailed C_{25} and C_{27}. This is due to the rapid transformation of low molecular weight petroleum alkanes. Even after the diesel fuel spill in Norilsk in May 2020, 2 months after the accident, despite the rather low Arctic temperatures, the composition of alkanes in the surface layer of bottom sediments did not match the composition of the spilled oil product [28].

The concentration of PAHs in sediments of the Murmansk shelf was also high—11,900–13,600 ng/g, but their content in C_{org} was 0.8–0.9% only, and in marine sediments, mostly less than 0.002% (Figure 9). For PAHs, it is more difficult to establish background concentrations, since their values depend on the amount of determined individual polyarenes. In the surface layer of sediments of the Barents Sea ($\sum 22$ PAHs), their content varied from 82 to 3076 ng/g with the highest values on the Svalbard shelf [29]. In the southwestern part of the Barents Sea, the PAH content varied from 10 to 1799 ng/g, and in the sediments of the Norwegian fjords they were predominantly of pyrogenic origin [30,31]. According to our 2020 data, the concentration of $\sum 19$ PAHs in the surface layer of bottom sediments varied from 2 to 2436 ng/g, which fits into the range of their determined earlier values in the sediments of the Barents Sea (Figure 4). Moreover, the highest concentrations are also found on the Svalbard shelf. Earlier, their increased values were also noted in the carbonaceous sediments of the shelf ([29–35] and others). Consequently, the specificity of this anomaly is stable. The erosion of carbonaceous deposits in the western part of Svalbard was regarded as the main source of PAHs. High concentrations of pyrogenic PAHs here are due to their natural formation in the sediment mass in low-temperature processes, since they were dominated by low-molecular compounds: naphthalene's and phenanthrene (Figure 8). At the same time, statistically significant differences in the concentrations and composition of PAHs in the sediments sampled in 1991–1998 and 2001–2005 [32,33], as well as in comparison with our data were absent.

In the bottom sediments of the Mohns and Knipovich Ridges, the characteristic structural elements are axial volcanic uplifts (AVP), which are confined to the rift valley [35–39]. Within the AVP, there are rifts/volcanic edifices with hydrothermal activity, and the release of gas/fluid creates hydro acoustic anomalies above these fields [40,41]. In this area, due to the coarsely dispersed composition of the sediments, in the surface layer, we have established rather low concentrations of AHCs, C_{org} and PAHs (Figures 2 and 3). At the same time, the average AHC concentrations in different years of research in the Mohns Ridge area were similar: 10–18 µg/g, with a rather low content of C_{org} in the composition of 0.13–0.28%. These results are consistent with the study of the hydrothermal plume of the Trollveggen field located east of the axial zone of the Mohns Ridge near the Jan Mayen hot spot [42]. The plume of this area was characterized by a moderate concentration of methane and a low concentration of suspended matter near the bottom, while the sediments were characterized by a rather low content of C_{org}—0.35% [43]. Nevertheless, with the maximum content at station 6131 (51 µg/g, Figure 2), the proportion of AHCs in the C_{org} composition increased to 1.9%, and in the composition of alkanes in the low-molecular-weight region there was a peak of n-C_{16} (Figure 3a) indicating the microbial nature of AHCs. In addition, in the sandy-silty-pelitic silts at station 6838, located in the area of ancient volcanic edifices, the AHC concentrations increased towards the lower horizon of the core (16–18 cm) to 36 µg/g, 1.2% in the C_{org} composition (Figure 9). At this station, boulder-sized basalt was raised, presumably taken from the side of a volcanic edifice.

In the Barents Sea, the nature of Holocene sediments is mainly marine terrigenous with a noticeable influence of alluvial facies in the coastal part of the shelf and ice-marine in the north of the water area [44]. Most of the sediments studied by us are represented by terrigenous carbonate-free aleurite and silty-pelitic oozes with an admixture of coarse-detrital material. Early diagenesis occurs under conditions of thermodynamic no equilibrium, and bioturbation complicates these processes [45].

On the slope of the Sturfiord trench at a depth of 392 m (station 6842, Figure 4), with the maximum AHC content (186 µg/g), their proportion in the C_{org} composition was also increased—1.18%. In this area, at station 6841, according to hydrophysical data, the most significant fluid flow was established. The gas torch rose above the bottom to a height of over 100 m [15]. The AHC content in the surface layer were only 37 µg/g (0.24% of C_{org}). However, the composition of alkanes in sediments at station 6841 sharply differed from their composition at station 6840, first of all, by the content of light homologues (Figure 6). In the northern part of the Medvezhinsky Trench, where there are craters formed as a result of the decomposition of gas hydrates [46,47], an increased AHC content was also found in 2017—up to 44 µg/g, with an increased proportion of low molecular weight homologues (L/H varied in the range 0.84–1.42) [43]. At the same time, during the transition from the oxidized to the reduced layer, the composition of alkanes became more "autochthonous" than in the surface horizon.

In the Franz-Victoria Trough, the change in the HC content in the sediment strata (Figure 9, station 6866) is most likely due to the presence of creep displacements here, which were found during the bathymetric survey.

The abnormal distribution and composition of HCs was established in the sediments of the Perseus Rise in 2017 (at depths of 107–200 m) [43,47]. Here, in the sedimentary strata, when passing from the 0–5 to the 5–10 cm layer, the AHC concentration increased by 53 times, and in the C_{org} composition—by 66 times (from 0.03 to 2.0%). Maximum AHC values at this station in terms of dry sediment (272 µg/g) and in the composition of C_{org}

(2.2%) were confined to the 15–20 cm horizon. Such changes in the sedimentary strata can occur during the transformation of seeping oil hydrocarbons [48], since the sediments of this region have a high petroleum and gas generation potential [16].

Fluid flows and their transformation in the surface layer of bottom sediments were considered as the main source of HCs in the study of bottom sediments in the area of the Stockman area [5,6]. The composition of alkanes in sediments had a mixed genesis: autochthonous homologues (n-C_{16}–C_{17}) dominated in the low-molecular-weight region and petroleum homologues in the high-molecular region; in the composition of PAHs—light polyarens [5]. It was assumed that rather low AHC concentrations in terms of dry weight (in the surface layer 4.4–18.6 µg/g and at a horizon of 10–20 cm—7.8–84.6 µg/g) and in the composition of C_{org} (on average $\leq$ 1%) in this area due to a decrease in the intensity of fluid flows in recent years. It should be borne in mind that the hydrocarbon deposits of the Stockman field are overlain by an impermeable stratum of predominantly clayey rocks [49].

Thus, in the Norwegian-Barents Sea basin, against the background of lateral variability of hydrocarbon molecular markers of bottom sediments, hydrocarbon anomalies of different genesis are distinguished. In the southwestern part of the Barents Sea, anomalies are associated with an anthropogenic component due to the influence of Atlantic waters, coastal runoff, coastal abrasion, aerosol flow and increased shipping. Therefore, in the sediments of the southwestern part of the Barents Sea, PAHs were found to contain compounds indicating air emissions from aluminum smelters when burning coal and wood [34].

Anomalies in the distribution of HCs in bottom sediments on Svalbard shelf, the Medvezhinsky and Sturfiord trenches suggest their natural formation in the sedimentary strata, which determines the specifics of their behavior. As a source of HCs, one can consider their input from the underlying horizons, since they dominate in most samples in the composition of C_{org} (Figure 9). Considering the high petroleum and gas potential of the Barents Sea and the features of the seabed surface (pockmark craters) make this assumption quite reasonable [50–54]. The existence of periods of rapid subsidence, as well as the existence of bituminous rocks, is a reliable indicator of the possible accumulation of a significant amount of HCs.

It is believed that low molecular HCs can move in fluid flows as a separate phase through the pores of sedimentary rocks and leave a geochemical trace in surface sediments due to accumulation, especially in places of gas discharge [55,56]. Low CPI values, indicating a low degree of alkane degradation, as well as the presence of low molecular weight PAHs, can serve as a confirmation of this assumption. Therefore, the sediments of the Barents Sea can be considered as a dynamic generating system that is a function of geological space and time [57].

5. Conclusions

The AHC concentration in the surface layer of bottom sediments in 2019 varied in the range of 6–64 µg/g with a maximum in the surface layer of bottom sediments of the Kaninsky Bank of the Barents Sea (11.7% in the composition of C_{org}), and in 2020, in the range 3–186 µg/g, with a maximum at Sturfiord on the Svalbard shelf (1.18% of C_{org}). A slight predominance of low molecular weight homologues (L/H $\leq$ 1) indicates the intensity of autochthonous processes in bottom sediments.

The PAH concentration in the surface layer of bottom sediments in 2019 varied in the range of 32–9934 ng/g with a maximum on the western shelf of Svalbard, and in 2020, 3–2430 ng/g, with a maximum on the eastern shelf of Svalbard. The PAHs were dominated by phenanthrene and naphthalene's, which is due to their natural formation in the sediment mass in low-temperature processes.

In the sedimentary strata, the lack of correlation in the distribution of HCs with the grain size type of sediments, the content of C_{org}, as well as changes in hydrocarbon molecular markers may indicate an endogenous effect in most of the studied areas (in particular, on the shelf of the Svalbard archipelago in Sturfiord and Medvezhinsky troughs, etc.). The enrichment of the sedimentary section with light alkanes and naphthalene's may be due to outbursts during point discharge of gas fluid from sedimentary rocks of the lower stratigraphic horizons.

In coastal areas on shipping lanes, the organic-geochemical background of bottom sediments is formed due to sedimentation processes, which leads to an increase in HC concentrations in the surface layer and in the composition of C_{org} (on the Kaninsky Bank up to 64 µg/g for AHCs and 600 ng/g for PAHs). However, the anthropogenic input into the bottom sediments of the Barents Sea is of subordinate importance in comparison with their natural input into fluid flows.

Supplementary Materials: The following are available online at https://www.mdpi.com/article/10.3390/fluids6120456/s1, Table S1: Characteristics of the surface layer of bottom sediments in the Norwegian and Barents Seas in different years of research, Table S2: Concentrations of organic compounds in bottom sediments at individual stations and distribution of markers in the composition of PAHs.

Author Contributions: I.A.N. collected samples in the 75 and 80 of the R/V Akademik Mstislav Keldysh and isolated hydrocarbons on board the vessel and interpreted the data obtained; A.V.K. determined hydro-carbons in laboratory conditions, analyzed alkanes, made a figure for the paper, All authors have read and agreed to the published version of the manuscript.

Funding: The expeditions were carried out within the framework of the state assignment of the Ministry of Education and Science of Russia (№ 0128-2021-0015), geochemical research and generalization of materials with the financial support of the Russian Science Foundation (project No. 19-17-00234).

Institutional Review Board Statement: Not applicable.

Informed Consent Statement: Not applicable.

Data Availability Statement: The data presented in this study are openly available in FigShare at.

Acknowledgments: The authors are grateful to Gordeev V.V. their valuable comments and advice when discussing the results; Halikov I., Popova M., Solomatina A., Chernov V. for assistance in conducting the analyses r.

Conflicts of Interest: The authors declare no conflict of interest.

References

1. Kaminskiy, V.D.; Suprunenko, O.I.; Smirnov, A.N.; Medvedeva, T.Y.; Chernykh, A.A.; Aleksandrova, A.G. The current resource state and prospects for the development of the mineral resource base of the shelf region of the Russian Arctic. *Explor. Conserv. Miner. Resour.* **2016**, *9*, 136–142. (In Russian)
2. Grigorenko, Y.N. Oil accumulation zones as an object of study and forecast Oil and gas geology. *Theory Pract.* **2016**, *11*, 1–11. Available online: http://www.ngtp.ru/rub/6/49_2016.pdf (accessed on 9 November 2021).
3. Tissot, B.; Welte, D. *Petroleum Formation and Occurrence*; Springer: Berlin, Germany, 1984; p. 699.
4. Pechora Sea. *System Research Experience*; RFFI: Moscow, Russia, 2003; p. 486. (In Russian)
5. Nemirovskaya, I.A. *Oil in the Ocean (Pollution and Natural Flows)*; Nauchnyy Mir: Moscow, Russia, 2013; p. 432. (In Russian)
6. Petrova, V.I.; Batova, G.I.; Kursheva, A.V.; Litvinenko, I.V.; Morgunova, I.P. Hydrocarbons in bottom sediments of the Shtokman area—Distribution, genesis, time trends. *Neftegazov. Geologiya. Theory Pract.* **2015**, *10*, 1–21.
7. AMAP (Arctic Monitoring and Assessment Programme). *Chapter 4. Sources, Inputs and Concentrations of Petroleum Hydrocarbons, Polycyclic Aromatic Hydrocarbons, and Other Contaminants Related to Oil and Gas Activities in the Arctic*; AMAP Assessment: Oslo, Norway, 2007; Volume 4, p. 87.
8. NAS (National Academy of Sciences). *Oil in the Sea III: Inputs, Fates, and Effects*; National Research Council; The National Academies Press: Washington, DC, USA, 2003; p. 265.
9. Yunker, M.B.; Macdonald, R.W.; Ross, P.S.; Sophia, C.; Johannessen, B.; Neil, D. Alkane and PAH provenance and potential bioavailability in coastal marine sediments subject to a gradient of anthropogenic sources in British Columbia, Canada. *Org. Geochem.* **2015**, *89–90*, 80–116. [CrossRef]
10. Lisitzyn, A.P. Dispersed sedimentary matter in the Earth's geospheres and in the White Sea system. White Sea. M. *Sci. World* **2012**, *2*, 19–47.
11. Kontorovich, A.E.; Borisova, L.S. Composition of asphaltenes as an indicator of the type of dispersed organic matter. *Geochemistry* **1994**, *11*, 1660–1667.

12. AMAP (Arctic Monitoring and Assessment Programme). *Chemicals of Emerging Arctic Concern*; AMAP Assessment: Oslo, Norway, 2017; p. 353.
13. Klyuvitkin, A.A.; Kravchishina, M.D.; Nemirovskaya, I.A.; Baranov, B.V.; Kochenkova, A.I.; Lisitzin, A.P. Studies of the sediment systems of the European Arctic during the cruise 75 of the R/V Akademik Mstislav Keldysh. *Oceanology* **2020**, *60*, 485–487. [CrossRef]
14. Klyuvitkin, A.A.; Politova, N.V.; Novigatsky, A.N.; Kravchishina, M.D. Studies of the European Arctic during the 80th Cruise of the Research Vessel Akademik Mstislav Keldysh. *Oceanology* **2021**, *61*, 139–141. [CrossRef]
15. Kravchishina, M.D.; Novigatsky, A.N.; Savvichev, A.S.; Pautova, L.A.; Lisicin, A.P. Investigation of sediment systems of the Barents Sea and the Norwegian-Greenland basin during the 68th voyage of the research vessel "Akademik Mstislav Keldysh". *Oceanology* **2019**, *59*, 173–176. [CrossRef]
16. Nemirovskaya, I.A. Hydrocarbons in the waters and bottom sediments of the Barents Sea during the period of ice cover variability. *Geokhimiya* **2020**, *65*, 679–692.
17. Korshenko, A. *Marine Water Pollution*; Annual Report 2019; Nauka: Moscow, Russia, 2020; p. 190.
18. Simard, R.; Hasegawa, J.; Bandaruk, W.; Headington, C.E. Infrared spectrometric determination of oil and phenol in water. *Anal. Chem.* **1951**, *23*, 1384–1387. [CrossRef]
19. *Monitoring of Hhazardous Substances in the White Sea and Pechora Sea*; Harmonisation with OSPAR's Coordinated Environmental Monitoring Programme (CEMP); Akvaplan-Niva: Tromsø, Norway, 2011; p. 71.
20. Nemirovskaya, I.A. Content and composition of hydrocarbons in water, suspended matter and bottom sediments of the Kara Sea. *Oceanology* **2010**, *50*, 717–729. [CrossRef]
21. Tarasov, G.A. Quaternary sedimentary cover of the western arctic shelf: Lithological structure, spatial distribution. *Vestnik KSC RAS* **2015**, *21*, 124–134. (In Russian)
22. Nemirovskaya, I.A. Oil units on the beaches of the Baltic Sea. *Water Resour.* **2011**, *3*, 315–323. [CrossRef]
23. Yamamoto, M.; Polyak, L. Changes in terrestrial organic matter input to the Mendeleev Ridge, western Arctic Ocean, during the Late Quaternary. *Glob. Planet. Chang.* **2009**, *68*, 30–37. [CrossRef]
24. Kashirtsev, V.A. *Organic Geochemistry of Naphthides in the East of the Siberian Platform*; YF. SO RAN: Yakutsk, Russia, 2003; p. 160.
25. Fernandes, M.B.; Sicre, M.A. The importance of terresrial organic carbon inputs on Kara Sea shelves as revealed by n-alkanes, OC and δ13C values. *Org. Geochem.* **2000**, *31*, 363–374. [CrossRef]
26. Tolosa, I.; Mora, S.; Sheikholeslami, M.R.; Villeneuve, J.; Bartocci, J.; Cattini, C. Aliphatic and Aromatic Hydrocarbons in coastal Caspian Sea sediments. *Mar. Pollut. Bul.* **2004**, *48*, 44–60. [CrossRef]
27. Gavshin, V.M.; Lapukhov, S.V.; Saraev, S.V. *Geochemistry of Lithogenesis under Conditions of Hydrogen Sulfide Contamination (Black Sea)*; Nauka: Moscow, Russia, 1988; p. 194. (In Russian)
28. Bouloubassi, I.; Saliot, A. Investigation of anthropogenic and natural organic inputs in estuarine sediments using hydrocarbon markers (NAN, LAB, PAH). *Oceanol. Acta* **1993**, *16*, 145–161.
29. Glyaznetsova, Y.S.; Nemirovskaya, I.A.; Flint, M.V. Study of the Effects of an Accidental Diesel Fuel Spill in Norilsk. *Dokl. Ross. Akad. Nauk. Nauk. O Zemle* **2021**, *501*, 113–118.
30. Eide, I.; Berg, T.; Thorvaldsen, B.; Christensen, G.N.; Savinov, V.; Larsen, J. Polycyclic aromatic hydrocarbons in dated freshwater and marine sediments along the Norwegian coast. *Water Air Soil Pollut.* **2011**, *218*, 387–398. [CrossRef]
31. Boitsov, S.; Jensen, H.K.B.; Klungsøyr, J. Geographical variations in hydrocarbon levels in sediments from the Western Barents Sea. *Nor. J. Geol.* **2009**, *89*, 91–100.
32. Boitsov, S.; Petrova, V.; Jensen, H.K.; Kursheva, A.; Litvinenko, I.; Klungsøyr, J. Sources of polycyclic aromatic hydrocarbons in marine sediments from southern and northern areas of the Norwegian continental shelf. *Mar. Environ. Res.* **2013**, *87*, 73–84. [CrossRef] [PubMed]
33. Dahle, S.; Savinov, V.; Petrova, V.; Klungsøyr, J.; Savinova, T.; Batova, G.; Kursheva, A. Polycyclic aromatic hydrocarbons (PAHs) in Norwegian and Russian Arctic marine sediments: Concentrations, geographical distribution and sources. *Nor. J. Geol.* **2006**, *86*, 41–50.
34. Boitsov, S.; Jensen, H.K.B.; Klungsøyr, J. Natural background and anthropogenic inputs of polycyclic aromatic hydrocarbons (PAH) in sediments of South-Western Barents Sea. *Mar. Environ. Res.* **2009**, *68*, 236–245. [CrossRef] [PubMed]
35. Zaborska, A.; Carroll, J.; Pazdro, K.; Pempkowiak, J. Spatio-temporal patterns of PAHs, PCBs and HCB in sediments of the western Barents Sea. *Oceanologia* **2011**, *53*, 1005–1026. [CrossRef]
36. Jiao, L.; Zheng, G.J.; Minh, T.B.; Richardson, B.; Chen, L.; Zhang, Y.; Yeung, L.W.; Lam, J.C.; Yang, X.; Lam, P.K. Persistent toxic substances in remote lake and coastal sediments from Svalbard, Norwegian Arctic: Levels, sources and fluxes. *Environ. Pollut.* **2009**, *157*, 1342–1351. [CrossRef]
37. Geli, L.; Renard, V.; Rommevaux, C. Ocean crust formation processes at very slow spreading centers: A model for the Mohns Ridge, near 72 N, based on magnetic, gravity, and seismic data. *JGR* **1994**, *99*, 2995–3013. [CrossRef]
38. Kokhan, A.V.; Dubinin, E.P.; Grokholsky, A.L. Geodynamic features of structure formation in the spreading ridges of the Arctic and Polar Atlantic. *Vestn. KRAUNC Earth Sci.* **2012**, *19*, 59–77.
39. Cherkashev, G.A.; Tamaki, K.; Baranov, B.V.; German, K.; Gusev, E.A.; Egorov, A.V.; Zhirnov, E.A.; Kerin, K.; Kurevits, D.; Okino, K.; et al. Exploration of the rift zone of the Knipovich Ridge: Expedition "Knipovich-2000". *Rep. Acad. Sci.* **2001**, *378*, 518–522.

40. Reimers, H. The Morphology of the Mohn's Ridge. *Nor. Univ. Sci. Technol.* **2017**, 1–114.
41. Pedersen, R.B.; Rapp, H.T.; Thorseth, I.H.; Lilley, M.D.; Barriga, F.J.; Baumberger, T.; Flesland, K.; Fonseca, R.; Früh-Green, G.L.; Jorgensen, S.L. Discovery of a black smoker vent field and vent fauna at the Arctic Mid-Ocean Ridge. *Nat. Commun.* **2010**, *1*, 126. Available online: www.nature.com/naturecommunications (accessed on 9 November 2021). [CrossRef] [PubMed]
42. Pedersen, R.B.; Thorseth, I.H.; Nygaard, T.E.; Lilley, M.D.; Kelley, D.S. Hydrothermal activity at the Arctic Mid-Ocean Ridge. In *Diversity of Hydrothermal Systems on Slow Spreading Ocean Ridges*; Rona, P., Devey, C., Murton, B., Eds.; Geophysical Monograph 188; American Geophysical Union: Washington, DC, USA, 2010; pp. 67–89.
43. Kravchishina, M.D.; Lein, A.Y.; Boev, A.G.; Prokofiev, V.Y.; Starodymova, D.P.; Dara, O.M.; Novigatsky, A.N.; Lisitzin, A.P. Hydrothermal Mineral Assemblages at 71° N of the Mid-Atlantic Ridge (First Results). *Oceanology* **2019**, *59*, 941–959. [CrossRef]
44. Gramberg, I.S. Barents shelf plate. *Leningr. Sci. Nedra* **1988**, 264.
45. Borisova, L.S. Geochemistry, composition, and structure of protoasphaltenes in organic matter of recent lacustrine sediments. *Russ. Geol. Geophys.* **2017**, *58*, 294–298. (In Russian) [CrossRef]
46. Tarasov, G.A.; Alekseev, V.V. Lithological-geological features of environment of the organisms. In *Paleogeography and Paleoecology of the Barents and White Seas in Quaternary*; Kola Scientific Center: Apatity, Russia, 1987; pp. 24–43. (In Russian)
47. Glyaznetsova, Y.S.; Nemirovskaya, I.A. Features of the distribution of bitumoids in the bottom sediments of the Barents Sea. *Oceanology* **2020**, *60*, 831–839. [CrossRef]
48. Andreassen, K.; Hubbard, A.; Winsborrow, M.; Patton, H.; Vadakkepuliyambatta, S.; PlazaFaverola, A.; Gudlaugsson, E.; Serov, P.; Deryabin, A.; Mattingsdal, R.; et al. Massive blow-out craters formed by hydrate-controlled methane expulsion from the Arctic seafloor. *Science* **2017**, *356*, 948–953. [CrossRef] [PubMed]
49. Ehrhardt, J.D. Negative-ion mass spectra of methylated diuretics. *Rapid Commun. Mass Spectrom.* **1992**, *6*, 349–351. [CrossRef]
50. Lein, A.Y.; Nemirovskaya, I.A.; Ivanov, M.V. Isotopic composition of organic and carbonate carbon in the surface horizons of bottom sediments in the area of the Shtokman field and on the "pockmark field" in the Barents Sea. *Doklady* **2012**, *446*, 67–70.
51. Judd, A.; Hovland, M. *Seabed Fluid Flow the Impact on Geology, Biology, and the Marine Environment*; Cambridge University Press: Cambridge, UK, 2007; p. 442.
52. Rise, L.; Bellec, V.K.; Chand, S.; Bøe, R. Pockmarks in the southwestern Barents Sea and Finnmark fjords. *Nor. J. Geol.* **2015**, *94*, 263–282. [CrossRef]
53. Plassen, L.; Knies, J. Fluid flow structures and processes; indications from the North Norwegian continental margin. *Nor. Geol. Tidsskr.* **2009**, *89*, 57–64.
54. Chand, S.; Thorsnes, T.; Rise, L.; Brunstad, H.; Stoddart, D. *Pockmarks in the SW Barents Sea and Their Links with Iceberg Ploughmarks*; Geological Society: London, UK, 2016; Volume 46, pp. 295–296.
55. Pau, M.; Hammer, Ø.; Chand, S. Constraints on the dynamics of pockmarks in the SW Barents Sea: Evidence from gravity coring and high-resolution, shallow seismic profiles. *Mar. Geol.* **2014**, *355*, 330–345. [CrossRef]
56. England, W.A.; MacKenzie, A.S.; Mann, D.M.; Quigley, T.M. The movement and entrapment of petroleum fluids in the subsurface. *J. Geol. Soc.* **1987**, *144*, 327–347. [CrossRef]
57. Petrova, V.I.; Batova, G.I.; Kursheva, A.V.; Litvinenko, I.V.; Morguva, I.P. Hydrocarbons in bottom sediments of the Shtokmanovskoe deposit: Distribution, genesis, and time series. *Neftegaz. Geol. Teor. Prakt.* **2015**, *10*. Available online: http://www.ngtp.ru/rub/1/35_2015.pdf (accessed on 9 November 2021). (In Russian)

Article

Functional Acrylic Surfaces Obtained by Scratching

Abraham Medina [1,*,†], Abel López-Villa [1,†] and Carlos A. Vargas [2,†]

[1] SEPI ESIME Azcapotzalco, Instituto Politecnico Nacional, Av. de las Granjas 682, Col. Sta. Catarina Azcapotzalco, Mexico City 02250, Mexico; abelvilla77@hotmail.com
[2] Departamento de Ciencias Básicas, Universidad Autónoma Metropolitana-Azcapotzalco, Av. San Pablo 180, Col. Reynosa Azcapotzalco, Mexico City 02200, Mexico; cvargas@azc.uam.mx
* Correspondence: amedinao@ipn.mx
† These authors contributed equally to this work.

Abstract: By using sandpaper of different grit, we have scratched up smooth sheets of acrylic to cover their surfaces with disordered but near parallel micro-grooves. This procedure allowed us to transform the acrylic surface into a functional surface; measuring the capillary rise of silicone oil up to an average height $\bar{h}$, we found that $\bar{h}$ evolves as a power law of the form $\bar{h} \sim t^n$, where t is the elapsed time from the start of the flow and n takes the values 0.40 or 0.50, depending on the different inclinations of the sheets. Such behavior can be understood alluding to the theoretical predictions for the capillary rise in very tight, open capillary wedges. We also explore other functionalities of such surfaces, as the loss of mass of water sessile droplets on them and the generic role of worn surfaces, in the short survival time of SARS-CoV-2, the virus that causes COVID-19.

Keywords: capillary rise; functional surfaces; human skyn; droplets; COVID-19

1. Introduction

Nowadays, research on the fabrication and physical behavior of functional surfaces, e.g., micro-structured surfaces with singular features able provide one or more functional properties, mainly those related to fluid transport, is a very relevant topic of study [1–8].

Commonly, the transport of liquids on functional surfaces is governed by the wetting, capillary action and the gravitational field, and therefore the details of the microstructured and nanostructured surfaces are essential to understand how liquid spreads in the preferred direction [2,5,8–12]. A feature of the wettable functional surfaces that has attracted much attention in recent years for its broad potential applications is that the spreading occurs without energy input, such as non-powered delivery systems, self-lubrication and microfluidic devices [8].

Depending on applications, the texture of many functional surfaces consists of grooves (open capillaries) of different height, width, and contact angle [2,10,11]. For example, in technology, micro V-grooves frequently appear, as in the case of grooved heat pipes where parallel micro grooves were drawn in copper plates to improve their energetic efficiency [2]. In nature, directional spreading of water is a feature that *Nepenthes alata*, a carnivorous pitcher plant whose slippery peristome remains completely wetted by water, has used as the source of an insect capture function [6]. Moreover, the structure of *Nepenthes* has been useful in the fabrication of bio-inspired surfaces through the replica molding method [3–8].

Incidentally, human skin bears similarities to a microfluidic system since the external layer forms a microchannel network comprised of large numbers of interconnected micro V-grooves. For instance, experiments on deposition of oil drops (moisturizer) on the middle of the forearm show that radial flows away from an initially placed drop occur through the grooves [12].

Similarly, functional acrylic surfaces are not new, for instance, acrylic sheets were micro-textured to generate fishbone architecture, to get blood repellent surfaces, by using continuous wave UV laser, which can generate patterned structures in range of 1–300 microns [13].

In the present work, we are interested in the fabrication and characterization of functional surfaces obtained through the raw scratching up of acrylic sheets, since the resulting irregular surface pattern dramatically modifies their surface wetting properties. Moreover, the issue of soft surfaces scratching is ubiquitous; however, we scarcely pay attention to it in our daily life, despite the fact that many surfaces at home, offices, hospitals, etc., turn into functional surfaces due to its use and necessary cleaning. Both facts produce finely and irregularly scratched surfaces.

To perform the study of the functionality of the scratched surfaces, we will texturize surfaces by following a simple protocol of longitudinal scratching up of acrylic sheets, which allows liquids to sprint uphill through micro V-grooves. In the experiments, we will harness such physical phenomenon, to understand the scope of a fluid mechanics model based on the capillary rise in V-grooved open capillaries having a small angle of aperture [14]. Additionally, the change of surface wettability will be characterized through the measurement of $\bar{h}$, the averaged front due to capillary rise, as a time function, when sandpaper of different grit is used.

Recently, the water droplets evaporation on functional surfaces has been studied, for instance, on micro-structured surfaces with hydrophilic and hydrophobic micropillars [15]. There, it has been observed that the rate of evaporation, $\dot{m}$, as a function of time t, of sessile droplets on both types of micropillar-structured surfaces obeys a relationship of the form $\dot{m} \propto t$, for most of the evaporation time. This can be attributed to the fact that when the droplets are sufficiently large (during the initial stages), evaporation is primarily governed by vapor diffusion at the liquid–vapor interface and heat conduction through the droplet.

The irregular texturing applied to acrylic surfaces also influences the temporal mass reduction of droplets, but the question is if evaporation or spreading dominates over one another. To quantify these processes, experiments of sessile droplets on scratched acrylic surfaces and on human skin will be also analyzed. In both of these latest cases, we will show that the droplets loss their mass at a high rate due to the capillary penetration into the V-grooves Consequently, evaporation seems to be a marginal phenomenon in the loss of mass of the droplets. The fast reduction of the mass of the water droplets on scratched surfaces, allows us to envisage that the virus SARS-CoV-2, which is carried by respiratory droplets to the functional surfaces must survive a fraction of time on scratched surfaces in comparison with its survival on smooth surfaces.

To reach our goals, in the next section we will revisit the main theoretical results of capillary action for two cases: the shape of a meniscus on a vertically standing plate and the capillary rise in open V-shaped capillaries making a small angle. Later, in Section 3, we propose a protocol for texturing acrylic plates, by scratching them up with different grit sandpaper. The surface characterization with electron microscopy of the coarse and fine scratches also is reported. Later on, we experiment how silicone oil forms menisci on smooth and textured vertical sheets and how it rises upwards on acrylic textured plates. We also compare our experimental results for the average front of rise $\bar{h}(t)$ for different types of scratches and tilt angles of the sheets with the formulas given in Section 2. In Section 4, we will perform experiments on the loss of mass of water sessile droplets on human skin and on scratched surfaces. Finally, in Section 5, we give the main conclusions of this work.

2. Equilibrium Profile and Dynamic Capillary Rise

2.1. Equilibrium Profile

When a vertical plane wall is in contact with a quiescent liquid of density ρ, dynamic viscosity μ and surface tension σ and it wets the wall, a meniscus is formed under the gravity action, see Figure 1. The height of the free surface on the wall can be determined through the capillary equilibrium condition given by the balance between the Laplace and the hydrostatic pressures, it yields the equilibrium height h given by [16].

$$h = l_c\sqrt{2(1 - \sin\theta)}, \qquad (1)$$

where the capillary constant, or *capillary length*, is $l_c = \sqrt{\sigma/\rho g}$, g is the acceleration due to gravity and the angle of contact is θ, which is a property of the liquid–solid contact, meaning that a perfect wetting yields $\theta = 0°$ and then $h = \sqrt{2}l_c$.

For water at 25 °C, $\sigma = 71.97$ mN/m, $\rho = 997.04$ kg/m^3, therefore $l_c = 2.71$ mm, consequently the maximum height of a water meniscus under perfect wetting is $h = 3.84$ mm.

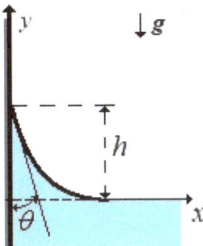

Figure 1. Schematic of the liquid meniscus formed, under the gravity action, on a vertical solid surface where θ is the angle of contact and h is the maximum height on the wall.

2.2. Capillary Rise

Later, we will experimentally show that the way by which the face was scratched of the acrylic sheet with sandpaper produces a micro-textured surface with V-grooves, distributed over the entire face. As mentioned earlier, several authors [2,5,8–12] agree that the spreading of liquid on functional surfaces with V-grooves can be modeled by assuming that each groove behaves as a capillary wedge, having a small angle of aperture α (two plates making an small angle), which can be termed as the Taylor–Hauksbee cell [14,17,18], as the one sketched in Figure 2.

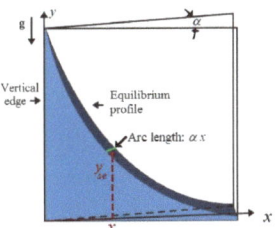

Figure 2. Depiction of a Taylor–Hauksbee cell, where two close together plates make a small aperture angle α. The equilibrium profile is the last stage of the temporal evolution of the liquid free surface. Notice that at a distance x from the edge, the arc length of the free surface, is approximately αx and there the local height of the equilibrium profile is $y_{se}(x)$.

In such a model, we assume that the lower part of a V-groove is brought into contact with a wetting liquid that will rise through the groove until it achieves a final equilibrium surface, which can be determined by knowing that the pressure jump (Laplace pressure [16]) across the surface at x is approximately $\Delta p_s = 2\sigma\cos\theta/\alpha x$, where αx is the arc length of the free surface just at a distance x from the edge. The equilibrium surface $y_{se}(x)$ of the meniscus is determined by the balance $\Delta p_s = \rho g y_{se}$ [14]. This balance yields the equilibrium surface

$$y_{se} = \frac{2\sigma\cos\theta}{\rho g \alpha x}, \qquad (2)$$

which is a rectangular hyperbola whose peculiarity is that very close to the vertical edge ($x = 0$) the liquid should reach, ideally, an infinite height; however, in actual experiments, the fluid always reaches the upper limit of the plates system.

Now, we must notice that the equilibrium surface will be achieved when the temporal evolution of the free surface $y_s(x,t)$ halts, i.e., $y_s(x,t) = y_{se}$. Through the use of the Reynolds lubrication equations and the free surface evolution equation of a slow, viscous flow under the gravity action [14,19] it was found that the free surface $y_s(x,t)$ evolves in a complex manner: if one follows the elevation of the meniscus y_s just at the edge, $x = 0$, for long times, the temporal evolution is as [14]:

$$y_s \approx \left(\frac{\sigma^2 \cos^2 \theta}{\mu \rho g} \right)^{1/3} t^{1/3}, \text{ at the vertical edge } (x = 0), \qquad (3)$$

meanwhile, if the measurement of y_s is made at $x \neq 0$, at intermediate times, the evolution of the free surface follows the power law [14]

$$y_s \approx \left(\frac{\sigma \alpha x \cos \theta}{3\mu} \right)^{1/2} t^{1/2}, \text{ at a small distance } x \text{ from the edge.} \qquad (4)$$

It is important to highlight that the power law $y_s \sim t^{1/3}$ closely fits experimental data when α is in the order of a few degrees and the capillary flow occurs in the edge, as schematically shown in Figure 2 [14,20]. The power law $y_s \sim t^{1/2}$, known as Lucas–Washburn law [21–23], is typical for capillary penetration in the absence of gravity into capillary tubes and porous media and in the cell it must be valid for capillary rise at small distances x from the edge. All these results will be useful later to understand the way how capillary rise in functional acrylic surfaces takes place in experiments.

3. Experiments

3.1. Equilibrium Height and Characterization of the Micro V-Grooves

In our experiments, we used transparent, 3 mm thick, 80 mm height and 48 mm wide acrylic sheets. Considering that we vertically dip the sheet in silicone oil (a nonvolatile liquid at room temperature) of nominal dynamic viscosity $\mu = 100$ cP, surface tension $\sigma = 0.021$ N/m and density $\rho = 960.0$ Kg/m^3, formula (1) lets us find that in this case $l_c = 1.49$ mm and the height of the meniscus on the acrylic wall is $h = 1.54$ mm. In Equation (1), we used the angle of contact $\theta = 28° \pm 0.5°$, which was measured on a clean and smooth sheet, as is shown in Figure 3.

Through measurements based on numerous pictures, such as Figure 3a, we found that $h = 1.47 \pm 0.05$ mm, which is close to the theoretical value, and the difference perhaps could be caused by the meniscus on the flange of the sheet.

The other pictures of the menisci of Figure 3 correspond to (b) an acrylic surface scratched up with 150 grit sandpaper and angle of contact $\theta = 28° \pm 0.5°$ and (c) an acrylic surface scratched up with 50 grit sandpaper $\theta = 29° \pm 0.5°$. All angles of contact were measured 5400 s after each sheet was dipped into the silicone oil reservoir and, consistently, the measured angles of contact are not appreciably affected by the raw scratching.

The scratching of the acrylic sheet was carried out to produce a series of V-grooves, moreover, the use of 150 grit and 50 grit sandpapers was intended to produce fine and coarse grooves, respectively, since 150 grit corresponds to an abrasive grain size of 92 μm and 50 grit is associated to 350 μm, in agreement to ANSI (American National Standards Institute). In experiments, we carried out the scratching up with *Fandeli* [24] "wet or dry" sandpaper made of abrasive faceted silicon carbide (SiC) grains. We use this sandpaper since acrylic has a hardness of 3 on the Mohs hardness scale from 1 to 10, meanwhile silicon carbide has a hardness of 9. Acrylic surfaces are relatively soft and easily scratched by SiC grits.

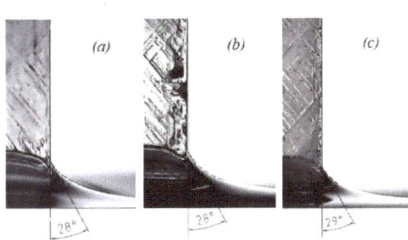

Figure 3. Representative pictures of menisci on vertically standing acrylic sheets, (**a**) without scratching, (**b**) scratched using 50 grit sandpaper and (**c**) scratched using 150 grit sandpaper. In all cases, the contact angles θ were measured by looking at the width of the sheet, frontally.

The protocol to produce the grooves on any acrylic face was the following: first, the face was cleaned up; then, it was uniformly and unidirectionally scratched three times (along the vertical edge); and finally, the face was cleaned up again with a fine hair brush. Following this procedure, we obtained a distribution of near parallel V-grooves. In Figure 4, we show scanning electron microscope (SEM) micrographs of specimens where the nearly parallel alignment, size and surface texture of the grooves on the acrylic face are observable. Typical width of grooves in the acrylic faces, since 150 grit sandpaper are between 10–70 µm and for 50 grit are between 100–200 µm.

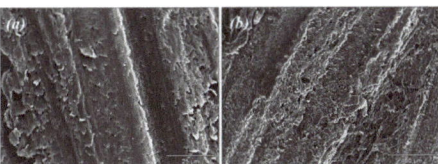

Figure 4. Micrographs of the near parallel grooves for (**a**) 150 grit and (**b**) 50 grit.

In Figure 5, SEM micrographs, near the edge of the sheets, where the capillary rise starts are shown for the cases where the scratching was performed with (a) 150 grit and (b) 50 grit sandpapers, respectively, notice that the V-grooves are best appreciated in the case of 150 grit, Figure 5a.

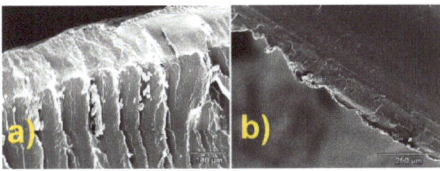

Figure 5. SEM micrographs near the edge of the sheets where the irregular V-grooves are observed: (**a**) 150 grit and (**b**) 50 grit.

3.2. Functional Surfaces: Capillary Rise

The scratching of the faces of acrylic sheets allowed us to texturize them with V-like grooves. To characterize the effect of the groove distribution on the acrylic sheets, we carried out several experiments of capillary rise. Our procedure to visualize the capillary flow along the faces consisted of dipping the sheet in a reservoir containing silicone oil 100 cP. In one case, we vertically dipped the sheet and in others tilted sheets were dipped with inclination angles of $\phi = 45°$ (counter clockwise direction) and $\phi = -45°$ (clockwise direction). Figure 6 depicts the corresponding ascending liquid films (blue thick lines) in each scratched face. In the vertical case, the flow rises opposite to gravity (Figure 6a), meanwhile in the case of $\phi = 45°$ (Figure 6b) the capillary flow surmounts the sheet, finally when $\phi = -45°$, the flow occurs at the lower face, when the liquid climbs a sloped ceiling configuration (Figure 6c). Interestingly, in these cases the V-grooves, with different orientation, take in the liquid in very contrasting ways.

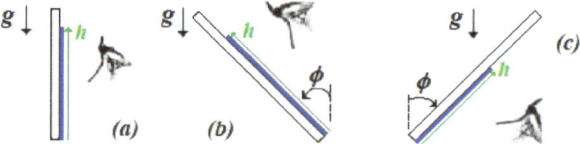

Figure 6. Depiction of the capillary rise of silicone oil on the functional surfaces scratched up: vertical ascent (**a**), ascent on the face tilted at $\phi = 45°$ (**b**) and ascent at $\phi = -45°$ (climbing flow at the lower face of the sheet) (**c**). The visualization of the thin films of liquid on the sheets and the measurement of respective fronts of rise is given by h.

In an early stage of experimentation, we dipped couples of sheets, each scratched with different grit and we video recorded, with a CCD camera, the ascent of liquid in the functional faces, as depicted in Figure 6. In Figure 7, snapshots of the capillary rise on the faces scratched with 50 grit (Figure 7a) and 150 grit (Figure 7b) are shown. On each face, the darkest lower regions correspond to well saturated zones and it is evident that the front of ascent is very irregular. To get a measure of the mean height of the front $\bar{h}$, at a given time t, in a second stage, we performed a simple procedure: after video recording the experiments, we obtained their single frames (digital pictures), each 1 s apart. Later, on each picture we traced ten evenly spaced parallel straight lines along the width of each face and on each line i ($i = 1, \ldots, 10$) we pointed out with arrows the local position of the front which allowed us to get a measured, in mm, of the local instantaneous positions of the front $h_i(t)$; in Figure 8, we show a series of time sequential images for a sheet scratched up with grit 50 in order to observe more accurately the corresponding measures of $h_i(t)$ (in mm).

Following that, for a given time t, we computed the arithmetic mean of data $h_i(t)$, obtaining the instantaneous mean position of the front $\bar{h}(t)$. Such a procedure was performed for all frames of a given video recording. We carried out four different experiments for each inclination and grit, always employing a new scratched sheet in a new experiment of capillary rise. At a final stage, we again averaged the four measurements of $\bar{h}(t)$, at a given time, for each inclination of the sheet, having a given grit.

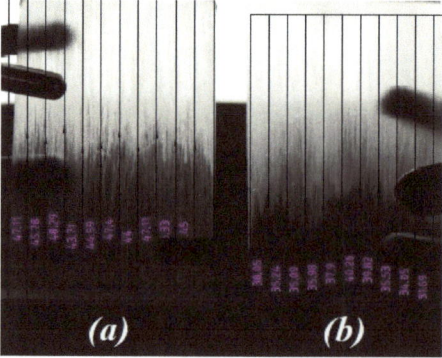

Figure 7. Snapshots of the capillary rise of silicone oil on a couple of vertical acrylic sheets with a face scratched up with (**a**) 50 grit and (**b**) 150 grit sandpapers. Arrowed lines, with numbers (in mm) aside, indicate the measure of the local heights, h_i. Notice that although the acrylic sheets were dipped at the same time, on average the height of rise is different due to the different grit on each plate.

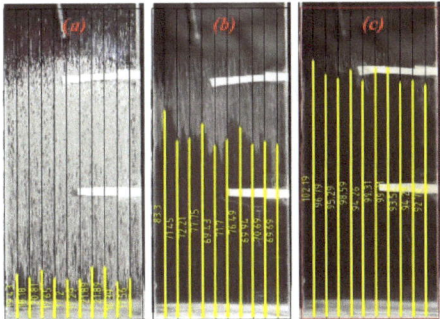

Figure 8. Time sequential images of the capillary rise in a vertical sheet scratched up with grit 50: (**a**) $t = 5400$ s, (**b**) $t = 170,220$ s and (**c**) $t = 360,360$ s. As in Figure 7, the numbers give the hight (in mm) of each yellow line.

In Figure 9, we show the log–log plot of the mean height $\overline{h}(t)$ on vertical sheets scratched with grits 50 and 150, respectively. There, we show that data fit the power law of the form $\overline{h} \approx at^n$, where the prefactor is $a = 0.69$ mm/s$^{0.39}$ and the exponent $n = 0.39 \pm 0.02$ for faces scratched with grit 50, and $a = 0.31$ mm/s$^{0.38}$ and $n = 0.38 \pm 0.02$ for faces with grit 150. In the inset of this figure, we show the plot of the mean speed of the fronts $d\overline{h}/dt$ (obtained from the power law fits) and this confirms that the liquid front is slightly faster in the face scratched up with 50 grit sandpaper. Is important to notice that in both cases, in the last few moments, the speed of the fronts remains nearly constant, which could be mainly attributed to flow occurring in the close vicinity of corners, since in this region the capillary pressure that drives the flow is large [14].

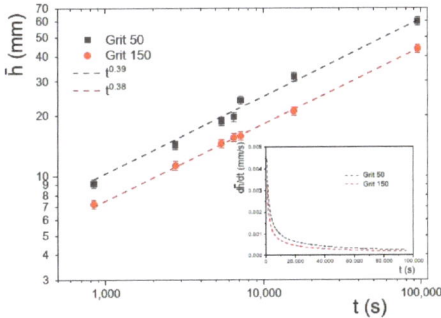

Figure 9. Log–log plot of the averaged height of capillary rise, $\overline{h}$, as a time function, on faces of acrylic with grit 50 and grit 150, respectively. In both cases, data fit power laws of the form $\overline{h} \sim t^n$, where $n = 0.38$ for grit 50 and $n = 0.39$ for grit 150.

With regard to the theoretical results of the capillary rise in very tight wedges discussed in Section 2, we highlight that for vertical sheets and tilted sheets with $\phi = 45°$, the average front obeys approximately that $\overline{h} \sim t^{0.40}$, which could be interpreted as an intermediate power law among that corresponding to the flow just at the edge $\overline{h} \sim t^{0.33}$ (Equation (2)) and the other corresponding to capillary rise at a small distance from the edge where $\overline{h} \sim t^{0.50}$ (Equation (3)). The same idea can be assimilated reasoning that due to our method of visualization of the fronts on the vertical sheet and on the 45° tilted sheet, we actually look at an depth-averaged flow composed of the edge flow and flow at a small distance from the edge. On the contrary, when we visualize the climbing flow on the sheet tilted at the angle $\phi = -45°$, the main flow occurs at a small distance from the edge, since gravity always tries to pull the liquid outside the V-grooves, thus in agreement with Equation (3) the exponent will have a value close to $n = 0.50$.

It can be seen that, in plots of Figures 9 and 10, there is a consistent difference between data for capillary rise in faces scratched with grit 50 and grit 150; however, they have nearly the same trend, given by the corresponding exponent n, consequently, the prefactor a must determine such a difference. Typically, the values of a for grit 50 are larger than those for grit 150. In order to explain this, we observe that the prefactor in Equation (4) contains the term ax (the arc length of the free surface of the liquid, located at distance x from the inner edge) which in general, for grit 50 is larger than for grit 150, since the V-grooves for grit 50 have a larger depth than for grit 150; additionally, we must remember from Figure 3 that no appreciable changes of the angles of contact θ were measured (by using the method of the meniscus under gravity) for smooth or scratched surfaces. However, we must notice that this latest condition will not be maintained for the sessile droplets on functional surfaces, as will be discussed in the next Section.

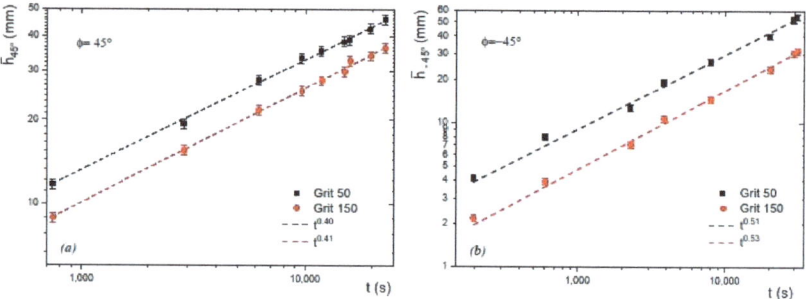

Figure 10. Log–log plot of the averaged front position, $\bar{h}$, as a time function, on faces tilted at (**a**) $\phi = 45°$ and (**b**) $\phi = -45°$, respectively. In both cases, we show data for 50 and 150 grit scratching. When $\phi = 45°$ data approximately fitted power laws of the form $\bar{h} \sim t^n$, where $n = 0.40$ for grit 50 and $n = 0.41$ for grit 150. Similarly for case $\phi = -45°$ data fitted power laws with $n = 0.51$ for grit 50 and $n = 0.53$ for grit 150.

All this quantitative characterization of the functional surfaces, specifically through the statics and dynamics effects of the capillary action, shows that our approximate models of fluid mechanics are very sensitive tools for these types of disordered or complex patterns. Now, the question is if a similar approach can be useful for the characterization of another phenomena on functional surfaces with irregular V-grooves, it will be studied experimentally later on.

4. Functionality of Scratched Surfaces in Other Cases

4.1. Droplets on Human Skin

We must notice that in nature and technology there are a myriad of functional surfaces following countless patterns. For instance, in Figure 11 we can observe that human skin has different microtextures on its different parts: in Figure 11a, we show the skin groove network on the forearm, in Figure 11b, the grooves on a fingertip are shown, whereas in Figure 11c, a water drop hanging from the forefinger can be visualized. In the case of spreading of liquid on a human forearm, the rate of spreading of a drop of a moisturizing oil was visualized through the fine surface V-grooves (characterized through replicas), obtaining that its advance front, L, follows a power law of the form $L \sim t^{0.5}$ [12]; to our knowledge the spreading of liquids on human forefingers is still an open problem, despite the well-known record of human fingerprints.

Similarly, the problem of the shape of a drop hanging from the fingertip depends on the random distribution of grooves and is awaiting for an accurate treatment. The skin is a live organ and is a tough outer protective layer that keeps water repellency, our proposal is that novel materials, using replicas, micromachining or directed sanding, may be useful to replicate geometrical configurations that nature has printed for eons to understand functionalities of our human integument.

Figure 11. Pictures of human skin groove networks of (**a**) forearm skin, (**b**) a fingertip and (**c**) a water drop hanging from the forefinger. Sample sizes in (**a**,**b**) are of a few millimeters.

Now, results interesting to follow the loss of mass of a water sessile droplet on skin; for instance, in Figure 12 we show three snapshots of a water droplet on the dorsal wrist skin, we observe that the droplet loss mass and we characterized this phenomenon by measuring the angle of contact θ as a function of the elapsed time, t. The plot of $\theta(t)$ in Figure 12d indicates that at short times the change of θ is very fast and at the last stage, the rate of change diminish. We will show later that such behavior is characteristic of sessile droplets on V-grooves.

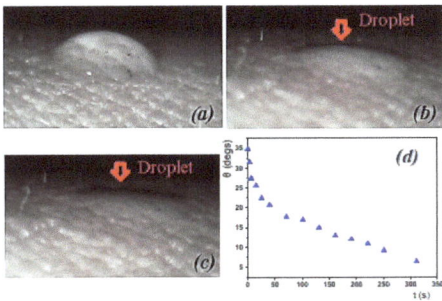

Figure 12. Time sequential images of the spreading of a water droplet on dorsal wrist skin: (**a**) $t = 0$ s, (**b**) $t = 163$ s, (**c**) $t = 198$ s and (**d**) plot of the change of the contact angle θ as a function of time, t.

4.2. Droplets on Scratched Acrylic Sheets

Finally, another functionality of acrylic was experimentally tested here: the loss of mass of water sessile droplets on scratched surfaces. This case is important due to its possible relevance on the surface transmission of SARS-CoV-2, the virus that causes COVID-19. The principal mode by which people are infected is through exposure to respiratory droplets carrying the infectious virus. SARS-CoV-2 is an enveloped virus, meaning that its genetic material is packed inside an outer layer (envelope) of proteins and lipids. The envelope contains structures (spike proteins) for attaching to human cells during infection. Since the droplet serves as a medium for virus survival, the infectivity of the virus is connected to an extent to the droplet lifetime.

Researchers have studied how long SARS-CoV-2 can survive on a variety of porous [25,26] and smooth surfaces [26,27]. On porous surfaces, studies report an inability to detect a viable virus within minutes to hours; on smooth surfaces, viable virus can be detected for days to weeks. The apparent, relatively faster inactivation of SARS-CoV-2 on porous compared with smooth surfaces might be attributable to capillary action within pores and faster aerosol droplet evaporation.

In Figure 13, we show the temporal evolution of a water droplet of approximately 1 mm diameter, whereas it evaporates on a smooth acrylic surface, we found that the time of evaporation under these conditions is of around 12 min. Similarly, we placed a water droplet of 1 mm on an acrylic surface with cross scratching (grit 100 sandpaper), see Figure 14. Now, the total loss of mass, on such a surface took place in about 3 min, in this case we observed that the main reason for it is the capillary penetration of liquid, into the horizontal micro V-grooves.

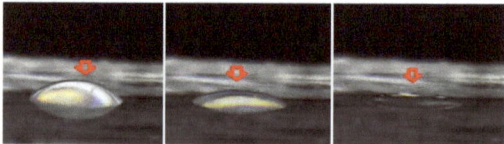

Figure 13. Snapshots of evaporation of a water drop (pointed out with arrows) on a smooth acrylic sheet. In this case evaporation took around 12 min.

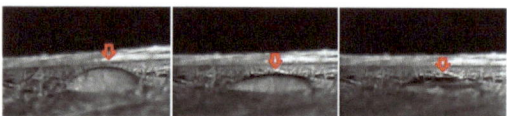

Figure 14. Snapshots of spreading of a water droplet (pointed out with arrows) on an acrylic sheet, which was cross scratched with a 100 grit sandpaper. In this case, the total loss of mass took around 3 min.

To highlight the importance of the type of scratching of the acrylic sheets, and its effect on the loss of mass of the droplets, in Figure 15 we show snapshots of the spreading of a water sessile droplet on a longitudinal scratching (grit 150), obtained by using the same protocol of Section 3. At time $t = 0$ the droplet is near circular, from the top view, after, at $t = 2$ s the droplet increases its diameter due to the high wettability between acrylic and water (the contact angle is $\theta_w = 36.5° \pm 1°$). A few seconds later, the mass of water flows spontaneously, due to the capillary action, along the V-grooves. Finally, there occurs the formation of very small drops, perhaps due to the instabilities in the fast flow, typical of the open V-shaped microchannels [28].

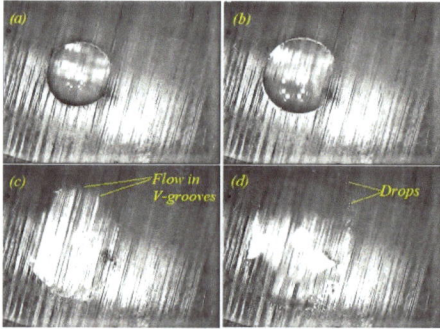

Figure 15. Snapshots of the top view of a water sessile water droplet of initial diameter $D \approx 1.5$ mm on an acrylic sheet with longitudinal scratching (grit 150): (**a**) $t = 0$ s, (**b**) $t = 2$ s, (**c**) $t = 34$ s and (**d**) $t = 87$ s. In (**c**), the capillary flow along the open V-grooves is appreciated, and in (**d**), the formation of very small drops there occurs.

The previous description details how a water droplet, on scratched surfaces, lose their mass, mainly due to the spontaneous capillary flow along the open microchannels. To reinforce these ideas, we also measured the change of the apparent contact angle, θ', of sessile droplets, as a function of time.

Studies of the droplet evaporation dynamics on hydrophobic or hydrophilic micropillared surfaces give plots of the contact angle of sessile water droplets [15]. In both cases, droplets evaporating on a structured surface predominantly exhibit the droplet contact angle decreasing slowly as the droplet evaporates and, after, the decreasing is faster. The first is attributed to the fact that when the droplets are sufficiently large (during the initial stages), evaporation is primarily governed by vapor diffusion at the liquid–vapor interface and heat conduction through the droplet.

To contrast with the previous case, in Figure 16, we show the plots of the apparent contact angles $\theta'(t)$, for water and silicone oil sessile droplets on longitudinal scratching acrylic sheets. It is appreciated that the change of these angles is very similar among them, despite silicone oil is a nonvolatile liquid at room temperature. Moreover, the contact angle of a water droplet on skin, given in Figure 12, changes in a similar manner. Contrary to cases of droplets on micro-pillared surfaces, all cases of loss of mass of droplets on V-grooved surfaces presents, during the initial stages, a strong loss of mass, quantified through the angles $\theta'(t)$ or $\theta(t)$, respectively. At the final stages, the slow loss of mass also occurs for all cases.

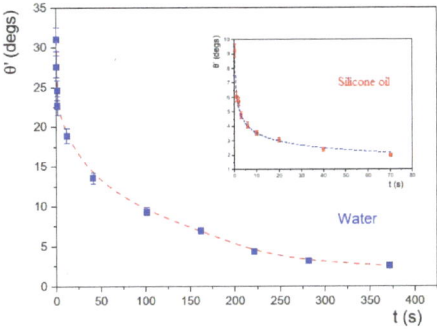

Figure 16. Plot of the temporal evolution of the apparent contact angle $\theta'(t)$, for a water sessile droplet and for a silicone oil sessile droplet (inset). The diameter of both droplets were $D \approx 1.2$ mm and it is below of the respective capillary lengths of water and silicone oil, see text.

We advice that measurements of θ', for times lower than 0.5 s, were not possible due to limitations of our microscope, however the static measurement of the contact angle for a sessile drop of silicone gave the value $\theta = 26° \pm 1°$, which is close to the contact angle measured through the method of the meniscus under gravity given in Section 2.

Our experimental observations on the spread of droplets on longitudinal scratching surfaces allows us to conclude that the fast loss of mass is due mainly to the capillary flow into the series of parallel open V-grooves which sustain the droplet, which also could be interconnected, meanwhile the slow loss of mass, at the later stages, it is imposed by the strong influence of the viscous stresses along the same series of V-grooves. The evaporation of the liquids in the horizontal V-grooves has also been reported [28,29].

In summary, in a general context, many solid surfaces suffer a profuse quantity of scratches, but in the context of COVID-19, when a droplet carrying the infectious virus is deposited on a scratched surface, it rapidly loses its envelope aqueous layer, and thus its survival time is very short. So, the scratching of surfaces brings up a beneficial functionality.

5. Conclusions

In this work, we experimentally studied the functionality of acrylic sheets when they were subjected to irregular scratching, nearly along a single direction. In a first stage, we revisited the classical theory of the capillary action by showing the formation of a liquid meniscus on vertically standing acrylic sheets. Afterwards, in the same context of the capillary action, we highlighted the various power laws, depending on the place of measurement, of capillary rise when a viscous and wetting liquid spontaneously penetrates into a Taylor–Hauksbee cell. Taking into account all these fundamental features of capillary action, we used 50 and 150 grit sandpapers in order to carve V-grooves on the face of the acrylic sheet and several of their geometrical characteristics were explored with the use of electronic microscopy.

An initial analysis of the functionality of the scratched surfaces was performed through the study of the capillary rise in vertical and tilted sheets, where measurements of the non-uniform fronts of silicone oil, for each case, were obtained. It is apparent that the scratching of surfaces with grit 50 produces larger grooves in comparison with those produced with grit 150. It was found that the averaged fronts of rise, $\bar{h}(t)$, obey power laws of the form $\bar{h}(t) \approx at^n$, with approximate values $n = 0.40$ for vertical and tilted sheets when $\phi = 45°$, meanwhile for the case when $\phi = -45°$ we found that approximately $n = 0.50$. Thus, the use of power laws given by Equation (2) or (3) is not direct since the size and orientations of the grooves affects the value of the exponent n, meanwhile the prefactor a depends more on the properties of the involved liquids and the grit size.

We also have argued about the possible utility of our procedure to study, as a geometric analog, the functionality of the human skin since our skin has different types of grooves at different parts of our body. It means that there are an infinity of functional characteristics in the human skin and it is possible that different types of scratches on acrylic (linear, circular, crossed, etc.), or types of micro patterning, can emulate simple micro flows on skin. Moreover, experiments of the loss of mass of sessile water droplets on skin allowed us to understand the importance of the V-grooves because they suck out the water from the droplets themselves. At the end of the current paper, we also gave evidence of the functionality of scratched surfaces for sessile water droplets because they loss mass very rapidly. We proved, for several fluids, that the loss of mass of sessile droplets is mainly due to the strong capillary flow in the open V-grooves. It allows us to conclude that in case that aqueous droplets containing SARS-CoV-2, fall on dry scratched surfaces, the time of loss of their aqueous layer is shorter than that of a smooth surface. As a consequence, the coronavirus remains viable a shorter time on scratched surfaces compared to smooth surfaces, a similar conclusion for many porous surfaces was found by other authors.

In broad context, wear of surfaces due to continuous use and cleaning, and specially when abrasives substances are used, provokes involuntary scratching that finally has a functionality as the one found in our controlled study, moreover, we hope that further studies will improve our comprehension of scratched surfaces.

Author Contributions: Experiments A.M., A.L-V., C.A.V.; modeling: A.M., A.L-V., C.A.V.; writing and revision: A.M., A.L-V., C.A.V. All authors have read and agreed to the published version of the manuscript.

Funding: This research received no external funding.

Data Availability Statement: The data presented in this study are available on request from the corresponding author. The data are not publicly available due to we still do not have a publicly accessible repository.

Acknowledgments: A.M. acknowledges F.J. Higuera, S. de Santiago, A. Jara, R. Diéz-Barroso, J. Casillas, D.A. Serrano and F. Hernández-Santiago their help in the different phases of this work.

Conflicts of Interest: The authors declare no conflict of interest.

References

1. Godi, A.; De Chiffre, L. Functional Surfaces in Mechanical Systems: Classification, Fabrication, and Characterization. In *Surface Engineering Techniques and Applications: Research Advancements*; Santo, L., Davim, J.P., Eds.; IGI Global: Hershey, PA, USA, 2014; pp. 1–44.
2. Deng, D.; Tang, Y.; Zeng, J.; Yang, S.; Shao, H. Characterization of capillary rise dynamics in parallel micro V-grooves. *Int. J. Heat Mass Transf.* **2014**, *77*, 311–320. [CrossRef]
3. Chen, H.; Zhang, P.; Zhang, L.; Liu, H.; Jiang,Y.; Zhang, D.; Han, Z.; Jiang, L. Continuous directional water transport on the peristome surface of Nepenthes alata. *Nature* **2016**, *532*, 85–89. [CrossRef]
4. Li, C.; Li, N.; Zhang, X.; Dong, Z.; Chen, H.; Jiang, L. Uni-Directional Transportation on Peristome-Mimetic Surfaces for Completely Wetting Liquids. *Angew. Chem. Int. Ed.*, **2016**, *55*, 14988. [CrossRef]
5. Chen, H.; Zhang, L.; Zhang, Y.; Zhang, P.; Zhang, D.; Jiang, L. Uni-directional liquid spreading control on a bio-inspired surface from the peristome of Nepenthes alata. *J. Mater. Chem. A* **2017**, *5*, 6914–6920 [CrossRef]
6. Zhang, P.; Zhang, L.; Chen, H.; Dong, Z.; Zhang, D. Surfaces inspired by the Nepenthes Peristome for unidirectional liquid transport. *Adv. Mater.* **2017**, *29*, 1702995. [CrossRef] [PubMed]
7. Zhou, S.; Yu, C.; Li, C.; Dong, Z.; Jiang, L. Programmable unidirectional liquid transport on peristome-mimetic surfaces under liquid environments. *J. Mater. Chem. A* **2019**, *7*, 18244–18248. [CrossRef]

8. Zhang, L.; Liu, G.; Chen, H.; Liu, X.; Ran, T.; Zhang, Y.; Gan, Y.; Zhang, D. Bioinspired unidirectional liquid transport micro-nano structures: A review. *J. Bionic Eng.* **2021**, *18*, 1–29. [CrossRef]
9. Tian, Y.; Jiang, Y.; Zhou, J.; Doi, M. Dynamics of Taylor rising. *Langmuir* **2019**, *35*, 5183–5190. [CrossRef]
10. Bamorovat Abadi, G.; Bahrami, M. A general form of capillary rise equation in micro-grooves. *Sci. Rep.* **2020**, *10*, 19709. [CrossRef]
11. Vorobyev, A.Y.; Guo, C. Metal pumps liquid uphill. *Appl. Phys. Lett.* **2009**, *94*, 224102. [CrossRef]
12. Dussaud, A.D.; Adler, P.M.; Lips, A. Liquid transport in the networked microchannels of the skin surface. *Langmuir* **2003**, *19*, 7341–7345. [CrossRef]
13. Paul, S.; Mukhopadhyay, S.; Mandal, S.; Agarwal, P.; Chanda, N. Fabrication of durable haemophobic surfaces on cast acrylic sheets using UV laser micromachining. *Micro Nano Lett.* **2019**, *14*, 1175–1177. [CrossRef]
14. Higuera F.J.; Medina, A.; Liñán A. Capillary rise of a liquid between two vertical plates making a small angle. *Phys. Fluids* **2008**, *20*, 102102. [CrossRef]
15. Günay, A.A.; Kim, M.-K.; Yan, X.; Miljkovic, N.; Sett, S. Droplet evaporation dynamics on microstructured biphilic, hydrophobic, and smooth surfaces. *Exp. Fluids* **2021**, *62*, 153. [CrossRef]
16. Landau, L.D.; Lifshits, I.M. *Fluid Mechanics (Volume 6 of Course of Theoretical Physics)*; Pergamon Press: Oxford, UK, 1987.
17. Taylor, B., IX. Part of a letter from Mr. Brook Taylor, F. R. S. to Dr. Hans Sloane R. S. Secr. concerning the ascent of water between two glass planes. *Philos. Trans. R. Soc.* **1712**, *27*, 538
18. Hauksbee, F.X. An Account of an experiment touching the ascent of water between two glass planes, in an Hyperbolick figure. *Philos. Trans. R. Soc. Lond.* **1712**, *27*, 539.
19. Sanchez, F.A.; Gutierrez, G.J.; Medina, A. Capillary penetration in cells with periodical corrugations. *Rev. Mex. Fis.* **2009**, *55*, 467–471.
20. Ponomarenko, A.; Quéré, D.; Clanet, C. A universal law for capillary rise in corners. *J. Fluid Mech.* **2011**, *666*, 146–154. [CrossRef]
21. Bell, J.M.; Cameron, F.K. The flow of liquids through capillary spaces. *J. Phys. Chem.* **1906**, *10*, 658–674. [CrossRef]
22. Lucas, R. Ueber das Zeitgesetz des kapillaren Aufstiegs von Flüssigkeiten. *Kolloid-Zeitschrift* **1918**, *23*, 15–22. [CrossRef]
23. Washburn, E.W. The dynamics of capillary flow. *Phys. Rev.* **1921**, *17*, 273–283. [CrossRef]
24. *Fandeli* Is a Manufacturer of Abrasives. Available online: https://fandeli.com/wp-content/uploads/2020/10/presentacion.pdf (accessed on 16 December 2021). (In Spanish)
25. Chatterjee, S.; Murallidharan, J.S.; Agrawal, A.; Bhardwaj, R. Why coronavirus survives longer on impermeable than porous surfaces. *Phys. Fluids* **2021**, *33*, 021701. [CrossRef] [PubMed]
26. Bhardwaja, R.; Agrawal, A. Tailoring surface wettability to reduce chances of infection of COVID-19 by a respiratory droplet and to improve the effectiveness of personal protection equipment. *Phys. Fluids* **2020**, *32*, 081702. [CrossRef]
27. Bhardwaja, R.; Agrawal, A. Likelihood of survival of coronavirus in a respiratory droplet deposited on a solid surface. *Phys. Fluids* **2020**, *32*, 061704. [CrossRef]
28. Tian, J.; Kannangara, D.; Li, X.; Shen, W. Capillary driven low-cost V-groove microfluidic device with high sample transport efficiency. *Lab Chip* **2010**, *10*, 2258–2264. [CrossRef] [PubMed]
29. Kolliopoulos, P.; Kumar, S. Capillary flow of liquids in open microchannels: Overview and recent advances. *npj Microgravity* **2021**, *7*, 51. [CrossRef]

Article

Crater Depth after the Impact of Multiple Drops into Deep Pools

Manfredo Guilizzoni [1,*] and Giuseppe Frontera [2]

[1] Department of Energy, Politecnico di Milano, Via Lambruschini 4, 20156 Milan, Italy
[2] Independent Researcher, 20852 Villasanta, Italy; giuseppe.frontera@outlook.com
* Correspondence: manfredo.guilizzoni@polimi.it; Tel.: +39-02-2399-3888

Abstract: Many studies have been devoted to single drop impacts onto liquid films and pools, while just a few are available about double drop or drop train impacts, despite the fact that the latter are more realistic situations. Thus, computational fluid dynamics with a volume-of-fluid approach was used here to simulate the impact of multiple drops into deep pools. The aim was to verify if multiple drop impacts significantly differ from single drops ones, and if the models available in the literature for the crater depth in the case of single impacts are reliable also for the multiple drop cases. After validation against experimental data for single and double drop impacts, simulations for four to 30 drops, with a diameter of 2.30 mm, impact velocities 1.0, 1.4, 1.8, and 2.2 m/s, and random initial positions in the domain were performed. The results showed that the time evolution of the crater depth for multiple impacts is similar to the single drop case during the inertial phase, while the following behavior is very different. Consequently, the available models for the maximum crater depth during single drop impacts can still predict the upper and lower bounds of the values of the crater depth during multiple drop impacts within 5% deviation.

Keywords: multiple drop impact; crater depth; computational fluid dynamics; numerical simulation; volume-of-fluid; interFoam; deep pool

Citation: Guilizzoni, M.; Frontera, G. Crater Depth after the Impact of Multiple Drops into Deep Pools. *Fluids* **2022**, *7*, 50. https://doi.org/10.3390/fluids7020050

Academic Editor: Mehrdad Massoudi

Received: 15 December 2021
Accepted: 19 January 2022
Published: 24 January 2022

Publisher's Note: MDPI stays neutral with regard to jurisdictional claims in published maps and institutional affiliations.

Copyright: © 2022 by the authors. Licensee MDPI, Basel, Switzerland. This article is an open access article distributed under the terms and conditions of the Creative Commons Attribution (CC BY) license (https://creativecommons.org/licenses/by/4.0/).

1. Introduction

Many phenomena, both of natural and technological interest, involve the interaction between liquid drops and an interface, in most cases between a solid and a gas or a liquid and a gas. The interaction may be mechanical, thermal, or chemical, and it may start with the generation of the drop on the interface (as during condensation) or with the impact of the drop onto the same. In the latter case, the impact velocity may be low (down to practically zero when the drop is gently deposed) or high, normal, or oblique. In most situations, a series of drops impact the interface at nearly the same time. A first example is obviously rain, with its effects on the Earth's water surfaces (oceans, seas, lakes, rivers, etc.) and on, e.g., fields, buildings, monuments, planes, and wind turbines. This also happens in many other scenarios, e.g., internal combustion engines, firefighting systems, surface cooling, spray painting, inkjet printing, pesticide distribution in agriculture, and blood sprays in crime scenes. Therefore, the importance of a deep understanding of this phenomenon is evident, which is far from being a simple one, up to the point that despite more than a century of investigation it is still not fully explained.

In the literature, many studies can be found, dealing either with the impact of a single drop (onto dry solid surfaces, onto surfaces wetted by liquid films of different thicknesses, still or moving, and onto gas–liquid interfaces alone in deep pools), or with the behavior of sprays, where the multitude of drops in the spray is considered in statistical terms. Such literature is so vast that it is impossible to cite all the relevant papers. Detailed introductory information and reviews about the single drop impact onto liquid surfaces can be found in [1,2], onto both solid and liquid surfaces in [3], and onto solid surfaces with the different possible scenarios in [4]. An in-depth analysis of sprays is described in [5]. Specifically on single drop impacts into deep pools, the work by Cole [6] is a valuable reference.

On the contrary, the studies about the impact of a limited number of drops, but not one, are quite scarce. Concerning drops in parallel with simultaneous or delayed impacts, the impact of two gemini drops onto a deep pool is analyzed in [7–9], of two gemini drops onto a thin liquid film in [10], and of three gemini drops onto a thin liquid film in [11,12]. Concerning drops in a series, the impact onto a dry surface is described in [13] and in [14] where heat transfer is also analyzed; the impact onto deep pools is described in [15,16] and in [17], in which the creation of a funnel is also reported.

To mitigate the lack of information about the impact of more than one drop, in the present work computational fluid dynamics (CFD) with a finite volume, volume-of-fluid approach was used to simulate the unsynchronous impact of multiple water drops into a deep pool of the same liquid and having the same temperature of the drops.

In fact, drop impact is a two-phase or three-phase (if the surface is of a solid) fluid dynamics phenomenon, in which inertial, viscous, and capillary forces merge their effects. It also becomes a multi-physics problem when heat transfer, mass transfer, or chemical reactions play a significant role (e.g., for hot or cold drops or surfaces, for drops impacting onto a chemically different liquid, and for reactive wetting). Therefore, the analytical solution of the resulting models is not viable, and very simplified models built on experimental results have been for a long time the only tool for analysis and prediction. In relatively recent years, CFD have also become commonly used. Many remarkable studies were developed, implementing the different CFD approaches: from the pioneering works [18], to developments with in-house software tools [19–21], to the most recent ones mainly using open-source packages [22–24], in addition to some of the previously cited papers. More specifically, Eulerian models using the finite volume method and the volume-of-fluid modelling technique [25] are the most used algorithms, followed by level-set [26], markers [27], interface capturing and tracking [28,29], combined volume-of-fluid and level-set [30,31] also with the ghost fluid method [32] and adaptive mesh refinement [33], Lattice Boltzmann [34,35], and molecular dynamics simulations (suitable up to the scale of nanodrops only) [36].

Specifically concerning multiple drop impacts—whose study is the major point of originality of the present work—onto deep pools, their outcomes depend on a multiplicity of factors: the fluids (drop, pool, surrounding atmosphere), the temperatures, the drop diameters, impact velocity, drop positions and mutual distances, and the time delay between the impacts of the single drops. Consequently, the number of simulations needed to uniformly cover all the space of the governing parameters would be extremely large. The approach here was therefore the following:

- A single fluid (water) at a fixed temperature was selected.
- Validation of the simulations against previously acquired experimental data [8,9] for single and double drop impacts was performed.
- Repeated simulations about the impact of four to 30 drops were performed, with the drops having the same diameter and impact velocity, but random positions in the domain, including the vertical distance from the pool, so that they hit the liquid–gas interface at slightly different time instants.

The aim was to statistically evaluate the effect of the mutual interactions between the craters formed by the single drops, particularly in terms of crater depth, to check if the latter is significantly affected by them, and consequently if the many models available in the literature for such a quantity during a single drop impact can be reliable also for the more realistic case of multiple drop impacts.

2. Materials and Methods

2.1. Experimental Setup

Validation of the numerical simulations was performed against the experimental results acquired in previous campaigns [8]. In such experiments, drop impacts were analyzed by means of high-speed videos of the drop-pool system seen in back illumination.

The experimental set-up and procedures, including the uncertainty analysis for the main parameters, were described in full detail by the authors of [8,9].

2.2. Numerical Simulations

The numerical simulations were performed using the *interFoam* solver of the OpenFOAM® open source CFD toolbox [37]. *interFoam* is a finite volume solver based on the volume-of-fluid (VOF) method and implementing the continuum surface force model to include the effects of surface tension at the interface [38]. OpenFOAM® was selected as it is free and open-source, and because of the many favorable reviews [39–42] and successful cases of use described in the literature both about drop impacts onto solid surfaces [43], normal impacts onto liquid surfaces for single drops [16,44,45], oblique impacts of single drops [46], drop trains [47], and other Eulerian-Eulerian two-phase fluid dynamics applications [48]. The model implemented in *interFoam* includes the continuity and momentum equations for a Newtonian and incompressible fluid [49,50], whose density and viscosity are calculated as a weighted average of the corresponding properties of the single phases, on the basis of an indicator function named volume fraction. The latter assumes a value of 0 for one phase, 1 for the other, and between 0 and 1 in the interfacial regions, and it is transported by the fluid velocity field. Volume tracking and interface reconstruction is then performed (typically as the isosurface at a volume fraction equal to 0.5), with no explicit interface tracking. With respect to the original VOF formulation, the *interFoam* model includes an additional term in the volume fraction equation, aiming at "compressing" the interface (even down to just 2–4 cells). In strict terms, such a term is a mass source, but both literature results [48] and verification by the authors proved that mass variation is completely negligible.

The complete system of equations solved by *interFoam* is constituted by the continuity equation (Equation (1)), the Navier-Stokes equation (Equation (2)) and the equation for the transport of the volume fraction (Equation (3)), as follows:

$$\nabla \cdot \mathbf{u} = 0 \qquad (1)$$

$$\frac{\partial(\rho \mathbf{u})}{\partial \tau} + \nabla \cdot (\rho \mathbf{u}\mathbf{u}) - \nabla \cdot (\mu \nabla \mathbf{u}) - (\nabla \mathbf{u}) \cdot \nabla \mu = -\nabla P_d - \mathbf{g} \cdot \mathbf{x} \nabla \rho + \sigma \kappa \nabla \gamma \qquad (2)$$

$$\frac{\partial \gamma}{\partial \tau} + \nabla \cdot (\bar{\mathbf{u}} \gamma) + \nabla \cdot [\mathbf{u}_r \gamma (1-\gamma)] = 0 \qquad (3)$$

where $\mathbf{u}$ is the flow velocity; $\mathbf{u}_r$ is the relative velocity at the interface ($\mathbf{u}_{liquid} - \mathbf{u}_{gas}$); $\mathbf{x}$ is the local coordinate vector; $\mathbf{g}$ is the gravity vector; σ is the interface tension between the phase (water–air surface tension in the present work); κ is the curvature of the interface; P_d is a modified pressure term, removing the hydrostatic contribution ($P_d = P - \rho \mathbf{g} \cdot \mathbf{x}$); $\bar{\mathbf{u}}$, ρ, and μ are the average velocity in the interfacial region and the fluid density and viscosity, respectively, all calculated as weighted averages of the single-phase quantities (even though this has a physical basis only for the density):

$$\bar{\mathbf{u}} = \mathbf{u}_{liquid} \gamma + \mathbf{u}_{gas}(1-\gamma) \qquad (4)$$

$$\rho = \rho_{liquid} \gamma + \rho_{gas}(1-\gamma) \qquad (5)$$

$$\mu = \mu_{liquid} \gamma + \mu_{gas}(1-\gamma) \qquad (6)$$

The advantage of the VOF method with respect to the two-fluid models is that a single set of equations must be solved. Further details about the *interFoam* solver and models can be found in [48,51], and in [52], where the source code can also be found. Very promising modified versions of *interFoam* were also presented in the literature [46,53], but the source code was not made publicly available, so the version included in the official OpenFOAM® distribution was used.

Concerning the discretization schemes and solution algorithms, the implicit Euler scheme (first-order accurate) was used for the time derivative, as it proved to offer better results in comparison with the second-order Crank-Nicholson discretization schemes that were tested during preliminary simulations, selecting different blending factors between 0.5 and 1. The conventional advection term was discretized using Gauss schemes: limited Van Leer for the volume fraction and limited linear for the velocity. For the latter, variations of the limiter parameter were tested, but the best results were obtained when keeping it equal to 1. Finally, the OpenFOAM® specific *interfaceCompression* scheme [54] was selected for the discretization of the compression term. In fact, the other possible scheme, *isoAdvector*, did not offer significant improvements, despite the good performances reported in the literature [55,56]. Adaptive time stepping was used, limiting the allowed Courant-Friedrichs-Lewy (CFL) number to 0.3, according to the recommendations for 3D cases [39,54]. Some volume fraction sub-cycles were also performed to further improve the accuracy.

Three-dimensional domains shaped as rectangular cuboids were used for the simulations, as shown in Figure 1.

Concerning the boundary conditions, for all the simulated cases:

- The bottom boundary was set as a wall, i.e., fixed value equal to 0 for the velocity, zero gradient for the pressure, volume fraction equal to 1 (water always present).
- The top boundary was set as an open boundary, i.e., zero gradient for the velocity, fixed value equal to 0 for the pressure, volume fraction equal to 0 (air always present).
- The side boundaries were set as symmetry boundaries for all the variables, to reduce the computational effort for single and double drop impacts, and to model the control volume as a "tile" of a larger physical domain for multiple drop impacts.

To reduce the height of the domain, the detachment of the drops from the generator and their fall towards the pool surface were not simulated: the drops were directly initialized as spheres near the free surface of the pool, with an initial velocity corresponding to the selected one. This approach has a point of weakness in the fact that drop oscillations after the detachment from the needle are not considered. In general, drop shape and oscillatory behavior may be an important influence on the impact outcomes, but in this case the shapes of the drops in the experiments were nearly spherical, so this approximation seemed acceptable. The mesh was purely structured, with hexahedral cells. Grading was used for the double drop simulations to better capture the "neck" between the two craters [9], while uniform meshes were used for the single and the multiple drop simulations. Adaptive remeshing is very slow in OpenFOAM® for 3D cases, so static meshes were used in all cases.

As the investigated impact velocities were low, laminar flow was assumed for both phases, in agreement with all the previously cited papers in this field. For each phase, the values of all the relevant thermophysical properties were taken at 28°C (average value from the experiments described in [8]).

At the beginning of the simulations, the pool height and the positions, diameters, and velocities of the drops within the domain were set using the *setFields* OpenFOAM® utility. Outside from the drops, the initial velocity was set to 0 for the whole domain. Given the assumption of incompressibility for both phases, pressure was initialized at 0. For the single and double drop impact cases, drop diameter and velocities were set according to the experimental values, in the range 2.27–2.32 mm and 1.0–2.0 m/s, respectively, as reported in [8,9]. For the multiple drop impacts, drop diameter was set to 2.30 mm for all cases, this value being the rounding of the average of the drop diameters in the experiments for single and double drop impacts. Drop impact velocities for the multiple impacts were set at 1.0, 1.4, 1.8, and 2.2 m/s, respectively. A total of 10 configurations were tested, 9 of them with a number of drops randomly chosen between 4 and 9, and 1 including 30 drops. For half of the simulations, the initial positions of the drops in the domain were randomly set independently for each impact velocity; for the other half (including the cases with 30 drops), the initial positions of the drops in the domain were randomly set for the cases at 1.0 m/s and then kept fixed for the other impact velocities. Two restrictions to the randomness of position were imposed: the distance from the domain boundaries cannot be less than 1.2 mm and the distance between two drop centers must be larger than 1.2 mm, so that at the beginning no drop touches the symmetry boundaries or another drop.

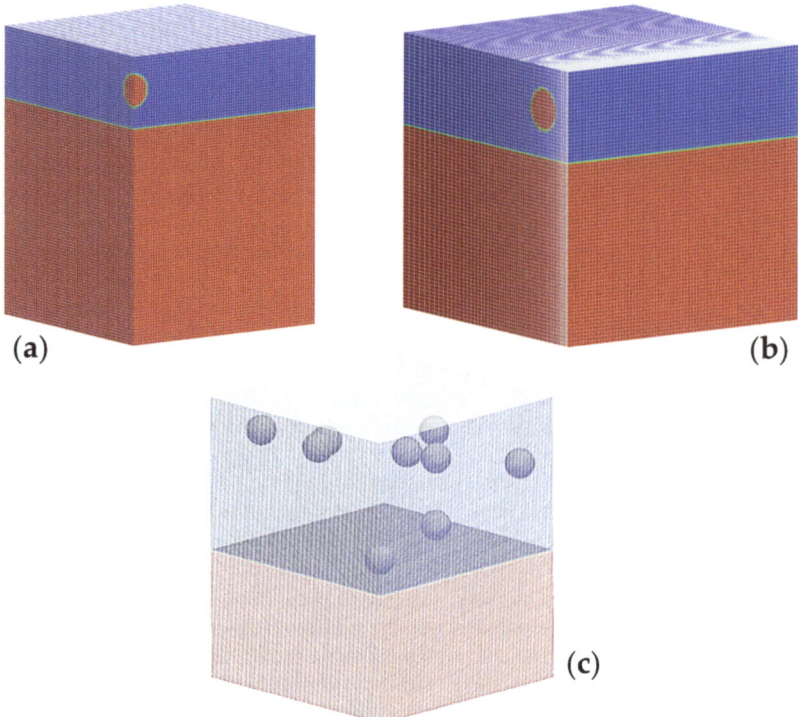

Figure 1. Domain and mesh (drawn at half the real refinement level, for better visualization) for the single (**a**), double (**b**), and multiple (**c**) impact simulations.

Figures 2–4 show some examples of the initial positions of the drops and of some frames extracted from the results of the numerical simulations (both as 3D views and as 2D views used to extract the crater depth), for a case of single drop impact and two cases of multiple drop impacts with 5 and 30 drops, respectively, impact velocity 1.4 m/s. Time $t = 0$ is set at the instant in which the first drop touches the pool free surface.

Table 1 reports the number and initial positions of the drops in all the multiple impact cases, by showing top views of the simulated domain. Cases from 5 to 9 are those with the same drop positions for all the velocities, while the cases from 1 to 4 are with drop positions different for each velocity. In some cases, the difference between the cases is only in the vertical positions of the drops, so it cannot be appreciated by the top view; still, it results in different crater interactions.

The dimensions of the domain selected after the preliminary simulations were 18×18 (horizontal) $\times 22$ (vertical) mm for the cases with 1 to 9 drops, and 36×36 (horizontal) $\times 22$ (vertical) mm for the cases with 30 drops; the cells were cubic with side 0.125 mm in all cases.

It is worth noting that in the VOF approach, regions of the same fluids coming into contact merge instantaneously, because the underlying model is not able to represent retarded coalescence (modified models would be needed, e.g., see [57]). On the other hand, real-world interfaces may resist even with direct contact—at least for a certain time—if the pressure of the contacting regions is not too different (that is why it is possible to have bouncing bubbles or drops [58]). These aspects cannot be reproduced in the simulations; for the present case this should influence the results only in a very limited number of cases during multiple drop impacts, in which two drops merge before impact, while in reality they may continue to fall touching each other, but without coalescence.

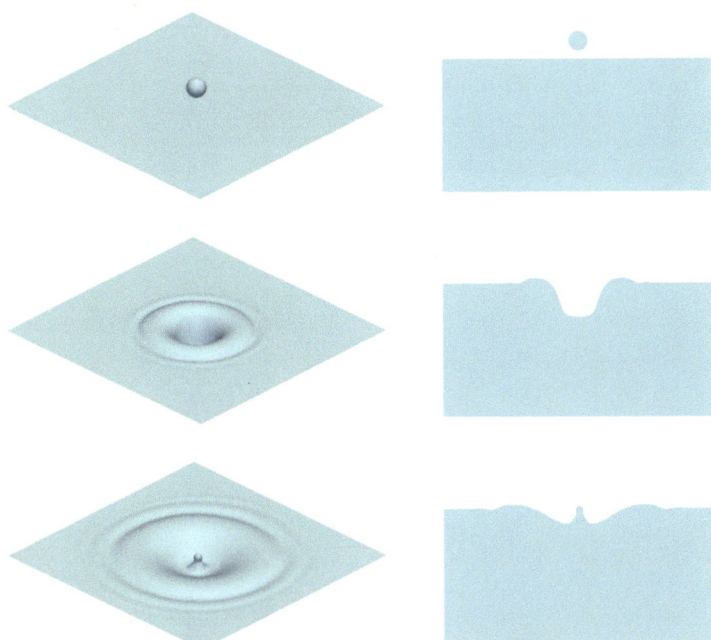

Figure 2. Examples of frames (3D and 2D used to extract the crater depth) showing the evolution of the craters and of the free surface of the pool for a case of single drop impact, impact velocity 1.4 m/s.

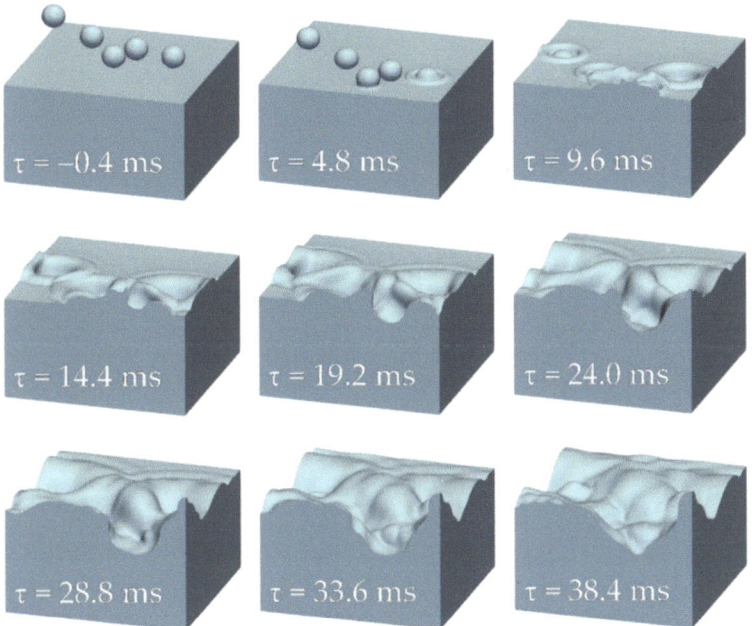

Figure 3. Examples of frames (3D and 2D used to extract the crater depth) showing the evolution of the craters and of the free surface of the pool for a case of multiple drop impacts with 5 drops, impact velocity 1.4 m/s.

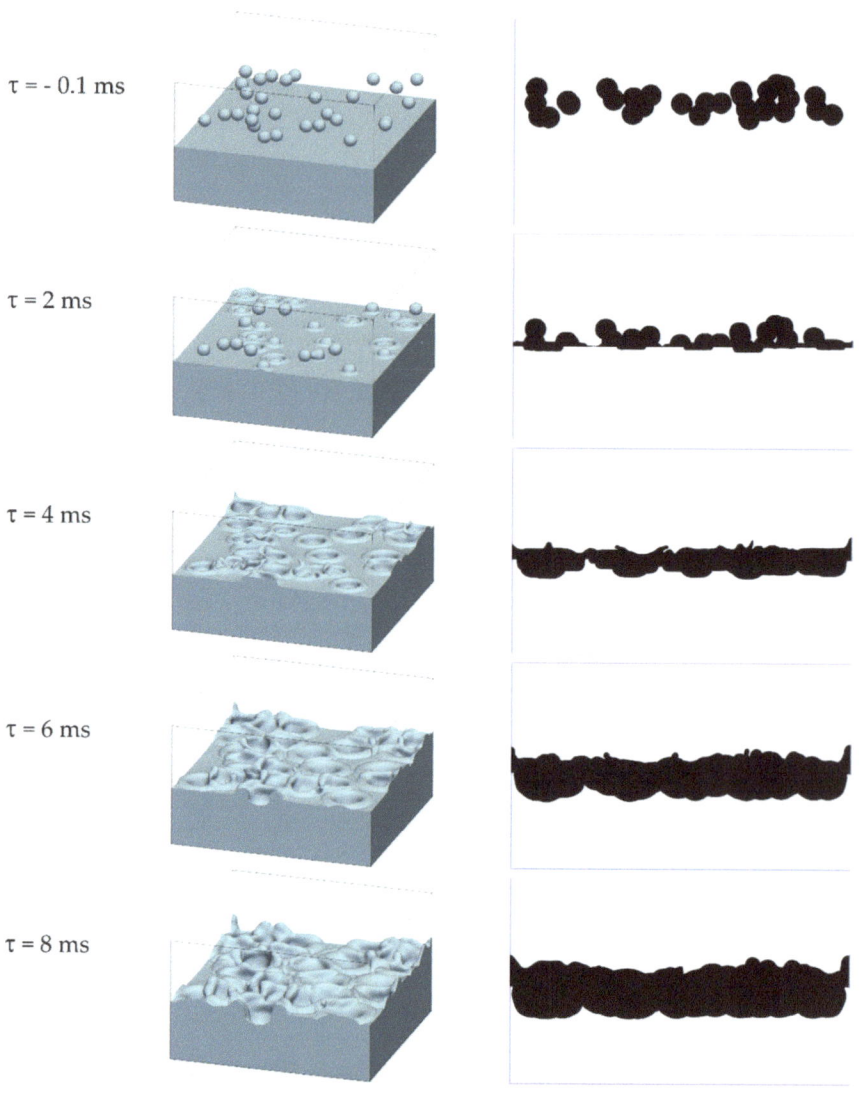

Figure 4. Examples of frames (3D and 2D used to extract the crater depth) showing the evolution of the craters and of the free surface of the pool for a case of multiple drop impacts with 30 drops, impact velocity 1.4 m/s.

Table 1. Top views of the simulated domain showing the initial positions of the drops in all the performed simulations about multiple impacts.

Test	w = 1.0 m/s	w = 1.4 m/s	w = 1.8 m/s	w = 2.2 m/s
1				
2				
3				
4				
5				
6				
7				
8				
9				

3. Results and Discussion

As already said, among the many parameters (depth, width, shape, capillary waves) that characterize the crater, expanding and then receding after the drop impact, the crater depth was selected as the quantity of interest for this study, even if it must be kept in account that the drop water was also transported towards much lower depths than the crater's bottom [8,9,59]). Therefore, crater depth evolution in time and maximum reached crater depth were analyzed. The aim was to evaluate the effect of the mutual interactions between the drops and the craters and particularly to check how they alter the maximum depth reached by the craters.

The results were also compared with the most credited models available in the literature, to quantify the discrepancies with respect to the simulated results and assess their reliability for multiple drop impacts.

3.1. Mesh Independence and Validation against Experimental Data

The validation of the numerical setup and procedure was performed by comparing the crater depth profiles as a function of time with respect to the experimental ones reported in [8]. Mesh independence was verified by performing numerical simulations with mesh sizes between 72,200 and 7,166,250 cells, as shown in Figure 5 for the case of the single drop impact at 1.4 m/s. The results with the different tested meshes and the *interfaceCompression* scheme are shown with continuous lines, while the results with the *isoAdvector* scheme are shown with dashed lines. The experimental data are evidenced with asterisk markers.

In Figure 5 and in all of the following charts, the crater depth is shown in its dimensionless version $z_{max\ dl}$, calculated as the ratio between the depth and the drop diameter before impact. In the same figure, the percent deviations in terms of maximum depth when using some of the tested meshes are also indicated. For single drop impacts, the percent deviations on the maximum crater depth were -10.33%, -0.78% and -9.87% for the velocities $w = 1.0$ m/s, 1.4 m/s, and 2.0 m/s, respectively; for double drop impacts, they were -6.10%, -1.70%, and -0.14% for the velocities $w = 1.0$ m/s, 1.4 m/s, and 2.0 m/s, respectively. Thus, the maximum crater depth was predicted within 10% accuracy, and less in the majority of cases; therefore, the agreement can be considered satisfactory.

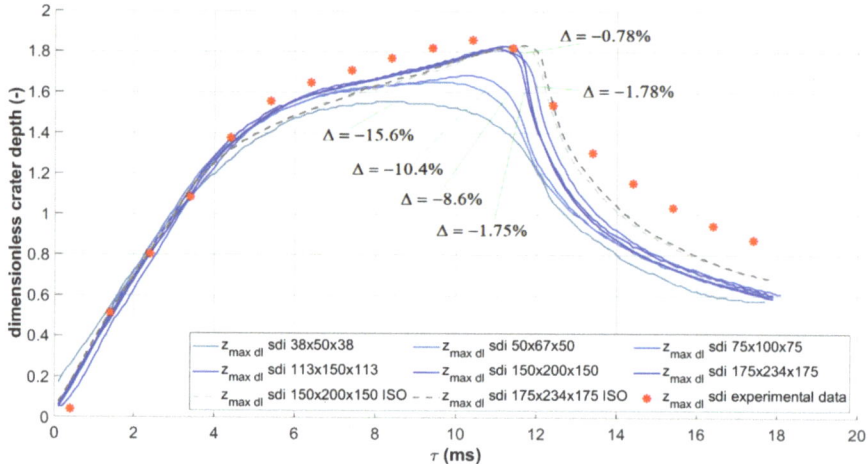

Figure 5. Dimensionless crater depth as a function of time for single drop impact at 1.4 m/s, simulations vs. the experiments reported in [8]. The results with different mesh resolutions are shown to evidence the mesh independence. The percent deviations between the simulated and experimental maximum crater depths with the selected meshes are also indicated.

During the inertial phase, in which the crater expands until it reaches the maximum depth, the mean absolute percentage difference between the simulation results with the $175 \times 234 \times 175$ mesh and the experimental data was 3.31% for the single drop impact cases and 3.79% for the double drop cases. The corresponding median values of the percentage deviation were 3.75% and 2.96%, respectively. As can be seen also in Figure 5, the agreement was much worse for the following capillary phase, in which the crater closes, with mean deviation 24.96% and median deviation 28.83% for the single impacts, and 16.30% and 14.72%, respectively, for the double impacts. No final explanation could be given for the significant difference in the performance of the simulations during the capillary phase between the single and double drop impacts.

On the basis of the simulations performed for validation, the domain was reduced to its already described final dimensions, and consequently the final mesh sizes selected for the simulations were $144 \times 144 \times 176$ (3.65 million cells) for the simulations with one to nine drops, and $288 \times 288 \times 176$ (14.6 million cells) for the simulations with 30 drops.

3.2. Results for Multiple Drop Impacts

Figures 6–9 report the results of the simulations in terms of time evolution of the dimensionless crater depth for the multiple drop impacts. In the same figures, the results for the single and double drop impacts and the prediction from some of the most credited literature models for the crater depth evolution and for the maximum crater depth are also shown. The equations and the references of the selected literature models are reported in Table 2.

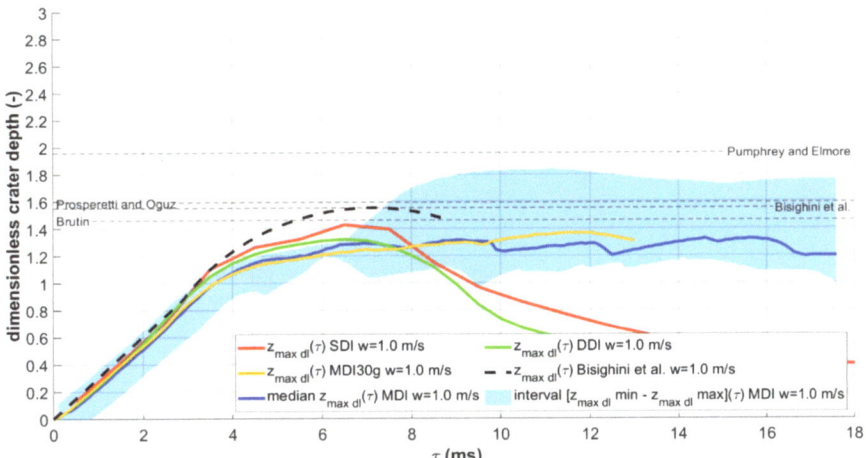

Figure 6. Dimensionless crater depth as a function of time for single, double, and multiple drop impacts at 1.0 m/s. Predictions from some of the most credited literature models are also shown.

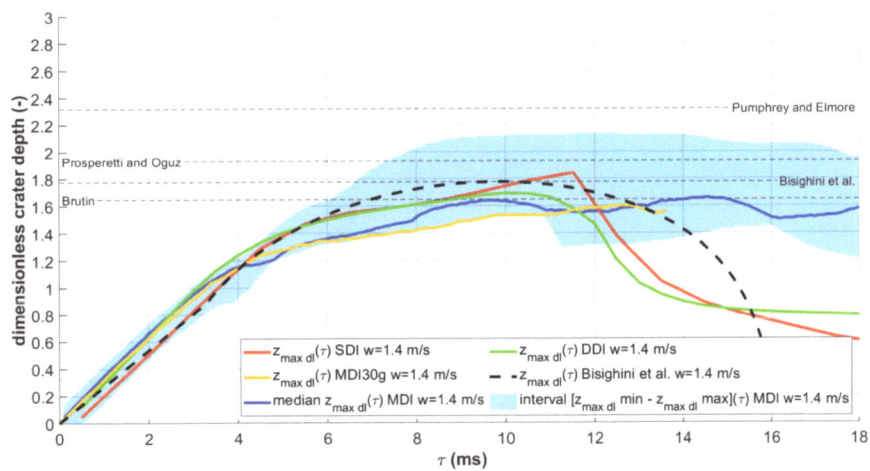

Figure 7. Dimensionless crater depth as a function of time for single, double, and multiple drop impacts at 1.4 m/s. Predictions from some of the most credited literature models are also shown.

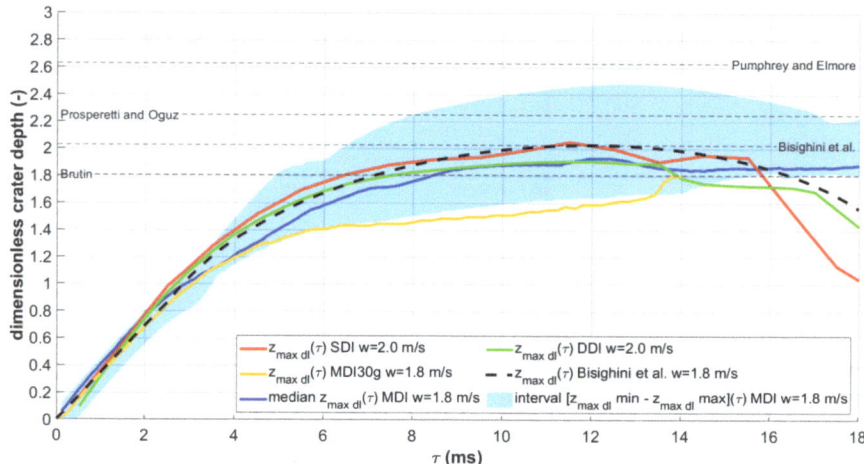

Figure 8. Dimensionless crater depth as a function of time for single, double, and multiple drop impacts at 1.8 m/s. Predictions from some of the most credited literature models are also shown.

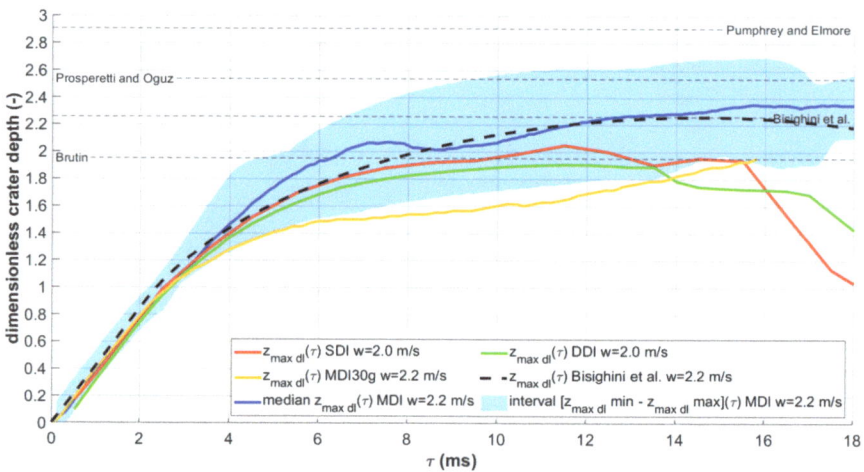

Figure 9. Dimensionless crater depth as a function of time for single, double, and multiple drop impacts at 2.2 m/s. Predictions from some of the most credited literature models are also shown.

The profiles in the multiple impact cases with four to nine drops enclosed a region, depending on the mutual interaction between the drops, that depended on the drop positions. Such a region is evidenced in the charts with a filled band, to better underline this aspect in comparison with the results of single and double drop impacts, that are drawn as continuous lines. The median of the crater depths for the multiple impact cases and the values for the 30 drops simulations are also shown as continuous lines, while the prediction from the model by Bisighini et al. is represented with a dashed line.

From the graphs it can be seen how, in the first part, the time histories of the crater depth for single, double, and multiple drop impacts are very similar. On the contrary, in the second part the profiles are different, particularly for the multiple drops where there is not the tendency of the crater to close as it does after single and double impacts. This can be seen from the last part of the plots, where the crater depth for multiple drop impacts tends to remain constant for a long time.

Table 2. Literature models for the dimensionless crater depth evolution and maximum crater depth.

Model	Equation[s]
Pumphrey and Elmore (1990) [60]	$z_{maxdl} = \left(\frac{Fr_D}{3}\right)^{\frac{1}{4}}$
Prosperetti and Oguz (1993) [61]	$z_{maxdl} = \left[2\left(\frac{2}{3}Fr_R + 4\frac{Fr_R}{We_R} + \frac{Fr_R^2}{We_R^2}\right)^{1/2} - 2\frac{Fr_R}{We_R}\right]^{1/2}$
Leng (2001) [62]	$z_{maxdl} = 0.727\left(\frac{Fr_D}{3}\right)^{\frac{1}{4}}$
Brutin (2003) [63]	$z_{maxdl} = (1+kWe)^{\frac{1}{5}}$ $k = \frac{20B_0^3}{24+3B_0^2} \quad B_0 = \frac{D_0}{a} \quad a = \sqrt{\frac{2\sigma}{\rho g}}$
Bisighini et al. (2010) [64]	$\ddot{\alpha} = -\frac{3}{2}\frac{\dot{\alpha}^2}{\alpha} - \frac{2}{\alpha^2 We} - \frac{1}{Fr}\frac{\zeta}{\alpha} + \frac{7}{4}\frac{\dot{\zeta}^2}{\alpha} - \frac{4\dot{\alpha}}{\alpha^2 Re}$ $\ddot{\zeta} = -3\frac{\dot{\alpha}\dot{\zeta}}{\alpha} - \frac{9}{2}\frac{\dot{\zeta}^2}{\alpha} - \frac{2}{Fr} - \frac{12\dot{\zeta}}{\alpha^2 Re}$ with initial conditions: $\dot{\alpha} = 0.17 \quad \alpha = \alpha_0 + 0.17\tau \quad \dot{\zeta} = 0.27 \quad \zeta = -\alpha_0 + 0.17\tau$

Figure 10 shows the maximum crater depths reached during multiple drop impacts as box plots, for the four investigated impact velocities.

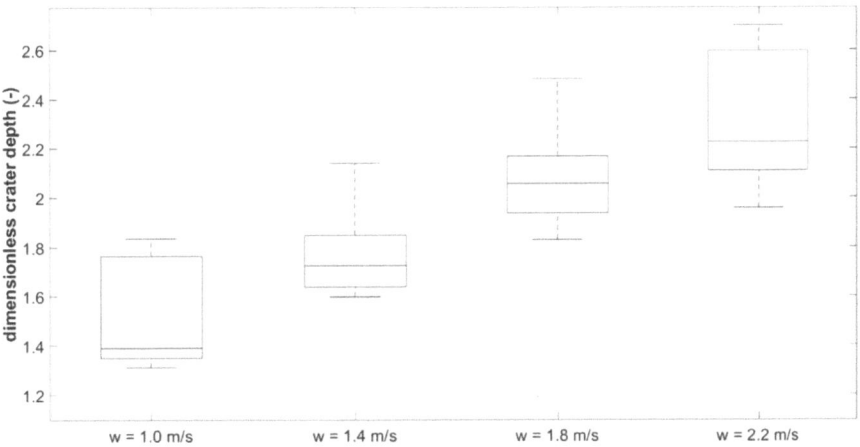

Figure 10. Box plots of the maximum dimensionless crater depth for multiple drop impacts, at the different impact velocities.

Tables 3–5 report the quantitative comparison between the simulation results and the predictions of the selected literature models, in terms of percent deviation of the model prediction with respect to the minimum, median, and maximum values of the maximum crater depth for each of the investigated impact velocities. Other literature models (e.g., those by Fedorchenko and Wang [65]) returned similar values.

Table 3. Deviation between the maximum dimensionless crater depths predicted from the literature models and the minimum values of the quantity calculated from the simulations at the different velocities.

Model	1.0 m/s	1.4 m/s	1.8 m/s	2.2 m/s
Pumphrey	49.49	41.68	43.84	38.54
Prosperetti and Oguz	22.22	18.03	23.00	20.99
Leng	8.68	3.00	4.57	0.72
Brutin	11.41	0.55	−1.10	−6.97
Bisighini	18.26	8.41	10.86	7.54

Table 4. Deviation between the maximum dimensionless crater depths predicted from the literature models and the median values of the quantity calculated from the simulations at the different velocities.

Model	1.0 m/s	1.4 m/s	1.8 m/s	2.2 m/s
Pumphrey	39.71	34.31	27.99	21.44
Prosperetti and Oguz	13.29	11.89	9.44	6.05
Leng	1.57	−2.35	−6.95	−11.72
Brutin	4.21	−4.67	−12.00	−18.45
Bisighini	10.52	2.78	−1.36	−5.74

Table 5. Deviation between the maximum dimensionless crater depths predicted from the literature models and the maximum values of the quantity calculated from the simulations at the different velocities.

Model	1.0 m/s	1.4 m/s	1.8 m/s	2.2 m/s
Pumphrey	6.78	8.47	5.96	7.78
Prosperetti and Oguz	−13.41	−9.64	−9.39	−5.87
Leng	−22.37	−21.14	−22.96	−21.64
Brutin	−20.42	−23.02	−27.14	−27.62
Bisighini	−15.53	−17.00	−18.33	−16.34

As can be seen from the tables, the models by Leng and by Brutin offer the best performance in predicting the median values of the maximum crater depths at the lowest impact velocities, while the model by Bisighini becomes the best at the highest ones.

The models by Leng and by Brutin are also the best in predicting the minimum values of the maximum crater depths at the different impact velocities. The model by Pumphrey and the model by Prosperetti and Oguz are overestimating the minimum and median values, particularly the first one, that then turns to give the best results in terms of the maximum values of the maximum crater depths at the lowest impact velocities.

Figures 11–13 show the maximum crater depths as a function of the dimensionless groups typically used to characterize drop impacts, the Reynolds Re_D, the Weber We_D, and the Froude Fr_D numbers:

$$Re_D = \frac{\rho\, w\, D}{\mu} \;;\; We_D = \frac{\rho\, w^2\, D}{\sigma} \;;\; Fr_D = \frac{w^2}{gD} \qquad (7)$$

The trends in the charts are obviously the same, but the latter helps in contextualizing the investigated conditions. The points corresponding to the simulations with between four and nine drops are shown with circle markers, while the triangle markers evidence the results from the 30 drop simulations. The cases in which the drop number and positions were the same in the simulations at the different impact velocities are connected by a line. In the last chart, some additional lines are also traced:

- the line corresponding to $Z_{max\,ad} = (Fr_D/3)^{1/4}$—that, as already said, is the "basis" of many of the literature models—is also shown as a dash-dot line.
- the two lines corresponding to $Z_{max\,ad} = 0.675\,(Fr_D/3)^{1/4}$ and $Z_{max\,ad} = 0.935\,(Fr_D/3)^{1/4}$, that can be used as "rounded" boundaries of the simulation results. A larger number of simulations would be needed to perform a significant fitting of the minimum, maximum, and median values of the maximum crater depths in order to propose a new model with reliable coefficients.

Concerning the influence of the distances between the drops, from the results it seems that the crater depth tends to reduce when increasing the number of drops, e.g., the results with 30 drops are those with the lowest values. Such a finding is not surprising, as when many drops are present and their mutual distance is relatively short, each crater "disturbs" the others during the inertial expansion phase. This is due both to the effect of pressure, as expanding craters displace water that is pushed towards the other craters, opposing their growth, and of surface tension when craters merge. To further verify this effect, an additional simulation was performed, with 16 drops uniformly distributed in the domain, synchronously impacting the pool at 1.4 m/s. This is an unrealistic situation, that was simulated as an extreme case. As can be seen from Figure 14, the craters arrive to "block" each other.

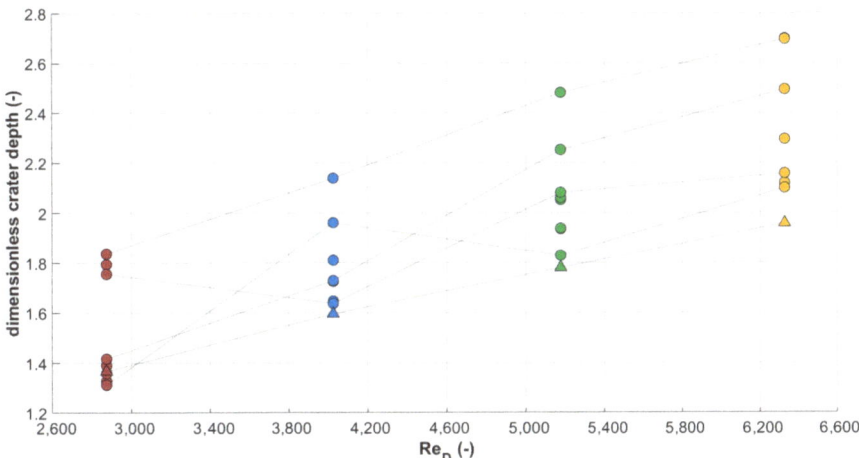

Figure 11. Maximum dimensionless crater depths for multiple drop impacts as a function of the Reynolds number. Circle markers are used for the points corresponding to the simulations with between four and nine drops, while the triangle markers for the results from the 30 drop simulations; the colors are related to the different impact velocities (brown 1.0 m/s, dark cyan 1.4 m/s, green 1.8 m/s, yellow 2.2 m/s).

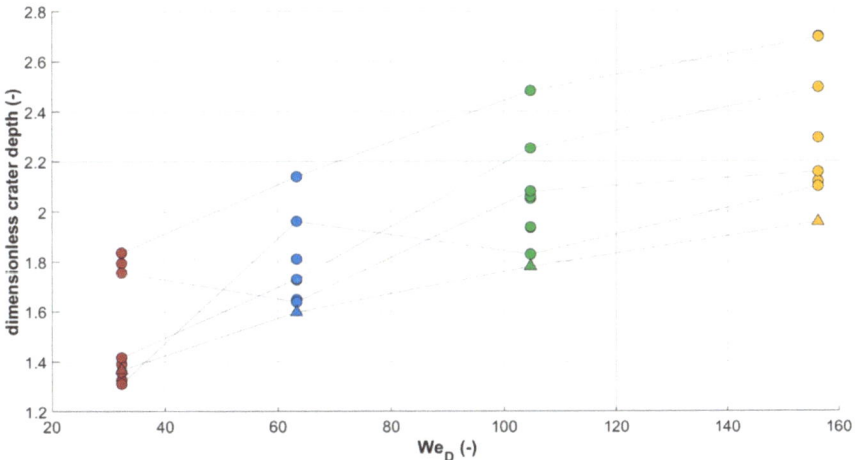

Figure 12. Maximum dimensionless crater depths for multiple drop impacts as a function of the Weber number. Circle markers are used for the points corresponding to the simulations with between four and nine drops, while the triangle markers for the results from the 30 drop simulations; the colors are related to the different impact velocities (brown 1.0 m/s, dark cyan 1.4 m/s, green 1.8 m/s, yellow 2.2 m/s).

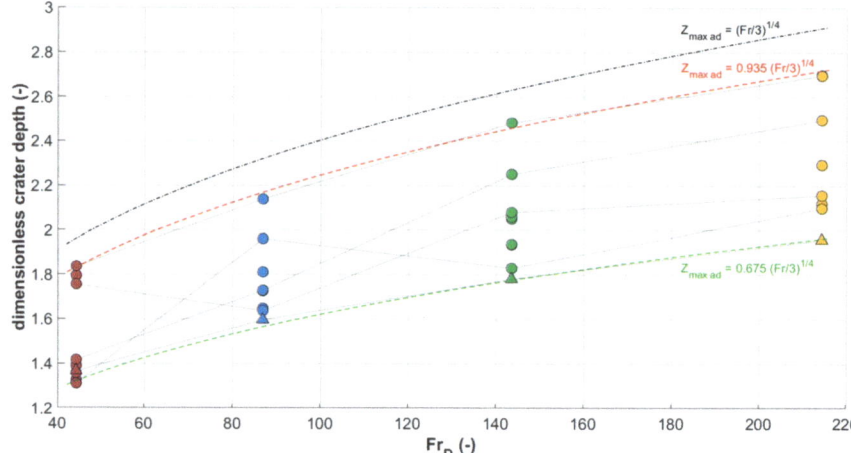

Figure 13. Maximum dimensionless crater depths for multiple drop impacts as a function of the Froude number. Circle markers are used for the points corresponding to the simulations with between four and nine drops, while the triangle markers for the results from the 30 drop simulations; the colors are related to the different impact velocities (brown 1.0 m/s, dark cyan 1.4 m/s, green 1.8 m/s, yellow 2.2 m/s).

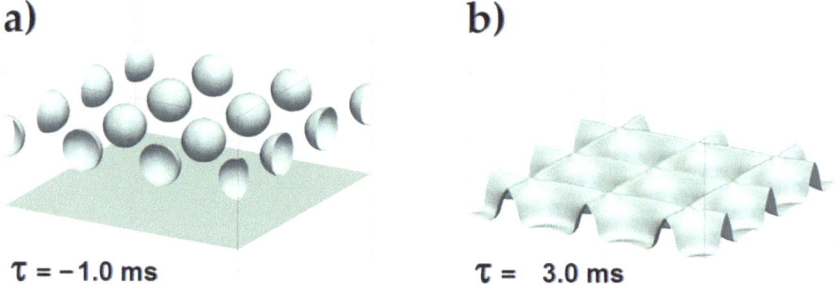

Figure 14. Three-dimensional view of the initial drop positions (**a**) 1 ms before the impact, and of the free surface configuration (**b**) 3 ms after the impact for a case with 16 drops uniformly distributed in the domain, impact velocity 1.4 m/s.

4. Conclusions

Numerical simulations of single, double, and multiple and not synchronous water drop impacts onto deep pools were performed in a finite volume framework, using the two-phase, incompressible *interFoam* solver of the OpenFOAM® open-source CFD package. After validation against the experimental data for single and double drop impacts, the focus was on the evolution in time and maximum value of the crater depth. Four impact velocities, namely 1.0 m/s, 1.4 m/s, 1.8 m/s, and 2.2 m/s were investigated, performing nine simulations for each of them with a random number (between four and 30) and positions of the drops in the domain.

The results were compared with some of the most credited literature models for the crater depth after single drop impacts, with the aim to assess the reliability of such models to predict such a quantity also for multiple impacts. The models by Leng, Brutin, and Bisighini offer very good performances in predicting the minimum value of the maximum crater depths at the different impact velocities, while the model by Pumphrey has the best agreement in terms of the maximum value of such a quantity. Thus, it can be concluded that some of the models available in the literature for the crater maximum depth can also be used as the upper and lower bounds of the values of the crater depth during multiple drop impacts. Moreover, the results show that in the presence of many drops the maximum depth is reduced, as the water displaced by each crater hampers the expansion of the others.

A larger number of simulations would be needed to perform a significant fitting and to provide a new model, e.g., for the median values of the maximum crater depth. This is foreseen as one of the future works for the research activity, included in a more extensive simulation campaign performed on the basis of a rigorous design of experiment, aimed at systematically exploring the effects of the separated variations of the single governing parameters. Additionally, the validation with experimental data directly acquired for multiple drop impacts and an extension of the investigation (both numerical and experimental) to drops having smaller diameters and higher velocities—to match more closely, e.g., those of rain drops—are needed to draw more in-depth conclusions about the differences between single and multiple drop impacts, in relation to the specific scenarios.

Author Contributions: Conceptualization, M.G.; formal analysis, G.F. and M.G.; investigation, G.F. and M.G.; methodology, M.G.; software, G.F. and M.G.; validation, G.F. and M.G.; visualization, G.F. and M.G.; writing—original draft preparation, G.F. and M.G.; writing—review and editing, G.F. and M.G. All authors have read and agreed to the published version of the manuscript.

Funding: This research received no external funding.

Data Availability Statement: The results of the numerical simulations are available from the authors upon reasonable request.

Conflicts of Interest: The authors declare no conflict of interest.

References

1. Worthington, A.M. On impact with a liquid surface. *Proc. R. Soc. Lond.* **1882**, *34*, 217–230. [CrossRef]
2. Worthington, A.M. *A Study of Splashes*; Longmans Green and Co.: London, UK, 1908.
3. Rein, M. Phenomena of liquid drop impact on solid and liquid surfaces. *Fluid Dyn. Res.* **1993**, *12*, 61–93. [CrossRef]
4. Yarin, A.L. Drop impact dynamics: Splashing, spreading, receding, bouncing. *Annu. Rev. Fluid Mech.* **2006**, *38*, 159–192. [CrossRef]
5. Ashgriz, N. (Ed.) *Handbook of Atomization and Sprays: Theory and Applications*; Springer Science & Business Media: Berlin, Germany, 2011; ISBN 9781441972644. [CrossRef]
6. Cole, D. The Splashing Morphology of Liquid-Liquid Impacts. Ph.D. Thesis, James Cook University, Townsville, Australia, 2007.
7. Bisighini, A.; Cossali, G.E. High-speed visualization of interface phenomena: Single and double drop impacts onto a deep liquid. *J. Vis.* **2011**, *14*, 103–110. [CrossRef]
8. Santini, M.; Fest-Santini, S.; Cossali, G.E. Experimental study of vortices and cavities from single and double drop impacts onto deep pools. *Eur. J. Mech. B Fluids* **2017**, *62*, 21–31. [CrossRef]
9. Guilizzoni, M.; Santini, M.; Fest-Santini, S. Synchronized Multiple Drop Impacts into a Deep Pool. *Fluids* **2019**, *4*, 141. [CrossRef]
10. Guo, Y.; Chen, G.; Shen, S.; Zhang, J. Double Droplets Simultaneous Impact on Liquid Film. *IOP Conf. Ser. Mater. Sci. Eng.* **2015**, *88*, 012016. [CrossRef]
11. Santini, M.; Cossali, G.E.; Marengo, M. Splashing characteristics of multiple and single drop impacts onto a thin liquid film. In Proceedings of the International Conference on Multiphase Flows, Leipzig, Germany, 9–13 July 2007.
12. Fest-Santini, S.; Steigerwald, J.; Santini, M.; Cossali, G.E.; Weigand, B. Multiple drops impact onto a liquid film: Direct numerical simulation and experimental validation. *Comput. Fluids* **2021**, *214*, 104761. [CrossRef]
13. Roisman, I.V.; Prunet-Foch, B.; Tropea, C.; Vignes-Adler, M. Multiple Drop Impact onto a Dry Solid Substrate. *J. Colloid Interface Sci.* **2002**, *256*, 396–410. [CrossRef]
14. Trujillo, M.F.; Lewis, S.R. Thermal boundary layer analysis corresponding to droplet train impingement. *Phys. Fluids* **2012**, *24*, 112102. [CrossRef]
15. Ray, B.; Biswas, G.; Sharma, A.; Welch, S.W.J. CLSVOF method to study consecutive drop impact on liquid pool. *Int. J. Numer. Methods Heat Fluid Flow* **2013**, *23*, 143–158. [CrossRef]
16. Bouwhuis, W.; Huang, X.; Chan, C.U.; Frommhold, P.E.; Ohl, C.-D.; Lohse, D.; Snoeijer, J.H.; van der Meer, D. Impact of a high-speed train of microdrops on a liquid pool. *J. Fluid Mech.* **2016**, *792*, 850–868. [CrossRef]
17. Lee, J.H.; Kim, S.; Kim, J.; Kim, H.; Kim, H.-Y. From an elongated cavity to funnel by the impact of a drop train. *J. Fluid Mech.* **2021**, *921*, A8. [CrossRef]
18. Oguz, H.N.; Prosperetti, A. Bubble entrainment by impact of a liquid drop on liquid surfaces. *J. Fluid Mech.* **1990**, *219*, 143–179. [CrossRef]
19. Rieber, M.; Frohn, A. A numerical study on the mechanism of splashing. *Int. J. Heat Fluid Flow* **1999**, *20*, 455–461. [CrossRef]

20. Bussmann, M.; Mostaghimi, J.; Chandra, S. On a three-dimensional volume tracking model of droplet impact. *Phys. Fluids* **1999**, *11*, 1406. [CrossRef]
21. Morton, D.; Rudman, M.; Liow, J.M. An investigation of the flow regimes resulting from splashing drops. *Phys. Fluids* **2000**, *12*, 747. [CrossRef]
22. Thoraval, M.-J.; Takehara, K.; Etoh, T.G.; Popinet, S.; Ray, P.; Josserand, C.; Zaleski, S.; Thoroddsen, S.T. von Karman Vortex Street within an Impacting Drop. *Phys. Rev. Lett.* **2012**, *108*, 264506. [CrossRef]
23. Thoraval, M.-J.; Li, Y.; Thoroddsen, S. Vortex-ring induced large bubble entrainment during drop impact. *Phys. Rev. E* **2016**, *93*, 033128. [CrossRef]
24. Reijers, S.A.; Liu, B.; Lohse, D.; Gelderblom, H. Oblique droplet impact onto a deep liquid pool. *arXiv* **2019**, arXiv:1903.08978.
25. Hirt, C.W.; Nichols, B.D. Volume of Fluid (VOF) Method for the Dynamics of Free Boundaries. *J. Comput. Phys.* **1981**, *39*, 201–225. [CrossRef]
26. Sussman, M.; Smereka, P.; Osher, S.J. A level set approach for computing solutions to incompressible two-phase flow. *J. Comput. Phys.* **1994**, *114*, 146–159. [CrossRef]
27. Popinet, S.; Zaleski, S. A front tracking algorithm for accurate representation of surface tension. *Int. J. Numer. Methods Fluid* **1999**, *30*, 775–793. [CrossRef]
28. Bonometti, T.; Magnaudet, J. An interface-capturing method for incompressible two-phase flows. Validation and application to bubble dynamics. *Int. J. Multiph. Flow* **2007**, *33*, 109–133. [CrossRef]
29. Tryggvason, G.; Bunner, B.; Esmaeeli, A.; Juric, D.; Al-Rawahi, N.; Tauber, W.; Han, J.; Nas, S.; Jan, Y.-J. A Front-Tracking Method for the Computations of Multiphase Flow. *J. Comput. Phys.* **2001**, *169*, 708–759. [CrossRef]
30. Albadawi, A.; Donoghue, D.B.; Robinson, A.J.; Murray, D.B.; Delauré, Y.M.C. Influence of surface tension implementation in Volume of Fluid and coupled Volume of Fluid with Level Set methods for bubble growth and detachment. *Int. J. Multiph. Flow* **2013**, *53*, 11–28. [CrossRef]
31. Taqieddin, A.; Liu, Y.; Alshawabkeh, A.N.; Allshouse, M.R. Computational modeling of bubbles growth using the coupled Level Set—Volume of Fluid method. *Fluids* **2020**, *5*, 120. [CrossRef] [PubMed]
32. Menard, T.; Tanguy, S.; Berlemont, A. Coupling level set/VOF/ghost fluid methods: Validation and application to 3D simulation of the primary break-up of a liquid jet. *Int. J. Multiph. Flow* **2007**, *33*, 510–524. [CrossRef]
33. Fuster, D.; Agbaglah, G.; Josserand, C.; Popinet, S.; Zaleski, S. Numerical simulation of droplets, bubbles and waves: State of the art. *Fluid Dyn. Res.* **2009**, *41*, 065001. [CrossRef]
34. Zheng, H.W.; Shu, C.; Chew, Y.T.; Sun, J.H. Three-dimensional lattice Boltzmann interface capturing method for incompressible flows. *Int. J. Numer. Methods Fluids* **2008**, *56*, 1653–1671. [CrossRef]
35. Huang, J.J.; Shu, C.; Chew, Y.T. Lattice Boltzmann study of bubble entrapment during droplet impact. *Int. J. Numer. Methods Fluids* **2011**, *65*, 655–682. [CrossRef]
36. Gentner, F.; Rioboo, R.; Baland, J.P.; De Coninck, J. Low Inertia Impact Dynamics for Nanodrops. *Langmuir* **2004**, *20*, 4748–4755. [CrossRef] [PubMed]
37. OpenFOAM®. Available online: https://openfoam.org (accessed on 5 December 2021).
38. Brackbill, J.U.; Khote, D.B.; Zemach, C. A Continuum Method for Modeling Surface Tension. *J. Comput. Phys.* **1992**, *100*, 335–354. [CrossRef]
39. Gopala, V.R.; van Wachem, B.G.M. Volume of fluid methods for immiscible-fluid and free-surface flows. *Chem. Eng. J.* **1998**, *141*, 204–211. [CrossRef]
40. Deshpande, S.S.; Anumolu, L.; Trujillo, M.F. Evaluating the performance of the two-phase flow solver interFoam. *Comput. Sci. Discov.* **2012**, *5*, 014016. [CrossRef]
41. Deshpande, S.S.; Trujillo, M.F. Distinguishing features of shallow angle plunging jets. *Phys. Fluids* **2013**, *25*, 082103. [CrossRef]
42. Costa, A.B.; Graham Cooks, R. Simulated splashes: Elucidating the mechanism of desorption electrospray ionization mass spectrometry. *Chem. Phys. Lett.* **2008**, *464*, 1–8. [CrossRef]
43. Roisman, I.V.; Weickgenannt, C.M.; Lembach, A.N.; Tropea, C. Drop impact close to a pore: Experimental and numerical investigations. In Proceedings of the ILASS—Europe 2010, 23rd Annual Conference on Liquid Atomization and Spray Systems, Brno, Czech Republic, 6–9 September 2010.
44. Berberovic, E.; van Hinsberg, N.P.; Jakirlic, S.; Roisman, I.V.; Tropea, C. Drop impact onto a liquid layer of finite thickness: Dynamics of the cavity evolution. *Phys. Rev. E* **2009**, *79*, 036306. [CrossRef]
45. Castillo-Orozco, E.; Davanlou, A.; Choudhury, P.K.; Kumar, R. Droplet impact on deep liquid pools: Rayleigh jet to formation of secondary droplets. *Phys. Rev. E* **2015**, *92*, 053022. [CrossRef]
46. Brambilla, P.; Guardone, A. Automatic tracking of corona propagation in three-dimensional simulations of non-normal drop impact on a liquid film. *Computing* **2013**, *95*, 415–424. [CrossRef]
47. Trujillo, M.F.; Alvarado, J.; Gehring, E.; Soriano, G.S. Numerical simulations and experimental characterization of heat transfer from a periodic impingement of droplets. *J. Heat Transf.* **2011**, *133*, 122201. [CrossRef]
48. Deshpande, S.S.; Trujillo, M.F.; Wu, X.; Chahine, G. Computational and experimental characterization of a liquid jet plunging into a quiescent pool at shallow inclination. *Int. J. Heat Fluid Flow* **2012**, *34*, 1–14. [CrossRef]
49. Kundu, P.K.; Cohen, I.M.; Dowling, D.R.; Tryggvason, G. *Fluid Mechanics*, 6th ed.; Academic Press: New York, NY, USA, 2015; ISBN 9780124059351. [CrossRef]

50. Ferziger, J.H.; Peric, M. *Computational Methods for Fluid Dynamics*; Springer: Berlin/Heidelberg, Germany, 2002; ISBN 9783540420743. [CrossRef]
51. Márquez Damián, S. Description and Utilization of InterFoam Multiphase Solver. Universidad Nacional del Litoral, Argentina. Available online: http://infofich.unl.edu.ar/upload/5e6dfd7ff282e2deabe7447979a16d49b0b1b675.pdf (accessed on 5 December 2021).
52. InterFoam. in the Unofficial OpenFOAM Wiki. Available online: https://openfoamwiki.net/index.php/InterFoam (accessed on 5 December 2021).
53. Raeini, A.Q.; Blunt, M.J.; Bijeljic, B. Modelling two-phase flow in porous media at the pore scale using the volume-of-fluid method. *J. Comput. Phys.* **2012**, *231*, 5653–5668. [CrossRef]
54. Jasak, H.; Weller, H.G. *Interface-Tracking Capabilities of the InterGamma Differencing Scheme*; Technical Report; Imperial College of Science, Technology and Medicine, University of London: London, UK, 1995.
55. Roenby, J.; Bredmose, H.; Jasak, H. A computational method for sharp interface advection. *R. Soc. Open Sci.* **2016**, *3*, 160405. [CrossRef] [PubMed]
56. Gamet, L.; Scala, M.; Roenby, J.; Scheufler, H.; Pierson, J.-L. Validation of volume-of-fluid OpenFOAM® isoAdvector solvers using single bubble benchmarks. *Comput. Fluids* **2020**, *213*, 104722. [CrossRef]
57. Krause, F.; Li, X.; Fritsching, U. Simulation of droplet-formation and -interaction in emulsification processes. *Eng. Appl. Comput. Fluid Mech.* **2011**, *5*, 406–415. [CrossRef]
58. Wu, Z.; Hao, J.; Lu, J.; Xu, L.; Hu, G.; Floryan, J.M. Small droplet bouncing on a deep pool. *Phys. Fluids* **2020**, *32*, 012107. [CrossRef]
59. Rodriguez, F.; Mesler, R. The penetration of drop-formed vortex rings into pools of liquid. *J. Colloid Interface Sci.* **1988**, *121*, 121–129. [CrossRef]
60. Pumphrey, H.C.; Elmore, P.A. The entrainment of bubbles by drop impacts. *J. Fluid Mech.* **1990**, *220*, 539–567. [CrossRef]
61. Prosperetti, A.; Oguz, H.N. The impact of drops on liquid surfaces and the underwater noise of rain. *Annu. Rev. Fluid Mech.* **1993**, *25*, 577–602. [CrossRef]
62. Leng, J.L. Splash formation by spherical drops. *J. Fluid Mech.* **2001**, *427*, 73–105. [CrossRef]
63. Brutin, D. Drop impingement on a deep liquid surface: Study of a crater's sinking dynamics. *C. R. Mec.* **2003**, *331*, 61–67. [CrossRef]
64. Bisighini, A.; Cossali, G.E.; Tropea, C.; Roisman, I.V. Crater evolution after the impact of a drop onto a semi-infinite liquid target. *Phys. Rev. E* **2010**, *82*, 036319. [CrossRef] [PubMed]
65. Fedorchenko, A.I.; Wang, A.-B. On some common features of drop impact on liquid surfaces. *Phys. Fluids* **2004**, *16*, 1349–1365. [CrossRef]

Article

Puncture of a Viscous Liquid Film Due to Droplet Falling

Viktor G. Grishaev [1,*], Ivan K. Bakulin [1], Alidad Amirfazli [2] and Iskander S. Akhatov [1]

[1] Center for Materials Technologies, Skolkovo Institute of Science and Technology, Bolshoy Boulevard 30, bld. 1, Moscow 121205, Russia; bakulin.ik@phystech.edu (I.K.B.); i.akhatov@skoltech.ru (I.S.A.)
[2] Department of Mechanical Engineering, York University, Toronto, ON M3J 1P3, Canada; alidad2@yorku.ca
* Correspondence: v.grishaev@skoltech.ru

Abstract: Droplet impact may rupture a liquid film on a non-wettable surface. The formation of a stable dry spot has only been studied in the inviscid case. Here, we examine the break-up of viscous films, and demonstrate the importance and role of the viscous dissipation in both film and droplet. A new model was therefore proposed to predict the necessary droplet energy to create a dry spot. It also showed that the dissipation contribution in film dominates when the ratio of the thicknesses to drop diameter is larger than 7/4.

Keywords: film rupture; drop impact; viscous dissipation

1. Introduction

Liquid films cover solid surfaces in many natural and industrial processes. Often films are exposed to falling droplets, for example, during raining, cooling, spraying or lubrication processes. To understand the interaction of droplets with films, many studies focus on drop impact outcomes (splashing, droplet deposition or rebound) or film deformations during these processes (see, e.g., [1–9]). Despite many works, less attention is paid to drop impact leading to film rupture, and more specifically the conditions at which a stable hole is formed. Understanding this latter point is important as it can affect the above applications, e.g., in lubrication, leading to areas that are not covered by the viscous film.

The rupture of a liquid film on a solid surface has also been considered in different aspects and conditions. For example, it is found that short- and long-range interfacial forces determine dewetting patterns of thin (<60 nm) liquid polymer films [10]. The motion of dry spots in thicker (1.5–4 mm) films of aqueous glycerol solution is described well by a lubrication model [11]. A dry spot may also appear in a dynamic liquid film forming under droplet impact onto a dry solid surface. In this case, the dry spot formation is determined by impact velocities, surface roughness and the size of defects presenting on solid surface [12–14]. Besides dry spot dynamics, it has been revealed that a stable dry spot exists in a liquid film, when its diameter is bigger than a critical value [15]. Although the various aspects of film rupture have been considered, the case due to droplet falling remains less studied.

Up to now, film rupture due to droplet falling has been considered for the cases when a dry spot growth is induced by a surface tension inhomogeneity or impact itself. If a droplet has a miscible liquid with a low surface tension, then its falling creates a local reduction of film surface tension. Such reduction is enough to puncture 1–100 μm thick water films spontaneously due to emerging Marangoni flows [16]. The rupture of thicker films has been considered for aqueous films and droplets. In this case, a stable dry spot forms if the impact dynamics leads to the appearance of a critical size crater [17] or hole [18]. As the impact-induced film rupture is observed at high Reynolds numbers (>>1), the fluid viscosity was neglected in the models developed to date.

Thus, the film rupture under drop impact has been studied only for the inviscid case. However, many applications involve fluids with viscosities considerably higher than water.

To understand the contribution of fluid viscosity, we experimentally studied the puncture of viscous films due to a falling drop made of the same fluid. Furthermore, we evaluate droplet energy required to form a stable dry spot.

2. Materials and Methods

The film rupture experiments were carried out with aqueous glycerol solutions. We also used the data for distilled water from [17]. The properties of the fluids are presented in Table 1. We choose the specific mass fractions of glycerol to have a suitable range of viscosities. Additionally, the specific fraction of glycerol was selected to be able to identify the threshold of the film rupture (see below).

The test substrate was aluminum (50 mm × 50 mm × 2 mm) covered with a superhydrophobic coating (NeverWet, Rust-Oleum, Vernon Hills, IL, USA). We used the superhydrophobic surface as the film rupture will be the most profoundly observed. The static contact angles of considered fluids on the substrates are given in Table 1.

Our experimental set-up consists of a syringe pump with a test liquid, a Petri dish with the test substrate glued to its floor, which was then covered by the test liquid, and a high-speed camera, as shown in Figure 1; further details can be found in [17]. The syringe pump dispenses the fluid slowly to form a droplet at the needle tip. The detached droplet falls on a liquid film covering the substrate in the Petri dish. The high-speed camera records the collision of the droplet with the film and subsequent formation (or not) of a stable dry spot.

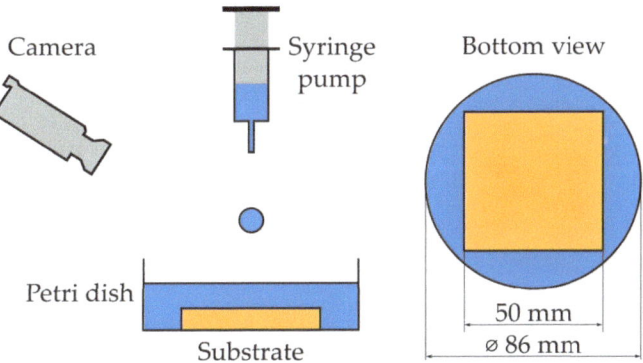

Figure 1. Schematic of the film rupture experiment.

The film thickness is set by forming a thick fluid layer on the substrate at the beginning and then removing the excessive volume. The addition and removal of liquid volumes are done with a digital pipette. The volume of removed liquid is determined using a calibration curve, that was obtained using a wet film thickness gauge (see details in [17]).

The droplet velocity is varied by changing the distance between the needle tip and the Petri dish. The velocity of a falling droplet and its diameter were determined from side view images captured by the high-speed camera. The details of the image processing can be found in [17,19]. The ranges of considered experimental conditions are presented in Table 2.

Table 1. Properties of experimental fluids (water and aqueous glycerol solutions) at 25 °C.

Glycerol, % Wt.	Viscosity [1], mPa·s	Density [1], kg·m^{-3}	Surface Tension, mN·m^{-1}	Static Contact Angle on Test Substrates, °
0	0.89 ± 0.02	1 ± 0.0002	72.0 ± 0.8	169 ± 2
68 ± 0.5	15 ± 0.9	1.17 ± 0.002	66.4 ± 0.9	167 ± 2
71 ± 0.5	20 ± 1.4	1.18 ± 0.002	65.1 ± 0.8	166 ± 2

[1] Values were taken from [20–22].

Table 2. Conditions of the film rupture experiments. Weber number, We, and Reynolds number, Re, were calculated by using film thickness as a characteristic length.

Glycerol, % Wt.	Droplet Diameter, mm	Droplet Velocity, m·s^{-1}	Film Thickness, mm	We	Re
0	2.0 ± 0.1	0.7–3.3	0.7–4.3	5–657	566–16,093
68 ± 0.5	1.9 ± 0.1	1.1–3.8	0.2–3.4	4–870	14–1013
71 ± 0.5	1.9 ± 0.1	1.2–5.1	0.7–3.8	25–1787	72–1137

3. Results

3.1. Experimental Observations

The drop impact leads to the formation of a crater in a liquid film; Figure 2 shows an example of an image sequence. At 16.7 ms, the thin film ruptures at the bottom of the crater, and a dry spot appears. Then, the rim of the crater diverged from the dry spot forming ripples. The left dry spot enlarged to an equilibrium size (it is not shown in the figure). For aqueous glycerol solutions, the rupture of a crater bottom always causes the formation of a stable dry hole at the end of the impact process.

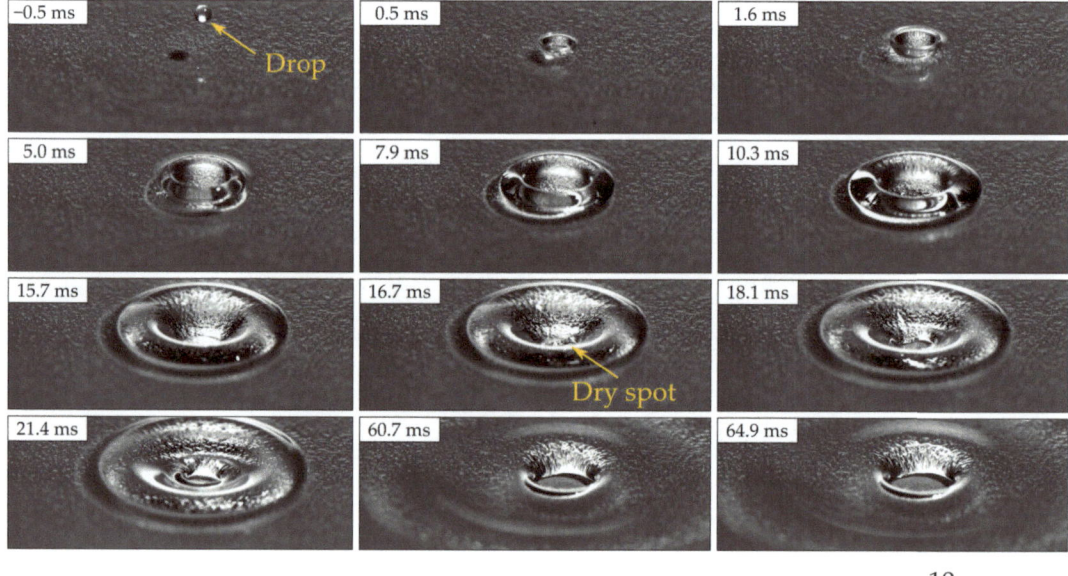

Figure 2. Dry spot formation in a liquid film on a superhydrophobic surface after droplet impact for the case of aqueous glycerol solution with viscosity $\mu = 15$ mPa·s. The droplet velocity—$U = 3.3$ m/s, diameter—$D = 1.9$ mm, and film thickness—$h = 3.1$ mm.

The behavior of the fluid films was studied for different droplet velocities and film thicknesses. For each case, the total energies (kinetic plus surface) of impacting droplets were calculated using measured droplet velocity U, diameter D and fluid properties (density, ρ, and surface tension, γ), i.e., as $\rho \frac{\pi}{12} D^3 U^2 + \gamma \pi D^2$. The calculated values leading or not to formation of a dry spot are shown in Figure 3. For comparison, we added a plot with data for pure water from [17]. Unlike water films, viscous films are broken up at higher droplet energies. Furthermore, they showed a new case when a dry spot may or may not appear, which depends on whether the thin film at the bottom of the crater ruptures or not (Figure 2). Note, water films rupture of the thin film at the bottom of the crater does not guarantee that a stable dry spot will be observed as the dry spot formed can close due to surface forces [17].

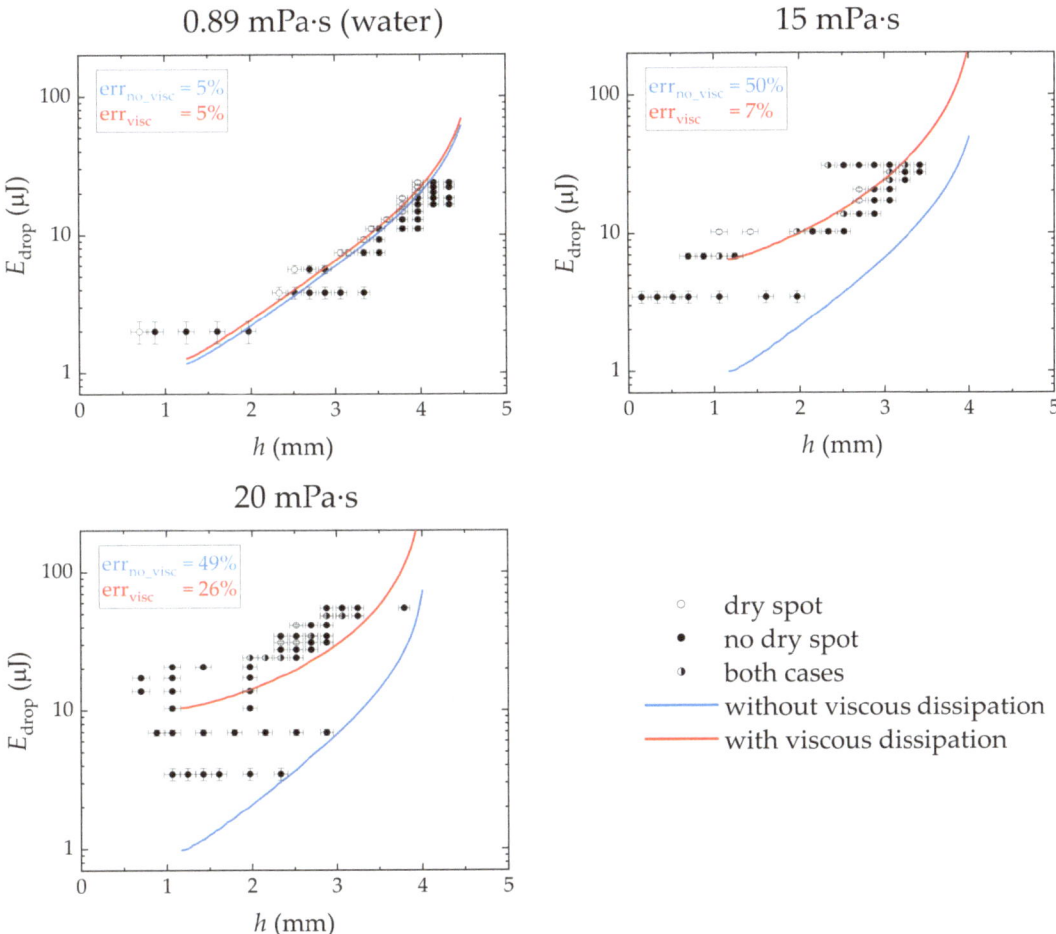

Figure 3. Energies of falling droplets, E_{drop}, leading or not to the formation of stable holes in the liquid films versus their thickness, h. Liquids are water (Data from [17]) and aqueous glycerol solutions. A joint legend outside the plots shows the notation of markers and lines. Legends inside of the plots present the separation errors of the crater model without and with the viscous dissipation of energy (blue and red fonts, respectively).

3.2. Evaluation of Required Droplet Energy

To describe the observed phenomena, as a first step, we used the approach proposed in [17] for the inviscid case. The approach in [17] assumes that a stable dry spot appears in a liquid film, if the droplet energy is sufficient to form a cylindrical crater with a critical size of a dry base (Figure 4). The critical area of the dry base, S_{crit}, was determined by calculation of the difference of free energies of the state of the film with and without the dry spot. The curve of the free energy difference versus the area of dry spot showed an energy barrier. If the film overcomes this barrier, then a state with a dry spot becomes favorable.

Therefore, the critical area of the crater base was set to be equal to the area corresponding to the peak of the barrier. The total change of surface and potential energies with the crater formation were calculated using volume conservation and the assumption that the width of the crater rim is equal to film thickness. This type of calculation, however, underestimated the values of droplet energies for aqueous glycerol solutions, i.e., viscous systems (Figure 3).

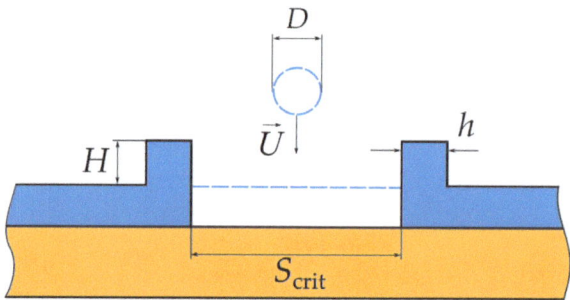

Figure 4. Model of the crater formation. The blue dashed lines show a droplet and film before impact.

To improve the model in [17], we considered an energy dissipation by viscous forces. For an incompressible Newtonian fluid, the viscous dissipation of energy is determined by the following integral over time, t, and fluid volume, V:

$$W = 2\mu \int_t \int_V e_{ij}^2 dV dt \quad (1)$$

where μ is the dynamic viscosity of the fluid and $e_{ij} = \frac{1}{2}\left(\frac{\partial u_i}{\partial x_j} + \frac{\partial u_j}{\partial x_i}\right)$ is the rate-of-strain tensor [23]. Solving of such integral is a complex task. Therefore, in drop impact, the evaluation of the viscous dissipation of energy is often done by using the following approximation:

$$W \sim \mu \left(\frac{U_c}{L_c}\right)^2 \Omega_c t_c \quad (2)$$

where U_c is the order of velocity change over the characteristic distance, L_c, for the fluid volume, Ω_c, during the time, t_c [24,25]. As the viscous dissipation of energy takes place during the droplet and film deformation, it can be written as the following summation with each term to be evaluated separately:

$$W = W_{\text{drop}} + W_{\text{film}} \quad (3)$$

For the deforming droplet, the parameters for the evaluation of the viscous dissipation of energy are:

$$U_c \sim U;\ L_c \sim D;\ \Omega_c \sim D^3;\ t_c \sim \frac{D}{U} \quad (4)$$

Then, the viscous dissipation of energy in a deforming droplet is proportional to

$$W_{\text{drop}} \sim \mu \left(\frac{U}{D}\right)^2 D^3 \frac{D}{U} \sim \mu U D^2 \quad (5)$$

In turn, the viscous dissipation of energy in the film happens due to formation of the crater, i.e., liquid being displaced from the bottom of the crater; as such, the relevant parameters are:

$$U_c \sim U;\ L_c \sim h;\ \Omega_c \sim h S_{\text{crit}};\ t_c \sim \frac{h}{U} \quad (6)$$

which leads to:

$$W_{\text{film}} \sim \mu \left(\frac{U}{h}\right)^2 h S_{\text{crit}} \frac{h}{U} \sim \mu U S_{\text{crit}} \quad (7)$$

Therefore, the total viscous dissipation of energy can be written as:

$$W = A\mu U D^2 + B\mu U S_{\text{crit}} \quad (8)$$

where A and B are non-dimensional constants.

Taking into account the viscous dissipation, the droplet energy for the film puncture is determined as:

$$E_{\text{drop}} = E_{\text{inviscid}} + W = E_{\text{inviscid}} + A\mu UD^2 + B\mu US_{\text{crit}} \quad (9)$$

where the term E_{inviscid} corresponds to the surface and potential energies arising from the critical size of the formed crater as in [17]. The constants A and B can be found by fitting Equation (9) to the experimental data.

The suitable values for constants A and B were found by varying them and searching for the minimum average separation error of Equation (9) in relation to all experimental points. As different numbers of experiments resulted in observing a dry spot or not, the separation error was calculated with the equalization on both classes as:

$$\text{separation error} = 0.5\frac{N_{\text{ds}}^-}{N_{\text{ds}}^{\text{all}}} + 0.5\frac{N_{\text{nds}}^-}{N_{\text{nds}}^{\text{all}}} \quad (10)$$

where N_{ds}^- is the number of experimental points for which dry spot was observed, but droplet energy was lower than the value of Equation (9), N_{nds}^- is the number of experimental points for which dry spot was not observed, but droplet energy was higher than the value of Equation (9); $N_{\text{ds}}^{\text{all}}$ and $N_{\text{nds}}^{\text{all}}$ are total number of experimental points with and without dry spot, respectively. The separation errors were averaged for all considered viscosities:

$$\text{average separation error} = \frac{1}{3}\sum_{j=1}^{3} \text{separation error}_j \quad (11)$$

Considering all test fluids, a minimum average separation error was seen for $A = 57.5$ and $B = 4.4$ (Figure 5).

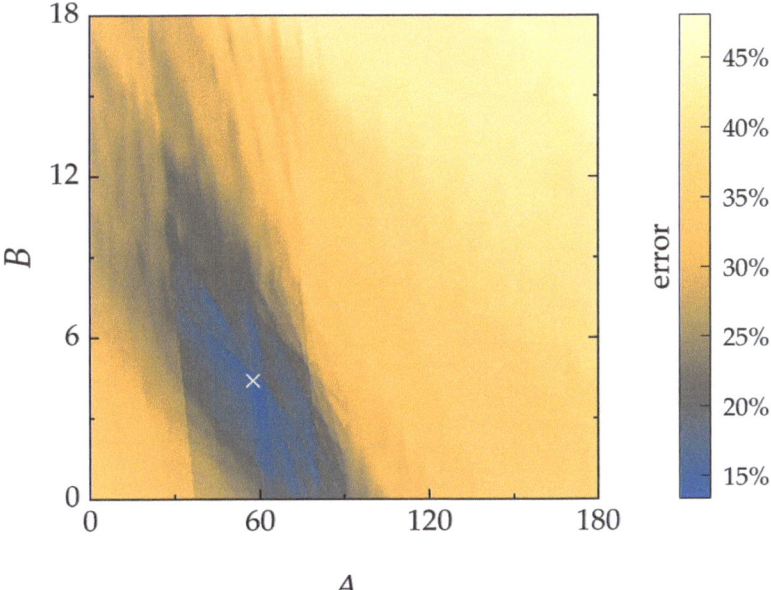

Figure 5. Distribution of the average separation error of the crater model with viscous dissipation of energy (Equation (9)) at different constants A and B. The white cross shows the constants (57.5; 4.4) at which the error has minimum value.

Using the above values for A and B, the plots of Equation (9) are shown by blue solid lines in Figure 3. The plotted curves separate the cases with and without dry spots much better than the model without the viscous dissipation of energy. To compare viscous dissipation in the film and in the drop, we plotted their ratios for all three fluids (see Figure 6).

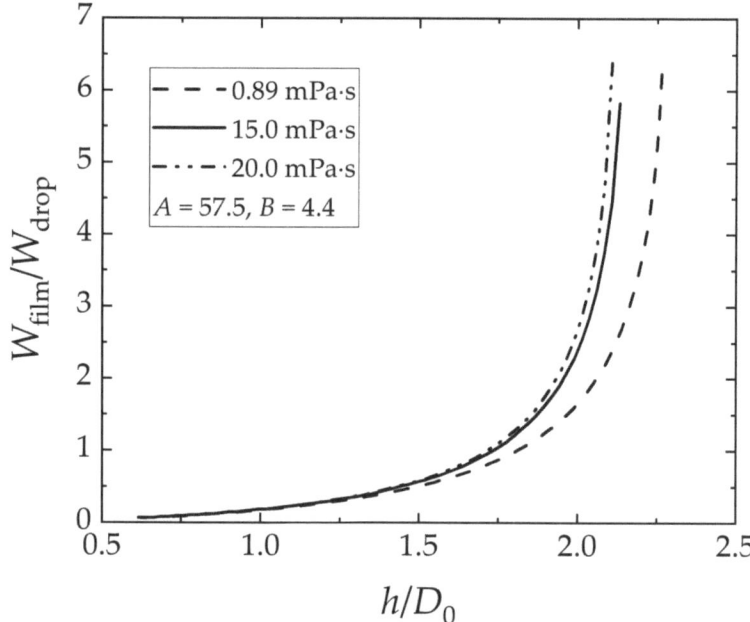

Figure 6. Ratio of viscous dissipation of energy in the film and the drop versus the film thickness. The plotted curves are generated using the values found for parameters A, B (Figure 5).

4. Discussion

The results show that the crater model of [17] underestimates the droplet energy for viscous films. The deviation of the thermodynamic model developed in [17] from experimental observation for the droplet energy needed to create a dry spot on a film increases as the viscosity of the film increases; for example, the underestimation for film viscosity of 20 mPa·s can be as large as an order of magnitude. Addition of the viscous dissipation term to the model developed in [17] substantially improved the prediction for the droplet energy needed to create a dry spot in a film.

The fitting of the developed model to experimental data allows us to evaluate the contribution of the viscous dissipation in a drop and a film (Figure 6). As the critical area of the crater bottom, S_{crit}, depends on the film thickness, the viscous dissipation in a film increased with increasing of the film thickness. The ratio of film to drop dissipation are not the same for the different viscosities, because of the difference in other fluid properties that influence the critical area. Nevertheless, the results in Figure 6 show that the contribution of viscous dissipation of energy in the film dominates for cases where film thicknesses to droplet diameter ratio are larger than 7/4.

The remaining error of the fitted model is likely due to the assumptions that are made in developing the model. For example, the real shape of the crater deviates from the cylindrical form assumed in the model, see Figure 4 (for model) and Figure 2 (the snapshots at 10.3 ms from experiments). Furthermore, the error may be due to the approximation of viscous dissipation. Another contributing factor can be a new mechanism of the film rupture (see below).

For the viscous fluids, we observed a new case when film rupture may happen or not at the same impact conditions. At such conditions, the falling droplet forms a crater with a thin film at the bottom. When the rim of the crater starts to settle, it not only diverges from an impact point but also starts to backfill the bottom of the crater (cf., snapshots from 10.3 to 16.7 ms in Figure 2). If a thin film at the bottom of the crater does not have sufficient time to break, then the diverging and settling crater rim fills the cavity. However, if the thin film has sufficient time to rupture before thickening, then the dry spot always prevents the filling of the cavity and the crater rim only moves outwards from the impact point with broadening and ripples observed (snapshots from 21.4 to 64.9 ms in Figure 2). Such peculiarity of the film rupture was not mentioned in the previous works [17,18], and it probably should be considered in further modelling.

5. Conclusions

It has been shown that to calculate the external energy needed to break a liquid film over a surface, one needs to consider viscous dissipation of energy, if liquids with a low viscosity, other than water, are used. As such, a new model was developed to allow prediction of the energy needed. Our semi-analytical model correctly predicts the order of magnitude of the needed droplet energy as the external stimuli to create a dry spot in a liquid film. Our experimental results show that droplet energy for the film rupture (the formation of a stable dry spot) increases with viscosity. This is correctly predicted by the new model developed here. Furthermore, using the model developed, we understood that the dissipation occurs in both the droplet and the film, but in non-equal terms. It is shown that the dissipation in films will be dominant for cases where the ratio of the thicknesses to drop diameter is larger than 7/4.

Author Contributions: Conceptualization, V.G.G., I.K.B. and A.A.; methodology, V.G.G., I.K.B. and A.A.; software, I.K.B.; validation, V.G.G. and I.K.B.; formal analysis, I.K.B.; investigation, I.K.B.; resources, I.S.A.; data curation, V.G.G. and I.K.B.; writing—original draft preparation, V.G.G.; writing—review and editing, A.A., I.K.B. and I.S.A.; visualization, I.K.B.; supervision, I.S.A.; project administration, V.G.G.; funding acquisition, V.G.G. All authors have read and agreed to the published version of the manuscript.

Funding: This research was funded by the Russian Science Foundation, grant number 19-79-10272.

Institutional Review Board Statement: Not applicable.

Informed Consent Statement: Not applicable.

Data Availability Statement: The data presented in this study are available on request from the corresponding author.

Acknowledgments: We thank Ivan Borodulin for fabricating the superhydrophobic substrates.

Conflicts of Interest: The authors declare no conflict of interest. The funders had no role in the design of the study; in the collection, analyses, or interpretation of data; in the writing of the manuscript, or in the decision to publish the results.

References

1. Tropea, C.; Marengo, M. The Impact of Drops on Walls and Films. *Multiph. Sci. Technol.* **1999**, *11*, 19–36. [CrossRef]
2. Cossali, G.E.; Brunello, G.; Coghe, A.; Marengo, M. Impact of a Single Drop on a Liquid Film: Experimental Analysis and Comparison With Empirical Models Experimental Set-Up. In Proceedings of the Italian Congress of Thermofluid Dynamics UIT, Ferrera, Italy, 30 June–2 July 1999.
3. Huang, Q.; Zhang, H. A Study of Different Fluid Droplets Impacting on a Liquid Film. *Pet. Sci.* **2008**, *5*, 62–66. [CrossRef]
4. van Hinsberg, N.P.; Budakli, M.; Göhler, S.; Berberović, E.; Roisman, I.V.; Gambaryan-Roisman, T.; Tropea, C.; Stephan, P. Dynamics of the Cavity and the Surface Film for Impingements of Single Drops on Liquid Films of Various Thicknesses. *J. Colloid Interface Sci.* **2010**, *350*, 336–343. [CrossRef] [PubMed]
5. Josserand, C.; Ray, P.; Zaleski, S. Droplet Impact on a Thin Liquid Film: Anatomy of the Splash. *J. Fluid Mech.* **2016**, *802*, 775–805. [CrossRef]

6. Kuhlman, J.M.; Hillen, N.L. Droplet Impact Cavity Film Thickness Measurements versus Time after Drop Impact and Cavity Radius for Thin Static Residual Liquid Layer Thicknesses. *Exp. Therm. Fluid Sci.* **2016**, *77*, 246–256. [CrossRef]
7. Chen, B.; Wang, B.; Mao, F.; Wen, J.; Tian, R.; Lu, C. Experimental Study of Droplet Impacting on Inclined Wetted Wall in Corrugated Plate Separator. *Ann. Nucl. Energy* **2020**, *137*, 107155. [CrossRef]
8. Ribeiro, D.F.S.; Silva, A.R.R.; Panão, M.R.O. Insights into Single Droplet Impact Models upon Liquid Films Using Alternative Fuels for Aero-Engines. *Appl. Sci.* **2020**, *10*, 6698. [CrossRef]
9. Lakshman, S.; Tewes, W.; Harth, K.; Snoeijer, J.H.; Lohse, D. Deformation and Relaxation of Viscous Thin Films under Bouncing Drops. *J. Fluid Mech.* **2021**, *920*, A3. [CrossRef]
10. Seemann, R.; Herminghaus, S.; Jacobs, K. Dewetting Patterns and Molecular Forces: A Reconciliation. *Phys. Rev. Lett.* **2001**, *86*, 5534–5537. [CrossRef]
11. Bankoff, S.G.; Johnson, M.F.G.; Miksis, M.J.; Schluter, R.A.; Lopez, P.G. Dynamics of a Dry Spot. *J. Fluid Mech.* **2003**, *486*, 239–259. [CrossRef]
12. Dhiman, R.; Chandra, S. Rupture of Thin Films Formed during Droplet Impact. *Proc. R. Soc. A Math. Phys. Eng. Sci.* **2010**, *466*, 1229–1245. [CrossRef]
13. Biance, A.L.; Pirat, C.; Ybert, C. Drop Fragmentation Due to Hole Formation during Leidenfrost Impact. *Phys. Fluids* **2011**, *23*, 022104. [CrossRef]
14. Rashidian, H.; Sellier, M.; Mandin, P. Dynamic Wetting of an Occlusion after Droplet Impact. *Int. J. Multiph. Flow* **2019**, *111*, 264–271. [CrossRef]
15. Lv, C.; Eigenbrod, M.; Hardt, S. Stability and Collapse of Holes in Liquid Layers. *J. Fluid Mech.* **2018**, *855*, 1130–1155. [CrossRef]
16. Neél, B.; Villermaux, E. The Spontaneous Puncture of Thick Liquid Films. *J. Fluid Mech.* **2018**, *838*, 192–221. [CrossRef]
17. Grishaev, V.; Bakulin, I.; Amirfazli, A.; Borodulin, I.; Akhatov, I. Energy of a Drop Required to Break a Liquid Film. *Langmuir* **2021**, *37*, 10433–10438. [CrossRef]
18. Ni, Z.; Chu, F.; Li, S.; Luo, J.; Wen, D. Impact-Induced Hole Growth and Liquid Film Dewetting on Superhydrophobic Surfaces. *Phys. Fluids* **2021**, *33*, 112113. [CrossRef]
19. Grishaev, V. Impact of Particle-Laden Drops on Substrates with Various Wettability. Ph.D. Thesis, Université Libre de Bruxelles, Bruxelle, Belgium, 2015.
20. Calculate Density and Viscosity of Glycerol/Water Mixtures. Available online: http://www.met.reading.ac.uk/~{}sws04cdw/viscosity_calc.html (accessed on 15 April 2022).
21. Cheng, N.-S. Formula for the Viscosity of a Glycerol–Water Mixture. *Ind. Eng. Chem. Res.* **2008**, *47*, 3285–3288. [CrossRef]
22. Volk, A.; Kähler, C.J. Density Model for Aqueous Glycerol Solutions. *Exp. Fluids* **2018**, *59*, 75. [CrossRef]
23. Acheson, D.J. Elementary Fluid Dynamics. In *Oxford Applied Mathematics and Computing Science Series*; Oxford University Press: Oxford, UK, 2005; ISBN 0198596790.
24. Chandra, S.; Avedisian, C.T. On the Collision of a Droplet with a Solid Surface. *Proc. R. Soc. A Math. Phys. Eng. Sci.* **1991**, *432*, 13–41. [CrossRef]
25. Huang, H.M.; Chen, X.P. Energetic Analysis of Drop's Maximum Spreading on Solid Surface with Low Impact Speed. *Phys. Fluids* **2018**, *30*, 022106. [CrossRef]

Article

Experimental Characterization of the Wettability of Coated and Uncoated Plates for Indirect Evaporative Cooling Systems

Roberta Caruana [1,*], Stefano De Antonellis [1], Luca Marocco [1], Paolo Liberati [2] and Manfredo Guilizzoni [1]

[1] Politecnico di Milano, Department of Energy, Via Lambruschini 4, 20156 Milano, Italy
[2] Recuperator S.p.A., Via Valfurva 13, 20027 Rescaldina, Italy
* Correspondence: roberta.caruana@polimi.it

Abstract: Indirect Evaporative Cooling (IEC) is a very promising technology to substitute and/or integrate traditional air conditioning systems, due to its ability to provide cooling capacity with limited power consumption. Literature studies proved that a higher wettability of the IEC plates corresponds to better performance of the system. In this work, wettability of three different surfaces used for IEC systems plates—uncoated aluminum alloy (AL), standard epoxy coating (STD), and a hydrophilic lacquer (HPHI)—is studied and characterized in terms of static and dynamic contact angles. The static contact angle resulted to be the lowest for the HPHI surface (average 69°), intermediate for the STD surface (average 75°), and the highest for the AL surface (average 89°). The analysis of the dynamic contact angles showed that their transient behavior is similar for all the surfaces, and the advancing and receding contact angles obtained are consistent with the results of the static analysis. These results will be useful as input parameters in models aimed at predicting the IEC system performance, also using computational fluid dynamics.

Keywords: wettability; contact angle; indirect evaporative cooling; coating; image processing

Citation: Caruana, R.; De Antonellis, S.; Marocco, L.; Liberati, P.; Guilizzoni, M. Experimental Characterization of the Wettability of Coated and Uncoated Plates for Indirect Evaporative Cooling Systems. *Fluids* 2023, 8, 122. https://doi.org/10.3390/fluids8040122

Academic Editors: Sourabh V. Apte and D. Andrew S. Rees

Received: 3 March 2023
Revised: 22 March 2023
Accepted: 31 March 2023
Published: 3 April 2023

Copyright: © 2023 by the authors. Licensee MDPI, Basel, Switzerland. This article is an open access article distributed under the terms and conditions of the Creative Commons Attribution (CC BY) license (https://creativecommons.org/licenses/by/4.0/).

1. Introduction and State of the Art

Nowadays, heating, ventilation, and cooling systems have become a necessity for people living in both developed and developing countries, thus representing a significant fraction of primary energy use on a world basis [1]. In particular, the energy consumption due to the cooling of indoor environments is becoming much more relevant than a few decades ago, both for the higher request of adequate comfort conditions, and for the increase of the average outdoor temperature due to the climate change.

In this area, evaporative cooling is of great interest, as it allows to cool the air through the evaporation of water [1], constituting the basis of technologies which can guarantee significant primary energy saving during summer air conditioning. In particular, Indirect Evaporative Cooling (IEC) is one of the most promising solutions to replace and/or integrate traditional air conditioning systems [2]. In fact, indirect systems, unlike the direct ones, are able to reduce the air temperature without increasing its humidity ratio, thus promoting preservation of comfort conditions in the built environment.

In an IEC system, the primary (product) air flows over the dry side of a plate, and the secondary (working) air flows over the opposite wet side of the same plate. The secondary air absorbs heat from the primary air through the aid of water evaporation on the wet surface of the plate, thus cooling down the primary air, which is the aim of the system [3]. The complete IEC system is made of a series of plates, constituting numerous channels in which primary and secondary air flow alternately.

In the last 25 years, researchers have been trying to model the behavior of IEC systems, by building analytical, phenomenological, and numerical models. In particular, Alonso et al. [4] developed the first relevant analytical model in this field, based on some simplifying assumptions which made the model quite easy to implement. This first model was the basis of the work of Chen et al. [5], who introduced the effects of condensation of fresh air, and of the study of Heidarinejad and Moshari [6], whose model takes into account also longitudinal heat conduction and the effects of the change of water temperature along the plates surfaces in cross-flow configuration. Then, in 2017, De Antonellis et al. [7] introduced a new phenomenological model which also considers the effect of the adiabatic cooling of the working air in the inlet plenum and the wettability of the plates. This model opens a new perspective, as it considers the wettability of the surfaces as a very important aspect to consider for the evaluation of the performance of IEC systems. Finally, in 2021, Adam et al. [8] proposed a numerical model of cross-flow IEC systems, which takes into account again the wettability of the surfaces.

Several literature papers have shown that performance of IEC systems is strongly related to the formation of uniform water layers on the plates. In particular, Chua et al. [9] proved that a high wetting condition significantly increases the IEC system performance by decreasing the primary air outlet temperature and thus increasing the wet bulb effectiveness of the system. Furthermore, as previously mentioned, De Antonellis et al. [7] highlighted the importance of taking into account the surface wettability factor in the modeling of IEC systems, also by evaluating its effect on the performance of the system. Finally, Guilizzoni et al. [10] analyzed the wettability of two types of plates, showing that an increase of wettability of the plates leads to an increase in the performance of the system. As a consequence, in order to improve the performance of a system, it is necessary to increase as much as possible the surface wettability of the plates.

It has been shown that covering the IEC plates with suitable coatings changes their surface characteristics, and thus the performance of the entire system [11]. In particular, using hydrophilic coatings turned out to be a very effective way to increase the surface wettability [10], namely to reduce both the static and dynamic contact angles that the water drops form when laying on the plates, thus favoring the surface wetting.

The scope of this work is to experimentally analyze the static and dynamic contact angles of water drops in contact with three different surfaces belonging to IEC systems plates, in order to fully characterize the wettability of these surfaces, and to provide the values of the contact angles to be used in models, both simplified and based on computational fluid dynamics, aimed at predicting the IEC system behavior and performance. This piece of information is extremely relevant because the only data currently available in literature about this kind of surfaces are those reported in [10], which refer to a much less extensive experimental campaign involving only static contact angles of coated surfaces. Therefore, further investigations were needed in this field.

The three surface that were considered in this work are: the aluminum uncoated surface (AL), the same surface covered with a standard epoxy coating (STD), and again the same surface covered with a hydrophilic lacquer (HPHI). Firstly, the static contact angle has been evaluated by deposing sessile drops on the three surfaces, taking macrophotography pictures of these drops, and measuring the contact angle by using the Axisymmetric Drop Shape Analysis (ADSA) technique [12]. Secondly, dynamic contact angles (advancing and receding) were estimated in two ways: for sessile drops on vertical surfaces, and by means of high-speed videos of the drop impacts on the three surfaces. Finally, the results have been related to the IEC systems performance measured in a previous experimental campaign [10], in order to lay the foundations for future wettability analyses.

2. Materials, Methods, and Motivation

In this section, the experimental procedure and setup used for each of the aforementioned analyses are described. Moreover it is highlighted the reason why all these analyses are needed for an in-depth characterization of the surface wettability, and how each of the analyzed angles can be related to the IEC system working processes.

2.1. Evaluation of the Static Contact Angle

The first quantity that is typically used to characterize the wettability of a surface is the static contact angle, measured for a sessile, namely gently deposed, drop [13]. For this reason, as previously mentioned, the static contact angle that the drops form with each of the three investigated surfaces has been evaluated.

The static or Young contact angle, θ, is defined as the angle formed by the tangent to the drop profile with the tangent to the solid surface profile in a plane where the normal vectors to both the liquid-gas and solid-gas interfaces are contained [14]. This angle is defined for horizontal, flat, and smooth surfaces, while if the surface does not fit in these requirements, different contact angle models are necessary [15]. For example, in order to predict the equilibrium contact angle of a drop which fully penetrates the asperities of a rough surface the Wenzel model is used [16], while the Cassie-Baxter model [17] is exploited for predicting the static contact angle of a drop when some air remains trapped between the drop and the asperities of the surface.

In order to evaluate the static contact angle, flat rectangular surface samples (with dimensions of few millimeters per side) were placed on suitable sample holders and prepared through a careful cleaning with alcohol, rinsing with distilled water, and complete drying. The samples were located on an anti-vibrating optical bench (Newport, *SA* Series, 1.2×0.80 m) with a carrying structure in aluminium alloy. Then, distilled water drops with volumes in the range 4–12 µL (to consider the dependence on the volume) were gently deposed on each sample. A high precision metering pump (Cole-Parmer Instrument Company, model *AD74900*) completed by suitable syringes (Hamilton) allowed to supply drops of controlled volume. Immediately after deposition, pictures of the drops were taken by using a Nikon *D90 SRL* digital camera equipped with a Nikkor 60 mm *F2.8 Micro* lens. Back illumination was provided by a 800 W halogen lamp equipped with a suitable diffusing screen. Finally, the contact angles were evaluated through the ADSA technique, whose details are thoroughly described in [12,18].

In short, the ADSA technique is based on the numerical fitting of the theoretical drop profile to the contour of experimental drops. The first is obtained by numerical integration of the classic Laplace-Young equation of capillarity [19]. The second is nowadays obtained by image processing of pictures of the drop-surface system.

Different versions of ADSA were developed along the years. In this work, the tests were conducted using the ADSA-P (perimeter) technique [12]. According to this method an objective function, to be minimized, is defined as the sum of the squares of the distances between the theoretical and the experimental drop profile points.

The experimental drop profile is extracted by a side view of the drop-surface sample system, using conventional edge detection operators. The numerical integration to obtain the theoretical drop profile, given by the Laplace-Young equation of capillarity, is usually done-under the assumption of drop axi-symmetry—in the peculiar arc length-turning angle coordinate system, that makes it very simple. From the best fit, the contact angle can be determined as the value of the turning angle at the intersection between the drop profile and the surface profile. The strategy employed in this work is to perform ADSA in the dimensionless version developed by Rotenberg et al. [20], with inclusion of some of the improvements by Cheng et al. [21], namely to fit the shape of the experimental drop to the theoretical drop profile using the Eötvös number as the adjustable parameter. The advantage of this version is that the only needed input is a side photograph of the drop surface system [14].

Practically, the measurement of contact angle consists of the following steps:

1. Pre-processing of the images (e.g., cropping if needed).
2. Image segmentation.
3. Extraction and smoothing of drop profile.
4. Fitting of the Laplace-Young equation to the experimental boundary.
5. Determination of the contact angle.

An a priori error propagation and uncertainty analysis of the ADSA technique is almost impossible to perform, as too many aspects are involved, from light and camera positions to camera resolution and sensor aspect ratio, to software parameters and operator's ability. However, the accuracy of the ADSA technique has already been validated many times in literature studies [12,21], and also specifically for the system used in this work by means of drops artificially generated by a calculator, with separate analysis of the effects of the parameters (including the ones of the software), and also on rendered drops, in order to replicate the problems of alignment and illumination [14]. Thus, the error made when using this technique results to be within 1.5° [18].

Figure 1 shows three example of sessile drops, one deposed on each of the investigated surfaces in horizontal orientation.

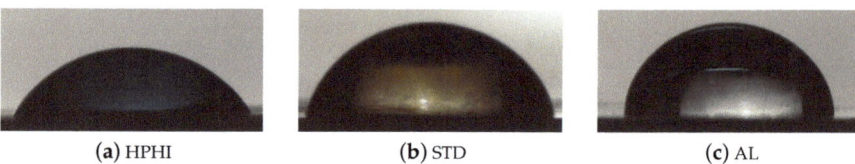

(a) HPHI   (b) STD   (c) AL

Figure 1. Examples of sessile drops on the three different surfaces investigated in horizontal orientation.

2.2. Evaluation of the Dynamic Contact Angles and Analysis of the Transient Behavior of the Drops

If a surface is not flat and horizontal, or if the drop is not gently deposed on the surface, it is not possible to evaluate the static or Young contact angle, so it is necessary to define the dynamic contact angles.

Focusing on the case of not gently deposed drops on a flat horizontal surface, it is possible to state that when a drop falls onto a surface, the contact angle changes with time, increasing and decreasing continuously until the drop reaches the equilibrium condition. In this scenario, two sets of dynamic contact angles can be defined: the advancing contact angles, θ_{adv}, which are the ones measured when the drop is spreading on the surface, and the receding contact angles, θ_{rec}, which are the ones measured when the drop is recoiling. The maximum and minimum values of these two sets are usually selected as the angles characterizing the dynamic behavior of the drop, and their difference $\theta_{adv} - \theta_{rec}$ is named contact angle hysteresis.

Dynamic angles can also be evaluated on sessile drops on vertical surfaces, even if the values obtained in this second way may differ from those obtained in the previous way.

For these reasons, the following step of the analysis was to evaluate the dynamic contact angles for the three investigated surfaces, both during drop impact and on vertical surfaces, and the transient behavior of these angles with time.

In order to perform the drop impact analysis, distilled water drops with the same range of volumes of the ones used for the static analysis were dropped from a fixed height on each of the three surfaces in horizontal orientation. Each sample was placed on sample holders and carefully cleaned as described in the previous section. The impact velocity was set to around 0.45 m s^{-1}. During the fall of each drop, a high-speed video of the side view of the drop-surface system was acquired, again using back illumination, through a Phantom Miro C110 camera, equipped with the same Nikkor 60 mm F2.8 Micro lens used for the static analysis. Then, the frames obtained from the acquired videos were processed to extract the drop contact angles. Further details about this procedure are described in [13].

Concerning the analysis of the dynamic contact angles on vertical surfaces, ten sessile distilled water drops (with the same volume range as before) were deposed on each of the investigated surfaces in vertical orientation, pictures of them were taken by using the same camera and lens of the static analysis, and their profiles were analyzed through image processing and polynomial interpolation (as the drops are no longer axisymmetric and ADSA cannot be used), in order to obtain the advancing and receding contact angles. The samples were again prepared as previously described.

Figure 2 shows three example of sessile drops, one deposed on each of the investigated surfaces in vertical orientation.

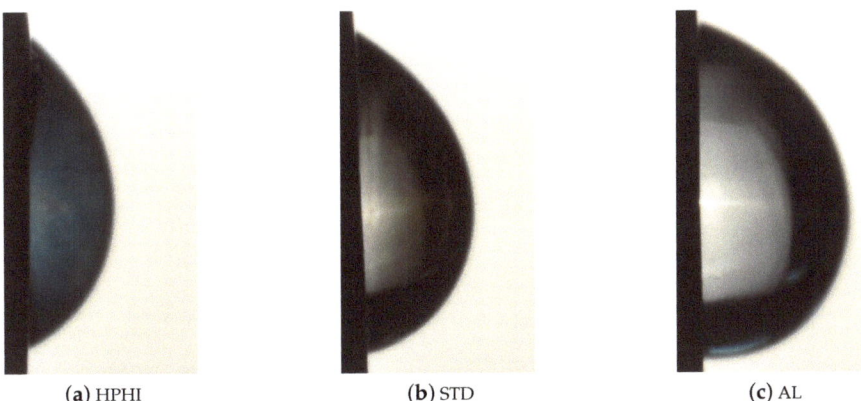

(a) HPHI   (b) STD   (c) AL

Figure 2. Examples of sessile drops on the three different surfaces investigated in vertical orientation.

3. Results and Discussion

In this section, the results of the previously described analyses are presented and discussed.

As previously mentioned, three surfaces were analyzed: uncoated aluminum alloy (AL), standard epoxy coating (STD), and a hydrophilic lacquer (HPHI). From the results of previous studies [10], the HPHI surface should be the one with the lowest contact angles, so highest wettability, and best performance.

3.1. Static Contact Angle

Due to manufacturing, the investigated surfaces are not perfectly smooth; on the contrary, their texture includes micro-grooves dominantly oriented in one direction. Such grooves may alter the contact angle uniformity along the triple line, due to the "pinning on sharp edges" (Gibbs effect) phenomenon. Therefore, the static contact angle was evaluated by taking pictures in two directions: with the lens parallel to these grooves and with the lens perpendicular to them. The drop deformation was evaluated from top views of the drops themselves and it can be considered small enough to still allow the use of the ADSA technique. For each of the 3 surfaces, 6 different plate samples were considered. On each sample, 20 drops for each orientation were deposed, for a total of 720 drops (240 for each surface).

Figure 3 shows the results for the static contact angle on the six samples of each surface, HPHI in Figure 3a, STD in Figure 3b, AL in Figure 3c, when the grooves are parallel to the lens (in blue), and when they are perpendicular to the lens (in red).

As usual in boxplots, the box represents the interval between the first and third quartiles of the data distribution, with the central line corresponding to the median. The external whiskers represent the maximum and minimum values of the data distribution, excluding the outliers data. The latter, represented by red crosses in Figure 3, are the values outside the box which are at a distance from the first and third quartiles greater than 1.5 times the size of the box itself. This representation allows about 99% coverage for normally distributed data.

From Figure 3, it is possible to notice that for all the surfaces there is a slight difference between the parallel and perpendicular orientations, and this difference strongly depends on the analyzed sample, so there is no "privileged" orientation.

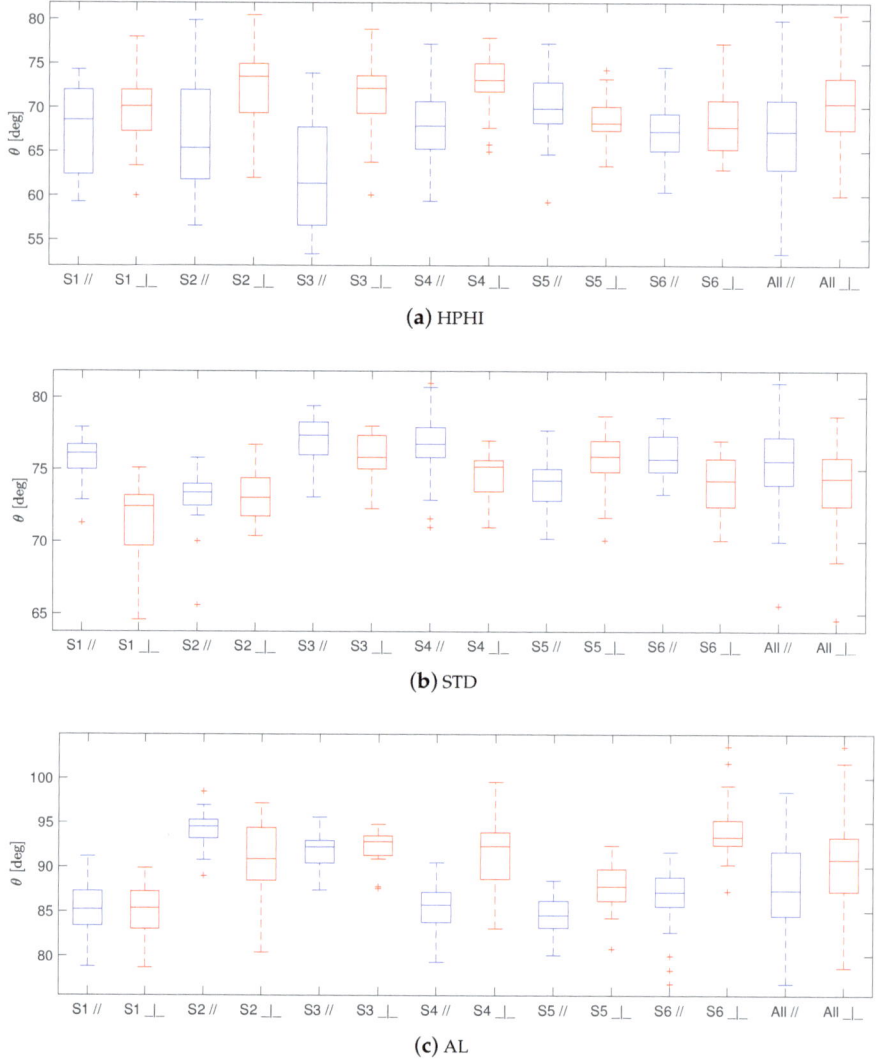

Figure 3. Boxplot representing the static contact angles obtained for the six samples of each of the three surfaces (S1, S2, S3, S4, S5, S6), and the overall results (All), with the camera lens parallel to the grooves (in blue) and perpendicular to them (in red).

Table 1 summarizes the overall results of the static contact angle analysis, in terms of average, median, and standard deviation values of the static contact angle for each of the three surfaces and for each of the two orientations, based on the "All" data of Figure 3a–c.

From this table, it is possible to notice that for all the surfaces, the difference between the parallel and perpendicular orientations in terms of average and median values is always less or at most equal to the corresponding standard deviation. Therefore, for subsequent analyses, the orientation of the grooves will be neglected for all the surfaces.

Figure 4 summarizes the overall results for the static contact angle, without considering the orientation of the grooves.

From this figure, it is possible to notice that the drops of the HPHI surface show the lowest static contact angle (global average: 69°, global median: 69°, global standard deviation: 5°), followed by the ones on the STD surface (global average: 75°, global median:

75°, global standard deviation: 3°), followed in turn by the ones on the AL surface, which shows the highest static contact angle (global average: 89°, global median: 89°, global standard deviation: 5°). These values are in good agreement with the results of previous analyses [10].

Table 1. Summary of the results of the static contact angle analysis.

Material	Orientation	Average	Median	Standard Deviation
HPHI	Parallel	67°	67°	6°
HPHI	Perpendicular	70°	70°	4°
STD	Parallel	75°	76°	2°
STD	Perpendicular	74°	74°	2°
AL	Parallel	88°	87°	5°
AL	Perpendicular	90°	91°	5°

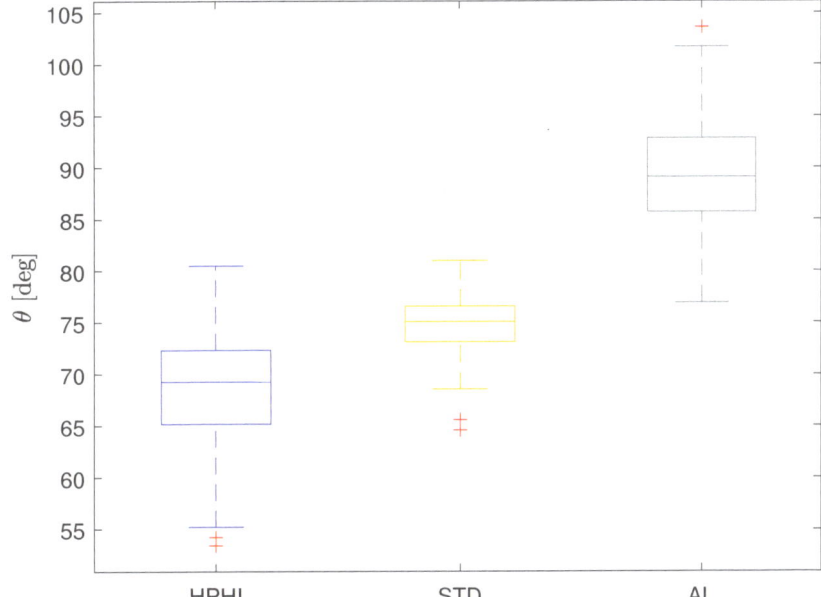

Figure 4. Boxplot representing the overall static contact angles for each of the three surfaces: HPHI in blue, STD in yellow, AL in grey.

The evolution of the drop profile in time was also evaluated by taking pictures of the drops on each of the three surfaces every 20 s for 3 min, confirming that the drop evaporation follows the well-known "constant contact area, varying contact angle" behavior, that is typical on wettable surfaces [22,23]. Therefore, the information obtained about the static contact angle can be considered exhaustive and reliable.

3.2. Dynamic Contact Angles and Transient Analysis

Figure 5 shows three sequences of frames obtained from the three high-speed videos of a drop falling on the HPHI surface, in Figure 5a, on the STD surface, in Figure 5b, and on the AL surface, in Figure 5c.

From this figure, it is possible to observe that the transient behavior of the drops falling on the three surfaces is almost the same, following the "usual" evolution for drop impacting surfaces with low-to-medium contact angles described in a large number of papers, e.g., [24–26]. Right after touching the surface, the drop takes on a particular shape, so that it seems to be divided into two or even three parts, and in this phase the contact angle is maximum (advancing contact angle). Then, the drop starts to recoil, and the contact angle decreases until reaching a minimum value (receding contact angle). After that, the contact angle increases and decreases continuously during drop oscillations (with contact angle hysteresis decreasing in time given the reduction of the liquid velocity, as expected), until reaching an equilibrium condition in which the contact angle is not changing anymore.

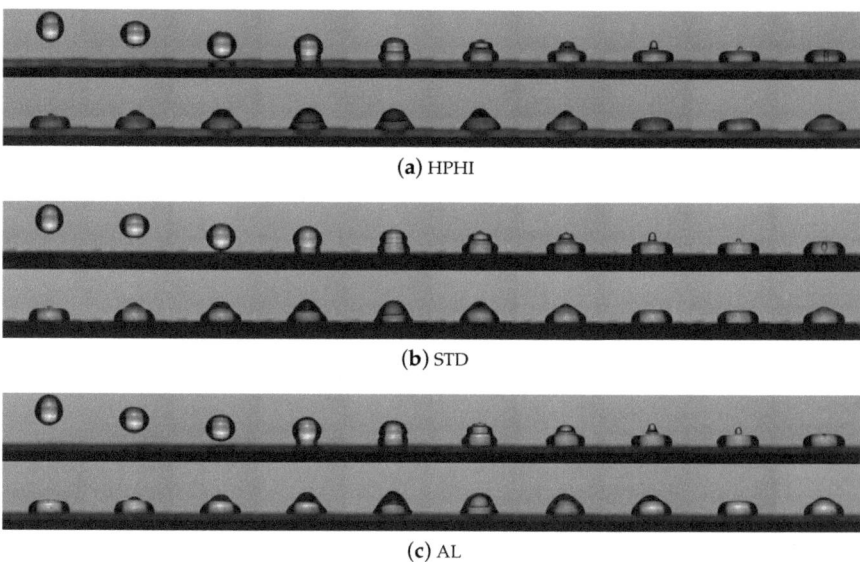

Figure 5. Frames from the high-speed video (at 909 fps) of a drop falling on each of the investigated surfaces.

In Figure 6 it is possible to observe an example of the trend of the contact angle over time for the drops falling on the three surfaces: HPHI in blue, STD in yellow, and AL in grey.

The first information that can be extracted from this chart concerns the total time of the transient, which is less than 1 s for all cases.

For each surface, it is also indicated the contact angle reached at the end of the transient, which is 62° for the surface covered with the HPHI coating, 72° for the surface covered with the STD coating, and 79° for the AL surface with no coating. These results are in a quite good agreement with those obtained from the static contact angle measurements.

The difference among the three surfaces is evident also for the dynamic contact angles: the advancing contact angles, represented by the values of the peaks of the curves, and the receding contact angles, represented by the values of the valleys of the curves, are again the lowest for the HPHI surface, intermediate for the STD surface, and the highest for the AL surface, consistently with the previous results.

Additionally, when considering the concave hull of the local minima and maxima of the contact angle profiles in time, it can be extracted that the contact angle hysteresis at the beginning of the transient is relatively similar between the three surfaces (in the range $60° \pm 3°$), then the reduction of the quantity shows a slightly different trend. Thus, the "ranking" of the surfaces in terms of contact angle hysteresis is the same as that according to the static contact angle.

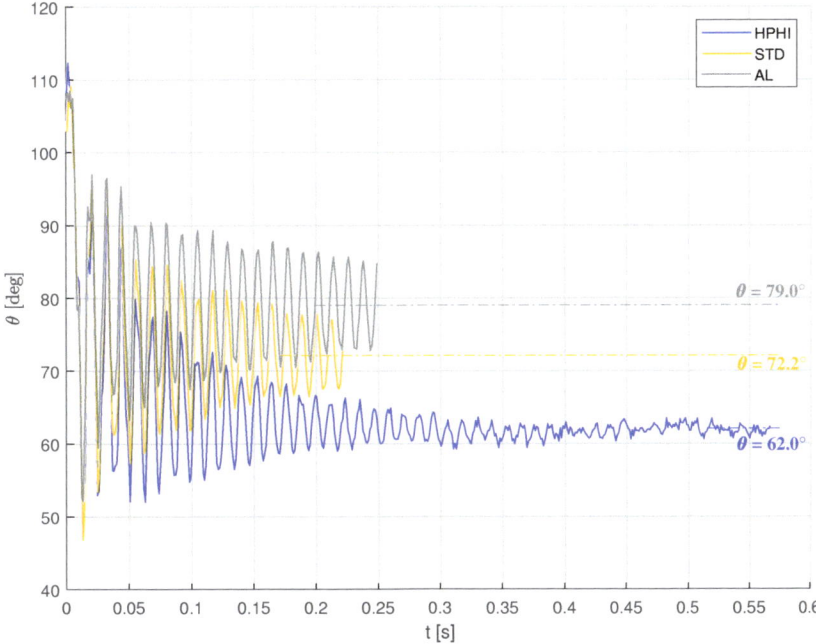

Figure 6. Transient behavior of the drops falling on the three surfaces: HPHI in blue, STD in yellow, AL in grey.

As previously mentioned, in order to evaluate the dynamic contact angles on the vertical surfaces, ten sessile drops were deposed on each surface. The results of this analysis can be seen in Figure 7, which shows the advancing and receding contact angles that the drops form with each surface.

From this Figure, it is possible to notice that again the HPHI surface shows the lowest contact angles, while the contact angles on the STD surface are intermediate, and the ones on the AL surface are the highest.

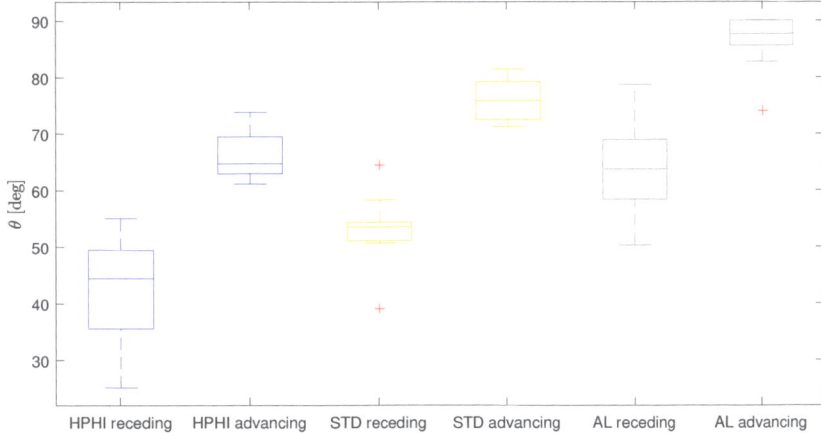

Figure 7. Boxplot representing the dynamic (advancing and receding) contact angles obtained by deposing sessile drops on each of the three surfaces in vertical orientation: HPHI in blue, STD in yellow, AL in grey.

Table 2 summarizes the results of the dynamic contact angle analysis obtained from sessile drops deposed on vertical surfaces, again in terms of average, median, and standard deviation.

Table 2. Summary of the results of the dynamic contact angle analysis obtained from sessile drops deposed on vertical surfaces.

Material	Dynamic Contact Angle Type	Average	Median	Standard Deviation
HPHI	Receding	42°	44°	9°
HPHI	Advancing	66°	65°	4°
STD	Receding	53°	54°	6°
STD	Advancing	76°	75°	4°
AL	Receding	64°	64°	8°
AL	Advancing	86°	88°	5°

To further assess the reliability of the obtained results, the relations between the equilibrium and dynamic contact angles were also compared with the predictions of the model by Tadmor [27], that allows to estimate the equilibrium contact angle from the dynamic ones. This model was selected among those available in literature as it combines simplicity and theoretical background. According to Tadmor's model, the static contact angle can be estimated as:

$$\theta = \arccos\left(\frac{\Gamma_{adv} \cos\theta_{adv} + \Gamma_{rec} \cos\theta_{rec}}{\Gamma_{adv} + \Gamma_{rec}}\right) \quad (1)$$

where:

$$\Gamma_{adv} = \left(\frac{\sin^3\theta_{adv}}{2 - 3\cos\theta_{adv} + \cos^3\theta_{adv}}\right)^{\frac{1}{3}} \quad (2)$$

and

$$\Gamma_{rec} = \left(\frac{\sin^3\theta_{rec}}{2 - 3\cos\theta_{rec} + \cos^3\theta_{rec}}\right)^{\frac{1}{3}} \quad (3)$$

When considering the dynamic angles evaluated on the vertical surfaces, the difference between the equilibrium contact angle predicted by the model and the experimental one was between 9° and 12°. On the contrary, when using the dynamic angles from the high-speed videos, the prediction was quite satisfactory for STD, with an error of 5°, while it was consistent with the results obtained with the vertical surfaces for HPHI and AL, with an error of 11° and 9°, respectively. Thus, the obtained results can be considered quite satisfactory.

3.3. Correlation of the Wettability Results with IEC System Performance

The performance of IEC systems using plates with the HPHI and STD coating has been previously studied [10], showing that the cooling capacity of the system using the plates with the HPHI coating, evaluated in terms of wet bulb effectiveness and fraction of evaporated water, are always better than the performance of the system using plates with the STD coating. These results suggest that there is a positive correlation between the wettability of the coating and the performance of the IEC system, namely that a system using plates with high wettability shows better performance than a system using plates with low wettability.

The results from the present work, also including the uncoated AL surface, will be the basis to improve the existing models aimed at predicting the behavior of IEC systems. In particular, the obtained wettability values will be included both in simplified models (e.g., by means of the spreading parameter), and in multiphase numerical simulations of the IEC systems, in order to properly set the boundary conditions on the plates enclosing the device channels in the virtual domain.

Furthermore, it will be necessary to study what happens to the wettability of the plates when the surfaces are not clean, but there is some fouling, e.g., due to limescale, as it is in IEC real operative conditions.

4. Conclusions

In this work, the static contact angle and dynamic contact angles that a drop forms with three different surfaces used for IEC systems plates have been analyzed.

Firstly, the static contact angle has been measured for 240 drops on each surface, and the result showed that the static contact angle is the lowest (average value: 69°) for the surface covered with the hydrophilic lacquer (HPHI), it is intermediate (average value: 75°) for the surface covered with the standard epoxy coating (STD), and it is the highest (average value: 89°) for the uncoated surface (AL). The obtained results regarding the HPHI and STD surfaces are in good agreement with the results of a previous experimental campaign [10].

Secondly, the dynamic contact angles (advancing and receding) have been studied for sessile drops deposed on the three surfaces in vertical orientation and for drops falling on the three surfaces in horizontal orientation. This last analysis was also useful to evaluate the transient behavior of the drops. The results showed that the transient behavior is similar for all the surfaces, with the classic oscillating behavior of drops impacting onto non-superhydrophobic surfaces. The values of the contact angles reached at the end of the transient are consistent with the results of the static analysis, as the advancing and receding contact angles. In particular, the dynamic contact angles were used to predict the corresponding static contact angles by using the model of Tadmor [27], and the predictions can be considered quite satisfactory.

In conclusion, the wettability of three surfaces belonging to IEC systems plates was fully characterized, thus the results obtained in this study will be of use to extend the correlation between plates wettability and IEC system performance parameters that was observed in previous works, and to provide input values for both simplified and computational fluid dynamics models.

Author Contributions: Conceptualization, R.C., S.D.A., L.M., P.L. and M.G.; methodology, R.C. and M.G.; resources, P.L. and M.G.; software, R.C. and M.G.; investigation, R.C. and M.G.; formal analysis, R.C. and M.G.; validation, all authors; writing—original draft preparation, R.C.; writing—review & Editing, all authors; visualization, R.C. All authors have read and agreed to the published version of the manuscript.

Funding: This research received no external funding.

Data Availability Statement: Data are available from the authors upon reasonable request.

Conflicts of Interest: The three surfaces used for IEC systems plates, namely HPHI, STD, and AL, have been supplied by Recuperator S.p.A.

Abbreviations

The following abbreviations are used in this manuscript:

ADSA	Axisymmetric Drop Shape Analysis
AL	Aluminum uncoated surface
HPHI	Aluminum surface coated with a hydrophilic lacquer
IEC	Indirect Evaporative Cooling
STD	Aluminum surface covered with a standard epoxy coating
V	Volume
deg	Degrees

fps	Frames per second
s	Seconds
t	Time
θ	Static contact angle
θ_{adv}	Advancing contact angle
θ_{rec}	Receding contact angle
Γ_{adv}	Advancing coefficient used in Tadmor model
Γ_{rec}	Receding coefficient used in Tadmor model

References

1. Sajjad, U.; Abbas, N.; Hamid, K.; Abbas, S.; Hussain, I.; Ammar, S.M.; Sultan, M.; Ali, H.M.; Hussain, M.; Wang, C.C.; et al. A review of recent advances in indirect evaporative cooling technology. *Int. Commun. Heat Mass Transf.* **2021**, *122*, 105140. [CrossRef]
2. Yang, H.; Shi, W.; Chen, Y.; Min, Y. Research development of indirect evaporative cooling technology: An updated review. *Renew. Sustain. Energy Rev.* **2021**, *145*, 111082. [CrossRef]
3. Duan, Z.; Zhan, C.; Zhang, X.; Mustafa, M.; Zhao, X.; Alimohammadisagvand, B.; Hasan, A. Indirect evaporative cooling: Past, present and future potentials. *Renew. Sustain. Energy Rev.* **2012**, *16*, 6823–6850. [CrossRef]
4. Alonso, J.S.J.; Martinez, F.R.; Gomez, E.V.; Plasencia, M.A.G. Simulation model of an indirect evaporative cooler. *Energy Build.* **1998**, *29*, 23–27. [CrossRef]
5. Chen, Y.; Luo, Y.; Yang, H. A simplified analytical model for indirect evaporative cooling considering condensation from fresh air: Development and application. *Energy Build.* **2015**, *108*, 387–400. [CrossRef]
6. Heidarinejad, G.; Moshari, S. Novel modeling of an indirect evaporative cooling system with cross-flow configuration. *Energy Build.* **2015**, *92*, 351–362. [CrossRef]
7. De Antonellis, S.; Joppolo, C.M.; Liberati, P.; Milani, S.; Romano, F. Modeling and experimental study of an indirect evaporative cooler. *Energy Build.* **2017**, *142*, 147–157. [CrossRef]
8. Adam, A.; Han, D.; He, W.; Chen, J. Numerical analysis of cross-flow plate type indirect evaporative cooler: Modeling and parametric analysis. *Appl. Therm. Eng.* **2021**, *185*, 116379. [CrossRef]
9. Chua, K.; Xu, J.; Cui, X.; Ng, K.; Islam, M. Numerical heat and mass transfer analysis of a cross-flow indirect evaporative cooler with plates and flat tubes. *Heat Mass Transf.* **2016**, *52*, 1765–1777. [CrossRef]
10. Guilizzoni, M.; Milani, S.; Liberati, P.; De Antonellis, S. Effect of plates coating on performance of an indirect evaporative cooling system. *Int. J. Refrig.* **2019**, *104*, 367–375. [CrossRef]
11. Zhao, X.; Liu, S.; Riffat, S.B. Comparative study of heat and mass exchanging materials for indirect evaporative cooling systems. *Build. Environ.* **2008**, *43*, 1902–1911. [CrossRef]
12. Del Río, O.; Neumann, A. Axisymmetric drop shape analysis: Computational methods for the measurement of interfacial properties from the shape and dimensions of pendant and sessile drops. *J. Colloid Interface Sci.* **1997**, *196*, 136–147.
13. Guilizzoni, M.; De Antonellis, S. Wettability analysis of desiccant beads for HVAC systems. *J. Phys. Conf. Ser.* **2019**, *1249*, 012005. [CrossRef]
14. Guilizzoni, M.; Sapienza, J. Axisymmetric Drop Shape Analysis using a low-cost home-made setup. *J. Phys. Conf. Ser.* **2021**, *1977*, 012003. [CrossRef]
15. Santini, M.; Guilizzoni, M.; Fest-Santini, S.; Lorenzi, M. Characterization of highly hydrophobic textiles by means of X-ray microtomography, wettability analysis and drop impact. *J. Phys. Conf. Ser.* **2017**, *923*, 012013. [CrossRef]
16. Wenzel, R.N. Resistance of solid surfaces to wetting by water. *Ind. Eng. Chem.* **1936**, *28*, 988–994. [CrossRef]
17. Cassie, A.; Baxter, S. Wettability of porous surfaces. *Trans. Faraday Soc.* **1944**, *40*, 546–551. [CrossRef]
18. Guilizzoni, M. Drop shape visualization and contact angle measurement on curved surfaces. *J. Colloid Interface Sci.* **2011**, *364*, 230–236. [CrossRef] [PubMed]
19. Laplace, P. *Traité de Mécanique Céleste*; Supplement au Dixième; Livre, Surl'Action Capillaire, Courcier: Paris, France, 1806.
20. Rotenberg, Y.; Boruvka, L.; Neumann, A. Determination of surface tension and contact angle from the shapes of axisymmetric fluid interfaces. *J. Colloid Interface Sci.* **1983**, *93*, 169–183. [CrossRef]
21. Cheng, P.; Li, D.; Boruvka, L.; Rotenberg, Y.; Neumann, A. Automation of axisymmetric drop shape analysis for measurements of interfacial tensions and contact angles. *Colloids Surf.* **1990**, *43*, 151–167. [CrossRef]
22. Birdi, K.; Vu, D. Wettability and the evaporation rates of fluids from solid surfaces. *J. Adhes. Sci. Technol.* **1993**, *7*, 485–493. [CrossRef]
23. Erbil, H.Y.; McHale, G.; Newton, M. Drop evaporation on solid surfaces: Constant contact angle mode. *Langmuir* **2002**, *18*, 2636–2641. [CrossRef]
24. Rein, M. Phenomena of liquid drop impact on solid and liquid surfaces. *Fluid Dyn. Res.* **1993**, *12*, 61–93. [CrossRef]
25. Rioboo, R.; Tropea, C.; Marengo, M. Outcomes from a drop impact on solid surfaces. *At. Sprays* **2001**, *11*. 155–166. [CrossRef]

26. Josserand, C.; Thoroddsen, S.T. Drop impact on a solid surface. *Annu. Rev. Fluid Mech.* **2016**, *48*, 365–391. [CrossRef]
27. Tadmor, R. Line energy and the relation between advancing, receding, and young contact angles. *Langmuir* **2004**, *20*, 7659–7664. [CrossRef]

Disclaimer/Publisher's Note: The statements, opinions and data contained in all publications are solely those of the individual author(s) and contributor(s) and not of MDPI and/or the editor(s). MDPI and/or the editor(s) disclaim responsibility for any injury to people or property resulting from any ideas, methods, instructions or products referred to in the content.

MDPI AG
Grosspeteranlage 5
4052 Basel
Switzerland
Tel.: +41 61 683 77 34

Fluids Editorial Office
E-mail: fluids@mdpi.com
www.mdpi.com/journal/fluids

Disclaimer/Publisher's Note: The title and front matter of this reprint are at the discretion of the Guest Editor. The publisher is not responsible for their content or any associated concerns. The statements, opinions and data contained in all individual articles are solely those of the individual Editor and contributors and not of MDPI. MDPI disclaims responsibility for any injury to people or property resulting from any ideas, methods, instructions or products referred to in the content.

www.ingramcontent.com/pod-product-compliance
Lightning Source LLC
LaVergne TN
LVHW072351090526
838202LV00019B/2518